Natural and Technological Hazards in Urban Areas: Assessment, Planning and Solutions

Natural and Technological Hazards in Urban Areas: Assessment, Planning and Solutions

Editors

George D. Bathrellos
Hariklia D. Skilodimou
Konstantinos G. Chousianitis
Charalampos Vasilatos

MDPI • Basel • Beijing • Wuhan • Barcelona • Belgrade • Manchester • Tokyo • Cluj • Tianjin

Editors
George D. Bathrellos
Department of Geology
University of Patras
Rio, Patras
Greece

Hariklia D. Skilodimou
Department of Geology
University of Patras
Rio, Patras
Greece

Konstantinos G. Chousianitis
Institute of Geodynamics
National Observatory
of Athens
Athens
Greece

Charalampos Vasilatos
Department of Geology
and Geoenvironment
National and Kapodistrian
University of Athens
Zografos, Athens
Greece

Editorial Office
MDPI
St. Alban-Anlage 66
4052 Basel, Switzerland

This is a reprint of articles from the Special Issue published online in the open access journal *Sustainability* (ISSN 2071-1050) (available at: www.mdpi.com/journal/sustainability/special_issues/Natural_Technological_Hazards).

For citation purposes, cite each article independently as indicated on the article page online and as indicated below:

LastName, A.A.; LastName, B.B.; LastName, C.C. Article Title. *Journal Name* **Year**, *Volume Number*, Page Range.

ISBN 978-3-0365-6397-8 (Hbk)
ISBN 978-3-0365-6396-1 (PDF)

Contents

About the Editors

George D. Bathrellos

Dr. George D. Bathrellos, Associate Professor at University of Patras, Faculty of Sciences, Department of Geology, Sector of General, Marine Geology & Geodynamics.

Research interests: geomorphology (pure and applied); natural hazards; urban planning; modeling; GIS. Dr. George Bathrellos has taught Geomorphology, Physical Geography, GIS, Urban Planning, Environmental Management and Monitoring, and Hydrology in graduate and post-graduate courses in four Greek universities. He has contributed to over 150 publications in international conferences and journals.

Hariklia D. Skilodimou

Dr. Hariklia D. Skilodimou, Ph.D. Researcher and Teaching Staff. Research interests: geomorphology (pure and applied); natural hazards; modeling; GIS. Dr. Skilodimou taught Geomorphology, Physical Geography, GIS, Urban Planning, Environmental Management and Monitoring, and Hydrology in in graduate and post-graduate courses in two Greek universities. She has contributed to over 150 publications in international conferences and journals.

Konstantinos G. Chousianitis

Dr. Konstantinos G. Chousianitis, PhD researcher. Research interests: tectonic geodesy; earthquake seismology; GMPEs; seismic hazard; GIS; earthquake-induced landslides; GPS measurements to constrain crustal deformation and strain rate patterns, investigate coseismic and postseismic processes.

Charalampos Vasilatos

Dr. Charalampos Vasilatos, Assistant Professor. Research interests: geochemistry; soil pollution control and remediation; hazardous waste, water and wastewater treatment; traditional and novel functional materials and environmental applications.

Preface to "Natural and Technological Hazards in Urban Areas: Assessment, Planning and Solutions"

This reprint contains fourteen (14) scientific papers related to floods, landslides, earthquakes, water and soil contamination, vegetation restoration, hazard and risk assessment. These contributions are republications of published papers in the Special Issue of the journal *Sustainability*, entitled "Natural and Technological Hazards in Urban Areas: Assessment, Planning and Solutions". This reprint presents the importance of studying natural and technological hazards in urban planning.

This reprint is directed to scientists, researchers, and readers who are interested in natural and technological hazards. The editors wish to thank the contributors and the *Sustainability* journal editorial staff for their support, whose professionalism and dedication have made this Special Issue possible. Finally, we kindly thank all authors for their participation and contribution to the volume, as well as all reviewers who have diligently reviewed the submitted manuscripts, and substantially contributed to the high quality of these published papers. We also highly appreciate the editorial support provided by the *Sustainability* journal.

George D. Bathrellos, Hariklia D. Skilodimou, Konstantinos G. Chousianitis, and Charalampos Vasilatos

Editors

Editorial

Natural and Technological Hazards in Urban Areas: Assessment, Planning and Solutions

Hariklia D. Skilodimou and George D. Bathrellos *

Department of Geology, University of Patras, 26504 Rio Patras, Greece; hskilodimou@upatras.gr
* Correspondence: gbathrellos@upatras.gr

Citation: Skilodimou, H.D.; Bathrellos, G.D. Natural and Technological Hazards in Urban Areas: Assessment, Planning and Solutions. *Sustainability* **2021**, *13*, 8301. https://doi.org/10.3390/su13158301

Received: 20 July 2021
Accepted: 23 July 2021
Published: 25 July 2021

Natural hazards are extreme natural phenomena whose associated consequences can lead to damage of both the natural and man-made environment. They occur worldwide, are rare at a particular place and time, and contribute to the progress of Earth's landscape [1–3]. Their cause, occurrence and evolution show important complexity and they differ in magnitude, frequency, speed and duration [4–6]. Moreover, their impact differs from place to place and when their consequences have a major impact on human life, they become natural disasters [7]. In this Special Issue, the term natural hazard refers to all atmospheric, hydrologic, geologic and geomorphologic hazardous natural phenomena that potentially affect human life and activities.

Generally, natural hazards and disasters take place more frequently in relation to our ability to restore the effects of past events [8]. On average over the past decade, about 60,000 people globally died from natural disasters each year. This number represents 0.1% of global deaths [9]. Additionally, global disasters produced $210 billion of losses in 2020 and were up 26.5% compared to 2019's cost of $166 billion [10].

On the other hand, a hazard of anthropogenic origin that can harm people, the environment or facilities is called a technological hazard. This type of hazard contains a wide range of modern issues and consequences of technology mismanagement and engineering mistakes [11]. Technological hazards such as desertification, water and soil pollution/degradation, land use changes, waste, hazardous materials incidents, pipelines and transportation are presented in this special issue.

Technological disasters account for about a third (36.4%) of all reported disasters worldwide since 1900 [12]. As technology becomes increasingly complex, technological hazards are likely to increase. The effects of technological disasters on humans may be long lasting. They are stressful, especially because they are random and unpredictable [13]. Technological hazards can bring on a crisis, menace the viability of a technological system, cause losses of life and property, and can put the social environment in which they occur at risk [14].

Natural hazards are physical phenomena active in geological time, while technological hazards result from actions or facilities created by human and occur in the recent past. Natural hazards are directly related to the natural environment and the use of technology may reduce their effects [15,16]. On the contrary, technological hazards can be mitigated by being aware of the natural environment [17,18].

Natural hazard events and technological accidents are separate causes of environmental impacts. In our time, combined natural and man-made hazards have been induced. Overpopulation and urban development in areas prone to natural hazards increase the impact of natural disasters worldwide [19,20]. Additionally, urban areas are frequently characterized by intense industrial activity and rapid but poorly planned growth that threaten environment and ecosystems, degrade quality of life and raise the likelihood of technological disasters.

To avoid the aforementioned effects, appropriate urban planning is crucial to minimize fatalities and reduce the environmental and economic impacts that accompany both natural and technological hazardous events. Thus, local authorities and countries should develop proper policies and strategies, aiming to effectively manage hazard risk, reducing the likelihood and consequences of disasters.

This Special Issue focuses on natural and technological hazards in urban areas. Additionally, hazard assessment, planning and solutions are included. Natural hazards such as floods, landslides, earthquakes, waterfall ice, fires, as well as technological hazards such as toxic elements, pipelines, waste and transportation, are presented.

Three works deal with the flood hazard assessment and one with the distribution of flash flood events. Joo et al. [21] assess a flood risk method in Korea. The leading indicators affecting flood damage were selected using factor analysis and principal component analysis. The analytic hierarchy process (AHP) [22], constant sum scale and entropy were used to assign the weights of the indicators. A relationship was examined between the elements and the indicators based on weights called the Integrated Index for flood risk assessment.

Liu et al. [23] estimate flood risk of urban areas in Kaohsiung city along the Dianbao River, in southern Taiwan. Floods and other extreme weather events occur continuously, threatening human lives and causing severe property losses [24]. A rainfall-runoff model (HEC-HMS) was adopted to simulate discharges in the watershed, and the simulated discharges were utilized as inputs for the inundation model (FLO-2D). A risk map was developed that compiled both flood hazards and social vulnerability levels.

Skilodimou et al. [25] propose a simple method to produce a flood hazard assessment map in burned and urban areas in a coastal part of the eastern Attica peninsula in central Greece. The Attica peninsula has suffered from wildfires and floods during the last decades [26]. Six factors were considered as the parameters most controlling runoff when it overdraws the drainage system's capacity. The analytical hierarchy process (AHP) method and geographical information system (GIS) were utilized to create a flood hazard assessment map of the study area.

Xiong et al. [27] measure the spatial and temporal distribution of flash floods and examine the relationship between flash floods and driving factors in Yunnan Province in China. Flash floods are one of the most serious natural disasters, and have a significant impact on economic development [28]. The results indicated that the number of flash floods occurring annually increased gradually from 1949 to 2015, and regions with a high quantity of flash floods were concentrated in Zhaotong, Qujing, Kunming, Yuxi, Chuxiong, Dali and Baoshan.

Another work focuses on landslide management in urbanized areas. Giordan et al. [29] show the importance of a dedicated monitoring solution and communication strategy for an effective management of complex active landslides in Italy. Large landslide emergencies are often managed by complex and multi-instrumental networks [30]. Since the management of these networks is often a complicated task, a new hybrid system focused on capturing and elaborating data-sets from monitored sites and on disseminating monitoring results to support decision makers was developed by the authors. It consists of an early warning application, which integrates a threshold-based approach, and a failure forecasting modeling.

Jia et al. [31] analyze and compare the importance of feature affecting earthquake fatalities in China mainland and establish a deep learning model to assess the potential fatalities based on the selected factors. Earthquakes cause significant damage to infrastructures and a large number of threats to the Chinese [32]. Thus, a proper rapid estimation of the number of casualties in an earthquake is important. In this paper, the random forest (RF) model, classification and regression tree (CART) model and AdaBoost model were used to assess the importance of nine features, and the analysis showed that the RF model was better than the other models.

Zhou et al. [33] examine the formation mechanism and influence factors of highway waterfall ice in northern China. Highways in the northern part of China have always been damaged by waterfall ice [34]. To explore the internal factors that lead to highway waterfall ice, gradation tests, penetration tests and freezing tests were conducted which revealed that coarse-grained particles can enhance the permeability of aquifers. Furthermore, to understand the formation mechanism of highway waterfall ice further, a mathematical model of saturated coarse-grained soil at the state of phase transition equilibrium was obtained.

Chen et al. [35] construct a natural hazard emergency relief alliance and analyze the mechanisms and dynamics of public participation. In China, the government plays an important role in emergency management and government supervision has a positive effect on environmental governance [36]. Using four different processes, namely participation proposals, negotiation interval, negotiation decision-making function and participation strategy, the authors comprehensively constructed an emergency relief alliance for natural hazards. In addition, the dynamic public interaction process was analyzed and a construction algorithm was given.

Additionally, two studies estimate the contamination of toxic elements in urban soils in Greece. Golia et al. [37] record the level of potentially toxic elements within the urban complex in the city of Volos, in central Greece. Urban and agricultural soils are often densely populated, and intensive human activities lead to large amounts of potentially toxic metals [38]. Soil pollution indices, such as the contamination factor (CF) and the geo-accumulation index (Igeo), were estimated regarding each of the metals of interest. The respective thematic maps were constructed, and the spatial variability of the contamination degree was displayed.

Alexakis et al. [39] investigate the soil quality of the Ioannina plain in western Greece concerning arsenic (As) and zinc (Zn), and delineate their origin as well as compare the As and Zn content in soil with criteria recorded in the literature. High concentration of toxic elements is reported in agricultural soil of Greece such as the Arta plain [40]. The geomorphologic settings, land use, and soil physicochemical properties were mapped and evaluated, including soil texture and concentrations of aqua-regia extractable As and Zn. The concentration of elements was spatially correlated with land use and the geology of the study area, while screening values were applied to assess land suitability.

Ali and Choi [41] review the existing methods for monitoring leakage in underground pipelines, the sinkholes caused by these leakages, and the viability of wireless sensor networking (WSN) for monitoring leakages and sinkholes. Leakage from underground pipe mains in urban areas may cause sudden ground subsidence or sinkholes [42]. The authors discussed the methods based on different objectives and their applicability via various approaches: (1) patent analysis; (2) web-of-science analysis; (3) WSN-based pipeline leakage and sinkhole monitoring.

Jia et al. [43] evaluate the vegetation restoration along an expressway in a cold, arid, and decertified area of China. Vegetation plays an important role in reducing soil erosion [44] and vegetation is significant in the restoration of expressways in the arid zone of China, although we still do not know which soil and vegetation types are most effective. The authors investigated soil particle size (SPZ), volume weight of the soil (VWS), soil water content (SWC), total porosity of soil (TP), soil organic matter (SOM), water erosion (WrE) and wind erosion (WdE) of eight sites, and evaluated them using the gray correlation method (GCM).

Kerpelis et al. [45] propose a theoretical approach for the estimation of seismic structural vulnerability of wastewater treatment plants. The assessment of seismic vulnerability is critical for lifelines such as wastewater treatment plants (WTPs) because failures may result in environmental degradation. For example, wildfires caused by earthquakes or urban fires can release toxic elements into the soil and water resources [46]. The authors tested and applied a rapid, simple methodology for assessing the seismic structural vulnerability (SSV) of WTPs (according to the qualitative method Rapid Visual Screening), using structural variables as indices of these infrastructures. An original new method involving the assessment of the SSV of 13 steps (four for a sample set of WTPs and nine for an individual one) was introduced following systematic literature retrieval.

Author Contributions: Conceptualization, G.D.B. and H.D.S.; writing—original draft preparation, G.D.B. and H.D.S.; writing—review and editing, G.D.B. and H.D.S.; visualization, G.D.B. and H.D.S. Both authors have read and agreed to the published version of the manuscript.

Funding: This research received no external funding.

Institutional Review Board Statement: Not applicable.

Informed Consent Statement: Not applicable.

Data Availability Statement: Not applicable.

Conflicts of Interest: The authors declare no conflict of interest.

References

1. Skilodimou, H.D.; Bathrellos, G.D.; Maroukian, H.; Gaki-Papanastassiou, K. Late Quaternary evolution of the lower reaches of Ziliana stream in south Mt. Olympus (Greece). *Geogr. Fis. Din. Quat.* **2014**, *37*, 43–50. [CrossRef]
2. Bathrellos, G.D.; Skilodimou, H.D.; Maroukian, H.; Gaki-Papanastassiou, K.; Kouli, K.; Tsourou, T.; Tsaparas, N. Pleistocene glacial and lacustrine activity in the southern part of Mount Olympus (central Greece). *Area* **2017**, *49*, 137–147. [CrossRef]
3. Bathrellos, G.D.; Skilodimou, H.D.; Zygouri, V.; Koukouvelas, I.K. Landslide: A recurrent phenomenon? Landslide hazard assessment in mountainous areas of central Greece. *Z. Geomorphol.* **2021**, *63*, 95–114. [CrossRef]
4. Youssef, A.M.; Abu-Abdullah, M.M.; AlFadail, E.A.; Skilodimou, H.D.; Bathrellos, G.D. The devastating flood in the arid region a consequence of rainfall and dam failure: Case study, Al-Lith flood on 23th November 2018, Kingdom of Saudi Arabia. *Z. Geomorphol.* **2021**, *63*, 115–136. [CrossRef]
5. Bathrellos, G.D.; Kalivas, D.P.; Skilodimou, H.D. Landslide Susceptibility Assessment Mapping: A Case Study in Central Greece. In *Remote Sensing of Hydrometeorological Hazards*; Petropoulos, G.P., Islam, T., Eds.; CRC Press, Taylor & Francis Group: London, UK, 2017; pp. 493–512, ISBN 978-1498777582.
6. Bathrellos, G.D.; Skilodimou, H.D.; Soukis, K.; Koskeridou, E. Temporal and spatial analysis of flood occurrences in drainage basin of Pinios River (Thessaly, central Greece). *Land* **2018**, *7*, 106. [CrossRef]
7. Alcántara-Ayala, I. Geomorphology, natural hazards, vulnerability and prevention of natural disasters in developing countries. *Geomorphology* **2002**, *47*, 107–124. [CrossRef]
8. Bathrellos, G.D.; Skilodimou, H.D. Land use planning for natural hazards. *Land* **2019**, *8*, 128. [CrossRef]
9. Ritchie, H.; Roser, M. Natural Disasters. Published online at OurWorldInData.org. 2014. Available online: https://ourworldindata.org/natural-disasters (accessed on 17 July 2021).
10. Pinchot, A.; Zhou, L.; Christianson, G. With Patchy Guidance, Companies May Have Climate Risk Blind Spots. Available online: https://www.wri.org/insights/patchy-guidance-companies-may-have-climate-risk-blind-spots (accessed on 17 July 2021).
11. Schmidt-Thomé, P.; Kallio, H.; Jarva, J.; Tarvainen, T. The spatial effects and management of natural and technological hazards in Europe. In *Final Report of the European Spatial Planning and Observation Network (ESPON) Project*; Geological Survey of Finland: Vuorimiehentie, Espo, Finland, 2005; Volume 1, pp. 1–197.
12. CRED. EM-DAT, CRED/UCLouvain, Brussels, Belgium. Available online: https://www.emdat.be/ (accessed on 17 July 2021).
13. Lindsey, A.B.; Donovan, M.; Smith, S.; Radunovich, H.; Gutter, M. Impacts of Technological Disasters. *Document FCS9265, Series of the Department of Family, Youth and Community Sciences, UF/IFAS Extension*. 2011. Available online: https://edis.ifas.ufl.edu/pdf/FY/FY123000.pdf (accessed on 17 July 2021).
14. Manion, M.; Evan, W.M. Technological catastrophes: Their causes and prevention. *Technol. Soc.* **2002**, *24*, 207–224. [CrossRef]
15. Skilodimou, H.D.; Bathrellos, G.D.; Chousianitis, K.; Youssef, A.M.; Pradhan, B. Multi-hazard assessment modeling via multi-criteria analysis and GIS: A case study. *Environ. Earth Sci.* **2019**, *78*, 47. [CrossRef]
16. Bathrellos, G.D.; Gaki-Papanastassiou, K.; Skilodimou, H.D.; Papanastassiou, D.; Chousianitis, K.G. Potential suitability for urban planning and industry development using natural hazard maps and geological–geomorphological parameters. *Environ. Earth Sci.* **2012**, *66*, 537–548. [CrossRef]
17. Makri, P.; Stathopoulou, E.; Hermides, D.; Kontakiotis, G.; Zarkogiannis, S.D.; Skilodimou, H.D.; Bathrellos, G.D.; Antonarakou, A.; Scoullos, M. The Environmental Impact of a Complex Hydrogeological System on Hydrocarbon-Pollutants' Natural Attenuation: The Case of the Coastal Aquifers in Eleusis, West Attica, Greece. *J. Mar. Sci. Eng.* **2020**, *8*, 1018. [CrossRef]
18. Alexakis, D.E.; Bathrellos, G.D.; Skilodimou, H.D.; Gamvroula, D.E. Land suitability mapping using geochemical and spatial analysis methods. *Appl. Sci.* **2021**, *11*, 5404. [CrossRef]
19. Bathrellos, G.D.; Skilodimou, H.D.; Chousianitis, K.; Youssef, A.M.; Pradhan, B. Suitability estimation for urban development using multi-hazard assessment map. *Sci. Total Environ.* **2017**, *575*, 119–134. [CrossRef]
20. Skilodimou, H.D.; Bathrellos, G.D.; Koskeridou, E.; Soukis, K.; Rozos, D. Physical and anthropogenic factors related to landslide activity in the Northern Peloponnese, Greece. *Land* **2018**, *7*, 85. [CrossRef]
21. Joo, H.; Choi, C.; Kim, J.; Kim, D.; Kim, S.; Kim, H.S. A Bayesian Network-Based Integrated for Flood Risk Assessment (InFRA). *Sustainability* **2019**, *11*, 3733. [CrossRef]

22. Joo, H.J.; Kim, S.J.; Lee, M.J.; Kim, H.S. A study on determination of investment priority of flood control considering flood vulnerability. *J. Korean Soc. Hazard Mitig.* **2017**, *18*, 417–429. [CrossRef]
23. Liu, W.-C.; Hsieh, T.-H.; Liu, H.-M. Flood Risk Assessment in Urban Areas of Southern Taiwan. *Sustainability* **2021**, *13*, 3180. [CrossRef]
24. Yang, T.-H.; Liu, W.-C. A General Overview of the Risk-Reduction Strategies for Floods and Droughts. *Sustainability* **2020**, *12*, 2687. [CrossRef]
25. Skilodimou, H.D.; Bathrellos, G.D.; Alexakis, D.E. Flood Hazard Assessment Mapping in Burned and Urban Areas. *Sustainability* **2021**, *13*, 4455. [CrossRef]
26. Bathrellos, G.D.; Karymbalis, E.; Skilodimou, H.D.; Gaki-Papanastassiou, K.; Baltas, E.A. Urban flood hazard assessment in the basin of Athens Metropolitan city, Greece. *Environ. Earth Sci.* **2016**, *75*, 319. [CrossRef]
27. Xiong, J.; Ye, C.; Cheng, W.; Guo, L.; Zhou, C.; Zhang, X. The Spatiotemporal Distribution of Flash Floods and Analysis of Partition Driving Forces in Yunnan Province. *Sustainability* **2019**, *11*, 2926. [CrossRef]
28. Xiong, J.; Wei, F.; Liu, Z. Hazard assessment of debris flow in Sichuan Province. *J. Geo-Inf. Sci.* **2017**, *19*, 1604–1612.
29. Giordan, D.; Wrzesniak, A.; Allasia, P. The Importance of a Dedicated Monitoring Solution and Communication Strategy for an Effective Management of Complex Active Landslides in Urbanized Areas. *Sustainability* **2019**, *11*, 946. [CrossRef]
30. Giordan, D.; Manconi, A.; Allasia, P.; Bertolo, D. Brief Communication: On the rapid and efficient monitoring esults dissemination in landslide emergency scenarios: The Mont de La Saxe case study. *Nat. Hazards Earth Syst. Sci.* **2015**, *15*, 2009–2017. [CrossRef]
31. Jia, H.; Lin, J.; Liu, J. An Earthquake Fatalities Assessment Method Based on Feature Importance with Deep Learning and Random Forest Models. *Sustainability* **2019**, *11*, 2727. [CrossRef]
32. Jia, H.; Lin, J.; Liu, J. Bridge Seismic Damage Assessment Model Applying Artificial Neural Networks and the Random Forest Algorithm. *Adv. Civ. Eng.* **2020**, *2020*, 6548682. [CrossRef]
33. Zhou, Z.; Lei, J.; Zhu, S.; Qiao, S.; Zhang, H. The Formation Mechanism and Influence Factors of Highway Waterfall Ice: A Preliminary Study. *Sustainability* **2019**, *11*, 4059. [CrossRef]
34. Zhou, Z.; Zhan, H.; Hu, J.; Ren, C. Characteristics of unloading creep of tuffaceous sandstone in east tianshan tunnel under freeze-thaw cycles. *Adv. Mater. Sci. Eng.* **2019**, *2019*, 7547564. [CrossRef]
35. Chen, Y.; Zhang, J.; Wang, Z.; Tadikamalla, P.R. Research on the Construction of a Natural Hazard Emergency Relief Alliance Based on the Public Participation Degree. *Sustainability* **2020**, *12*, 2604. [CrossRef]
36. Chen, Y.X.; Zhang, J.; Tadikamalla, P.R.; Gao, X.T. The Relationship among Government, Enterprise and Public in Environmental Governance from the Perspective of Multi-player Evolutionary Game. *Int. J. Environ. Res. Public Health* **2019**, *16*, 3351. [CrossRef]
37. Golia, E.E.; Papadimou, S.G.; Cavalaris, C.; Tsiropoulos, N.G. Level of Contamination Assessment of Potentially Toxic Elements in the Urban Soils of Volos City (Central Greece). *Sustainability* **2021**, *13*, 2029. [CrossRef]
38. Golia, E.E.; Tsiropoulos, N.G.; Dimirkou, A.; Mitsios, I. Distribution of heavy metals of agricultural soils of central Greece using the modified BCR sequential extraction method. *Int. J. Environ. Anal. Chem.* **2007**, *87*, 1053–1063. [CrossRef]
39. Alexakis, D.E.; Bathrellos, G.D.; Skilodimou, H.D.; Gamvroula, D.E. Spatial Distribution and Evaluation of Arsenic and Zinc Content in the Soil of a Karst Landscape. *Sustainability* **2021**, *13*, 6976. [CrossRef]
40. Papadopoulou-Vrynioti, K.; Alexakis, D.; Bathrellos, G.; Skilodimou, H.; Vryniotis, D.; Vassiliades, E. Environmental research and evaluation of agricultural soil of the Arta plain, western Hellas. *J. Geochem. Explor.* **2014**, *136*, 84–92. [CrossRef]
41. Ali, H.; Choi, J.-h. A Review of Underground Pipeline Leakage and Sinkhole Monitoring Methods Based on Wireless Sensor Networking. *Sustainability* **2019**, *11*, 4007. [CrossRef]
42. Ali, H.; Choi, J.-h. Risk Prediction of Sinkhole Occurrence for Different Subsurface Soil Profiles due to Leakage from Underground Sewer and Water Pipelines. *Sustainability* **2020**, *12*, 310. [CrossRef]
43. Jia, C.; Sun, B.-p.; Yu, X.; Yang, X. Evaluation of Vegetation Restoration along an Expressway in a Cold, Arid, and Desertified Area of China. *Sustainability* **2019**, *11*, 2313. [CrossRef]
44. Jia, C.; Sun, B.; Yu, X.; Yang, X. Analysis of Runoff and Sediment Losses from a Sloped Roadbed under Variable Rainfall Intensities and Vegetation Conditions. *Sustainability* **2020**, *12*, 2077. [CrossRef]
45. Kerpelis, P.N.; Golfinopoulos, S.K.; Alexakis, D.E. A Proposed Theoretical Approach for the Estimation of Seismic Structural Vulnerability of Wastewater Treatment Plants. *Sustainability* **2021**, *13*, 4835. [CrossRef]
46. Alexakis, D.E. Suburban areas in flames: Dispersion of potentially toxic elements from burned vegetation and buildings. Estimation of the associated ecological and human health risk. *Environ. Res.* **2020**, *183*, 109153. [CrossRef] [PubMed]

Article

Flood Risk Assessment in Urban Areas of Southern Taiwan

Wen-Cheng Liu *, Tien-Hsiang Hsieh and Hong-Ming Liu

Department of Civil and Disaster Prevention Engineering, National United University, Miao-Li 36063, Taiwan; leo2601673@gmail.com (T.-H.H.); dslhmd@gmail.com (H.-M.L.)
* Correspondence: wcliu@nuu.edu.tw; Tel.: +886-37-382357

Abstract: A flood risk assessment of urban areas in Kaohsiung city along the Dianbao River was performed based on flood hazards and social vulnerability. In terms of hazard analysis, a rainfall-runoff model (HEC-HMS) was adopted to simulate discharges in the watershed, and the simulated discharges were utilized as inputs for the inundation model (FLO-2D). Comparisons between the observed and simulated discharges at the Wulilin Bridge flow station during Typhoon Kongrey (2013) and Typhoon Megi (2016) were used for the HEC-HMS model calibration and validation, respectively. The observed water levels at the Changrun Bridge station during Typhoon Kongrey and Typhoon Megi were utilized for the FLO-2D model calibration and validation, respectively. The results indicated that the simulated discharges and water levels reasonably reproduced the observations. The validated model was then applied to predict the inundation depths and extents under 50-, 100-, and 200-year rainfall return periods to form hazard maps. For social vulnerability, the fuzzy Delphi method and the analytic hierarchy process were employed to select the main factors affecting social vulnerability and to yield the weight of each social vulnerability factor. Subsequently, a social vulnerability map was built. A risk map was developed that compiled both flood hazards and social vulnerability levels. Based on the risk map, flood mitigation strategies with structural and nonstructural measures were proposed for consideration by decision-makers.

Keywords: flood risk; hazard; social vulnerability; hydrological and hydrodynamic model; analytic hierarchy process; fuzzy Delphi

Citation: Liu, W.-C.; Hsieh, T.-H.; Liu, H.-M. Flood Risk Assessment in Urban Areas of Southern Taiwan. *Sustainability* **2021**, *13*, 3180. https://doi.org/10.3390/su13063180

Academic Editors: Marc A. Rosen and George D. Bathrellos

Received: 31 January 2021
Accepted: 12 March 2021
Published: 14 March 2021

1. Introduction

In recent years, the economy has developed rapidly in various parts of the world, and countries have used many fuels, such as oil and coal, that cause holes in the ozone layer and produce a large amount of greenhouse gases such as carbon dioxide and methane, resulting in a continuous rise in temperature. Droughts, floods, and other extreme weather events occur continuously, and extreme weather causes natural disasters that continue to threaten human lives and cause severe property losses [1–5]. According to statistics from the United Nations Disaster Database (www.emdat.be, (accessed on 30 January 2021)), since 1950, the number of major natural disasters has increased year by year, and after 1990, it has exceeded 200 annually. The total number of disasters that occurred in 2018 was 282. In the seven continents in the world, the number of disasters that occurred in Asia in 2018 was 129, accounting for 46% of the total. Among the types of natural disasters that occurred in 2018, flood-related disasters accounted for 74% of all disasters; floods made up 39% of the total number of disasters, storms made up 30%, and slope disasters made up 5%. The rest of the disaster types account for approximately 2% to 9% each. Therefore, quantifying flood risk is an important way to develop adaptation and management strategies for decision-making [6,7]. In general, flood risk can be defined as the product of flood hazard and vulnerability [8–10].

Based on a literature review, several approaches have been utilized to perform flood risk assessments. These methods include historical disaster flood risk assessments [11–17], index systems for flood risk assessments [18–23], scenario simulation approaches [24–30], and remote sensing and geographic information systems [31,32]. Different approaches display corresponding advantages and disadvantages [26]. Cai et al. [26] employed a one-dimensional pipe network and a two-dimensional surface-coupled model and a multi-index fuzzy comprehensive evaluation model to build flood disaster risk. Wang et al. [23] utilized a fuzzy synthetic evaluation method based on combined weight to generate flood risk maps in the Beijing-Tianjin-Hebi urban area. Geng et al. [33] proposed a coupled one- and two-dimensional hydrodynamic model to simulate inundation as a hazard indicator, and social and economic characteristics were analyzed as vulnerability indicators. The analytic hierarchy process (AHP) was used to compute the weights of the assessment indices. Afterwards, a fuzzy comprehensive evaluation approach was adopted to assess urban flood risk. However, these studies did not clearly describe that how the social vulnerability was selected for further analysis.

Taiwan is located on the western edge of the Pacific Ocean, surrounded by the sea; it is an island country. The river slopes are steep and winding. It is often exposed to extreme rains and typhoons during the rainy season and in summer and autumn. In the wet season, southwesterly winds and typhoons bring frequent heavy rains, resulting in failures of regional drainage systems in the middle and lower reaches of rivers; these failures cause levees to break or overflow, resulting in flooding and threatening the lives of the people in the middle and lower reaches of the rivers. The property and economic losses associated with these floods are serious. In the future, rainfall patterns will change as a result of climate change. There may be more rainfall during the wet season, and flooding situations in urban regions and low-lying areas will become more serious. Therefore, the government must take preventive measures with flood control projects such as constructing urban drainage systems, riverbanks, and rainwater sewers as early as possible to ensure the safety of people and property.

In addition to establishing flood warning systems and improving existing drainage systems, governments are now gradually moving toward flood prevention and disaster mitigation planning; these goals are realized through the formulation of laws and regulations, the implementation of disaster preventative measures, and mitigation education and training aimed at achieving effective risk reduction. Flood risk assessments are composed of hazards and vulnerability. Among these, hazards are associated with occurrences of natural disasters that humans cannot eliminate or change through existing technology. To effectively reduce risk, we must decrease the level of vulnerability.

The objective of the present study is to utilize the hydrological model HEC-HMS and the inundation model FLO-2D to explore the areas that are prone to flooding in the urban region of the Dianbao River watershed in southern Taiwan and draw a flood potential map using the degree of hazard. The fuzzy Delphi method and the hierarchical analysis approach were utilized to analyze and calculate the degree of vulnerability. By combining the degree of hazard and the degree of vulnerability, an urban area flood disaster risk map was produced. Based on the risk map, adaptation strategies were proposed for disaster prevention and mitigation planning.

The hypotheses of this study include that (1) the minimum unit for flood risk assessment is village. It means that flood risk level in a village is same; (2) no engineering works is executed to reduce the flood hazard; and (3) the social vulnerability for different rainfall return periods is also same.

2. Data Collection

This study collected rainfall data at eight rain gauge stations, flow data at the Wulilin Bridge flow station, and water level data at the Changruu water level station (see Figure 1a). The collection period of the rainfall, flow, and water level data was from August 2012 to March 2018.

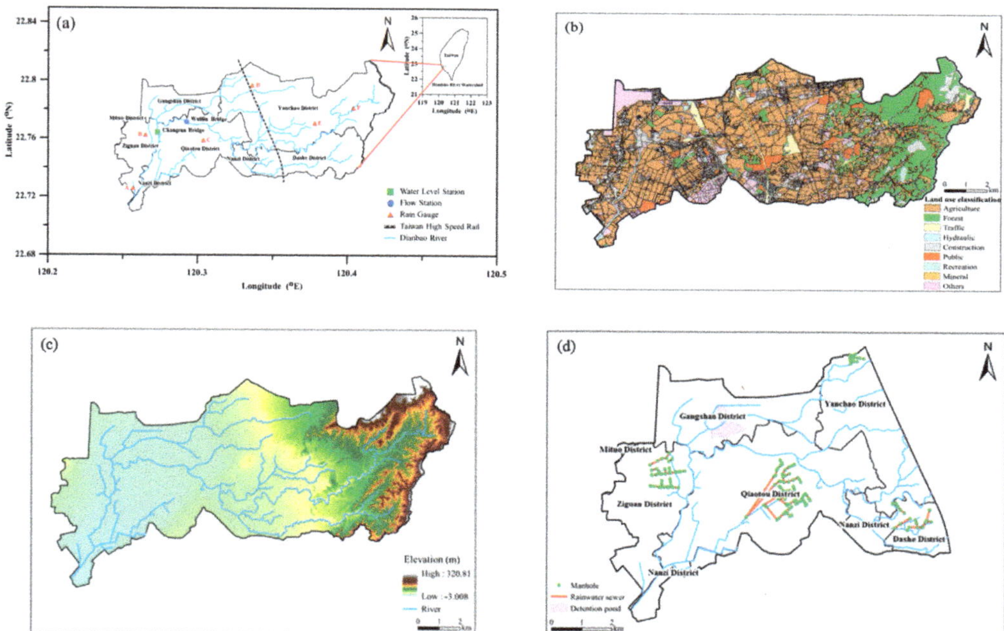

Figure 1. (**a**) The study area in the Dianbao River watershed and the urban region, (**b**) land use classification, (**c**) elevation, and (**d**) rainwater sewer system and detention ponds.

The land use data were obtained from the Taiwan Experimental Watershed Information Platform established by the Taiwan Typhoon Flood Research Institute and the Water Resources Agency. The land use types can be classified as agriculture, forest, traffic, hydraulic, construction, public, recreation, mineral, or others (Figure 1b).

For the elevation data of the Dianbao River watershed, 20-m DTM data released from the information disclosure platform of the Ministry of the Interior were adopted (Figure 1c). A model was utilized to generate grids using DTM data to obtain the numerical elevation of the terrain, which was further used to determine the accuracy of the model simulations. The cross-sectional river data were collected from the Taiwan Experimental Watershed Information Platform.

In urban areas, cities are frequently flooded due to heavy rainfall events with short durations and the rapid collection of water resulting from urban development. Therefore, the construction of rainwater sewer systems and flood detention ponds would help urban areas drain excessive rainwater quickly. In the Dianbao River watershed, there are flood detention ponds in zones A and B, located on both banks of the Daliao drainage area. The area of the flood detention pond in zone A is 17 hectares, and the flood detention volume of this pond is approximately 425,000 m^3, while the area of the flood detention pond in Area B is 42 hectares and its flood detention volume is approximately 1.05 million m^3. In this study, the data of rainwater sewers and flood detention ponds in the middle and lower reaches of the Dianbao River were taken from the Taiwan Experimental Watershed Information Platform (Figure 1d).

Apart from the data collection used for the rainfall-runoff model and inundation model, data related to social vulnerability were also collected for further analysis.

3. Materials and Methods

3.1. Study Area

The boundaries of the Dianbao River watershed in Kaohsiung city border the Agongdian River watershed in the north, the Houjin River drainage basin in the south, the hilly area of the Central Mountain Range in the east, and Mituo District in the west. The basin originates from the top of Wushan Mountain in Yanchao District, Kaohsiung city. The elevation is approximately 320 m, and the river flows westward through Yanchao District, Dashe District, Nanzi District, Qiaotou District and Gangshan District and finally into the Taiwan Strait on the south side of the Oziliao fishing port in Ziguan District [34]. Most of the Dianbao River watershed consists of plains. The main roads passing through the area are high-speed highways, longitudinal railways, other highways and roads, such as county roads. The road network is quite dense, and the traffic is convenient for users. This study uses the Taiwan High Speed Rail as a boundary to simulate rainfall-runoff with the HEC-HMS model in the upstream watershed area in the east and to simulate the flooding area with the FLO-2D in the west (Figure 1a).

The average annual rainfall in the watershed of the Dianbao River is approximately 1825 mm [35]. The wet season is mainly from June to August, and the dry season is from October of a given year to March of the following year.

3.2. Methodological Outline

The procedure followed in this research includes two parts: hazards and vulnerability. Before starting the research, fundamental data, including terrain and land classification, urban drainage networks, basin geography, meteorological and hydrological observations, population, social economy, rescue equipment, and special agency data, were gathered. In terms of hazards, the research steps were to establish the hydrological and inundation models, perform model calibration and validation, predict the flooding extents and depths under different rainfall return periods, and finally complete the hazard map. In terms of social vulnerability, the research steps were to select social vulnerability factors, design questionnaires, apply the fuzzy Delphi method to determine the main factors affecting social vulnerability, use the AHP to determine the weights of the main social vulnerability factors, calculate the social vulnerability index (SVI), and finally establish a social vulnerability map. According to these two maps, the risk map was built. The research outline is illustrated in Figure 2.

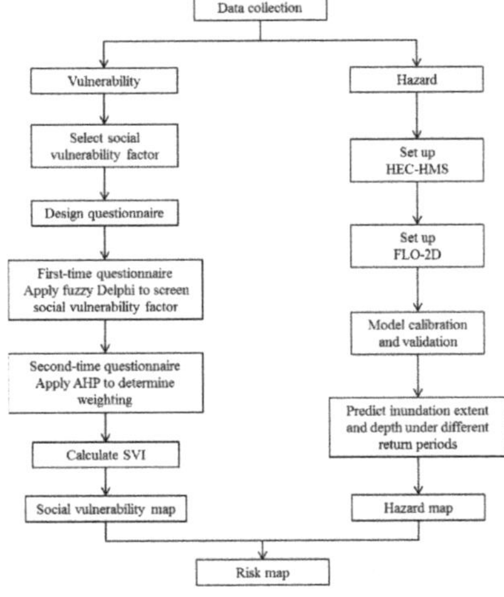

Figure 2. Outline of the flood risk assessment methodology.

3.3. Hazard Assessment with Hydrological and Hydrodynamic Models

In recent years, the impacts of climate change have caused extreme rainfall events, which have resulted in urban areas being unable to discharge internal waters, leading to flooding. This study uses the hydrological model HEC-HMS and the inundation model FLO-2D to simulate an inundation map of urban areas, serving as the hazard map.

3.3.1. Hydrological Model (HEC-HMS)

The rainfall-runoff model used in this study, HEC-HMS software, was developed by the Hydrological Engineering Center of US Army Corps Engineers. The HEC-HMS model, which was modified from HEC-1, is a new version used for hydrological simulations [36]. In this study, HEC-HMS version 4.3 was employed to calculate the rainfall-runoff discharge in the watershed [36]. The model was designed to be applied to different geographic areas, such as in studies of river basin flood hydrology and large and small watershed/urban runoff [37–43]. The HEC-HMS includes three modules: the basin model, meteorological model, and control specification. In the basin model, the US Soil Conservation Service (SCS) curve number (CN) was used to calculate precipitation loss. The input data include the initial loss, CN, and impermeability. The HEC-GeoHMS toolbox can be adopted to calculate the average CN for each subwatershed. The SCS synthetic unit hydrograph method was applied to convert rainfall to runoff discharge. Two parameters, the time of concentration (T_C) and storage coefficient (R), were used in the Clark unit hydrograph method. The time of concentration (T_C) and the regional value (RV), which is related to the time of concentration and the storage coefficient, are expressed as follows [44,45]:

$$T_C = 2.27 \times 10^{-4} \frac{L^{0.8}}{S^{0.5}} \left(\frac{1000}{CN} - 9\right)^{0.7} \tag{1}$$

$$RV = \frac{R}{R + T_C} \tag{2}$$

where L denotes the length of the main channel (m) and S expresses the slope of the watershed (%).

The time lag (T_{lag}) then can be determined as follows:

$$T_{lag} = 0.6 * T_C \tag{3}$$

The recession base flow method, which uses an exponentially decayed algorithm, was employed to separate the base flow. In a rainfall event, the recession base flow strongly affects the volume and timing of the base flow. In this study, the recession constant and ratio to the peak were set in the ranges of 0.2–0.35 and 0.01–0.1, respectively.

The rainfall method was used to establish a meteorological model in this study. In the module, the time-series rainfall data of each subcatchment and the total rainfall of each subcatchment are given. The main parameters to be set in the control specification are the start time and end time of the simulation and the time step. In this study, the computational time step was set to one hour.

3.3.2. Inundation Model (FLO-2D)

FLO-2D model is generally adopted for studies on flood routing. The model is currently certified by the Federal Emergency Management Agency. It is mainly used in urban flood simulations, flood plain management, engineering risk designs, and debris flow simulations. FLO-2D is a combination of a one-dimensional river channel unsteady flow model and a two-dimensional horizontal overland flow model [46]. It is used to simulate the depth and velocity of a two-dimensional fluid flow and to estimate the reasonable disaster range of a fluid. The detailed governing equations can be found in Dimitriadis et al. [47] and Hu et al. [48].

The FLO-2D model divides the DEM (digital elevation model) into fixed-size square grids. Each grid can be input with a surface elevation value, Manning roughness coefficient, infiltration coefficient, area reduction factor, outflow, inflow, and other hydrological and geological data to simulate the real flood flow in the study area. This model starts with basic two-dimensional surface overland flow, and many data parameters can be added; users can freely choose to turn on or off the additional functions, including the presence of river channels, dikes, roads, streets, urban area reduction, rainfall, debris flow, infiltration, and rainwater sewers.

3.3.3. Establishment of HEC-HMS and FLO-2D Models

To save computational time, the simulation domain of the FLO-2D model is smaller than that of the HEC-HMS model; therefore, connecting these two modeling domains is necessary. Figure 3a illustrates the connection of the HEC-HMS and FLO-2D models. It shows that the whole domain of the Dianbao River watershed is used for HEC-HMS, while the urban area is adopted for FLO-2D.

Using geomorphological analysis with the input function of the HEC-HMS model and the output function of HEC-GeoHMS can achieve the purpose of automatically building watersheds. The Dianbao River watershed was built for the HEC-HMS model as shown in Figure 3b and includes 7 subcatchment units, 16 channel units, and 16 confluence point units.

To simulate the inundation extent and depth of urban areas, a numerical grid was established in the FLO-2D model. The DTM data were imported into the FLO-2D model, and using the grid generation module, a grid with a size of 20 m × 20 m in the horizontal plane was generated. The new version of FLO-2D has been enhanced and can be combined with the EPA-SWMM model. Therefore, the grid system calculations can be used to simulate the drainage of rainwater sewer nodes and the overflow of full pipes. Figure 3c displays the grid generation of the FLO-2D model, which contains a total of 142,608 grids.

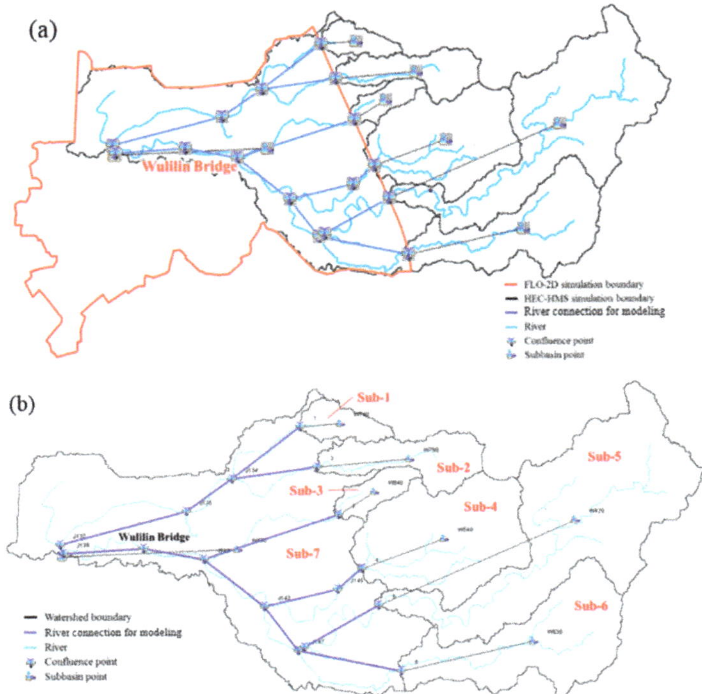

Figure 3. *Cont.*

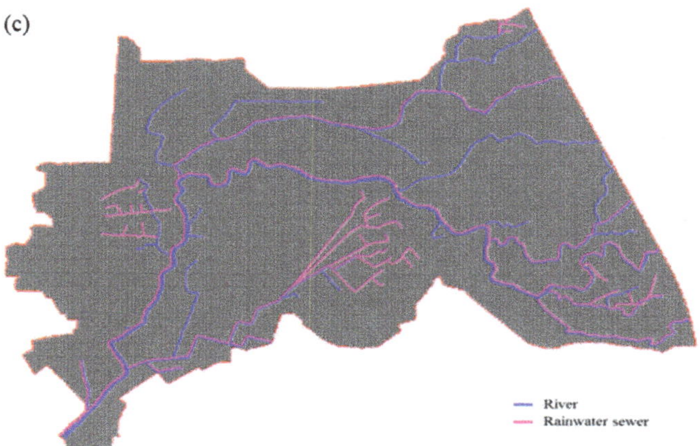

Figure 3. (**a**) Modeling connection for HEC-HMS and FLO-2D in the study area, (**b**) model setup for HEC-HMS, and (**c**) grid established for the FLO-2D model simulations.

3.4. Fuzzy Delphi (FD)

The Delphi method proposed by Dalkey and Helmer [49] is a procedural method used to express expert opinions in a systematic manner. However, the Delphi method has some shortcomings; it is time-consuming, and the cost of collecting expert opinions can be high. The so-called consistency of expert opinions only falls within a certain range of expert opinions but does not determine the point of impact. Therefore, this law implies ambiguity, so the final results cannot truly reflect the opinions of experts. To resolve these shortcomings, the fuzzy Delphi method was developed. Murray et al. [50] first combined the fuzzy set with Delphi, and then Ishikawa [51] proposed the max-min method and the fuzzy integration method to integrate the opinions of experts into fuzzy numbers, which is called the fuzzy Delphi (FD) method. In the first stage of the expert questionnaire, the method proposed by Jeng [52] was adopted to select important social vulnerability factors.

The expert questionnaire is mainly filled out based on the judgment of each expert's professional literacy and the importance of each evaluation factor. The evaluation scale is divided into levels from 0–10. The higher the evaluation score is, the more representative the factor is of the research trend. An option column was added at the end of the questionnaire. If the experts felt that there were deficiencies, their opinion can be supplemented, making the questionnaire more complete and in line with the meaning of the fuzzy Delphi questionnaire. The consensus value can be yielded by triangular fuzzy number method [50]. The results using FD to obtain important social vulnerability factors can be found in Section 4.4.1.

3.5. Analytic Hierarchy Process (AHP)

Saaty [53] proposed the analytic hierarchical process (AHP) method. Through the establishment of a hierarchical structure with mutual influence, the AHP can be used to solve multiple evaluation criteria and decision-making problems associated with uncertainty; through quantitative judgments and comprehensive evaluation, the AHP can provide decision-makers with sufficient information to choose appropriate options and reduce the risk of making mistakes. The purpose of the development of the hierarchical analysis method is to systematize complex issues, provide hierarchical decomposition from different levels, organize the levels in a quantitative manner, and comprehensively evaluate them to provide decision-makers with information with which they can choose appropriate solutions.

The AHP builds complex multi-objective decision-making problems into hierarchical systems with dendritic structures; each layer is composed of different elements. The complex problems are changed from high-level problems to low-level problems as they are gradually decomposed, and each level in the framework only affects one other level and, at the same time, is only affected by one other level. Before establishing the AHP framework, one thing must be confirmed: each factor must be independent of the other factors. The hierarchical structure method can be used to confirm a hierarchical relationship, but there is no specific construction method in the actual construction process. The AHP uses the characteristic vector method to calculate the weights of the elements and obtains the priority value of each scheme. The larger the value is, the higher the priority of the adopted scheme is.

Therefore, this study adopts the AHP method to design expert questionnaires, which are completed by experts in related fields, and the weights of each vulnerability factor are determined based on the results of the questionnaire. The results using AHP to obtain the weight of social vulnerability factors can be found in Section 4.4.2.

3.6. Social Vulnerability Index (SVI) and Normalization

This study uses the FD method to select and screen out the vulnerability factors and then uses the AHP method to determine the weight of each vulnerability factor. In terms of calculating vulnerability, this research refers to the social vulnerability index proposed by Shaw et al. [54] to identify the vulnerability of a certain area. Then, the vulnerability data are standardized based on this index. Since the values of the variables are different and the units are also different, to compare different variables with each other, it is necessary to standardize all the statistical data obtained before the vulnerability index is calculated. The index is calculated in an average manner. The standardized formula is expressed as follows:

$$z - score = Z = \alpha \frac{\sum_{i=1}^{n} (x_i - M)}{SD} \tag{4}$$

$$SD = \sqrt{\sum_{i=1}^{n} \frac{(x_i - M)^2}{n-1}} \tag{5}$$

where x denotes the statistics of different variables, M represents the mean value $(=\frac{1}{n}\sum_{i=1}^{n} x_i)$, n expresses the total number of variables, SD indicates the standard deviation, and α is the correction factor of each vulnerability factor (=+1 or −1). The correction factor ($\alpha = -1$) decreases the level of social vulnerability, representing a positive effect on social vulnerability, whereas the corrector factor ($\alpha = +1$) increases the level of social vulnerability, denoting a negative effect on social vulnerability. The social vulnerability index (SVI) can be expressed as follows:

$$SVI = \sum_{j=1}^{N} k_j Z_j \tag{6}$$

where k is the weight value, Z denotes the standardized value, and N indicates the total number of vulnerability factors.

The next step is to normalize the SVI value and convert it to a value between 0 and 1 ($=I_{SVI}$). When the I_{SVI} value is closer to 1, the vulnerability is higher; in contrast, when the I_{SVI} value is closer to 0, the vulnerability is lower. The normalization formula can be represented as follows:

$$I_{SVI} = \frac{SVI - SVI_m}{D} \tag{7}$$

where SVI_m represents the minimum SVI value and D denotes the full range of SVI values, which is the maximum SVI value minus the minimum SVI value.

After calculating the I_{SVI} value and dividing the social vulnerability into 5 grades, including extremely low, low, medium, high, and extremely high (see Table 1) based on every 0.2 increment [55], ArcGIS software was utilized to draw the social vulnerability map.

Table 1. Grades and levels of the social vulnerability index (*SVI*) normalization (I_{SVI}) values, inundation depths, and risk values.

Degree of Vulnerability/Hazard/Risk	Social Vulnerability Level/Hazard Level/Risk Level	I_{SVI} Value	Inundation Depth (m)	Risk Value
Extreme low	1	0.0–0.2	<0.3	1, 2, 3, 4
Low	2	0.2–0.4	0.3–0.5	5, 6, 8, 9
Medium	3	0.4–0.6	0.5–1.0	10, 12
High	4	0.6–0.8	1.0–3.0	15, 16
Extreme high	5	0.8–1.0	>3.0	20, 25

3.7. Hazard Assessment

Urban flooding disasters are mainly caused by heavy rainfall. When rainfall exceeds the rainfall volume that the drainage system of an area can load, flooding disasters occur, causing casualties and property losses and indirectly leading to the suspension of commercial activities and classes in schools. In this study, the simulated inundation depth is used as the hazard assessment index, and the inundation depth is divided into 5 grades: below 0.3 m, 0.3 m to 0.5 m, 0.5 m to 1.0 m, 1.0 m to 3.0 m and above 3.0 m (see Table 1). When the inundation depth is 0.3 m, the water depth is approximately the height of a child's knees, so it is difficult for children to escape. When the inundation depth is 0.5 m, the water depth is approximately the height of an adult's knees, which causes adults to have difficulty walking. When the inundation depth is 1.0 m, the water depth is approximately the height of the waist of an adult, so it indeed causes difficulties for adults walking. When the inundation depth is 3.0 m, the water depth is as high as a building is tall, which threatens human lives.

3.8. Risk Assessment

Different definitions of risk have been documented by various researchers [56,57]. Generally, risk can be defined as the product of hazard (*H*) and social vulnerability (*V*) in Taiwan [58]. The simple equation to describe the risk is expressed as follows:

$$Risk = H \times V \tag{8}$$

Because hazards and social vulnerability are each graded in five levels, the approach of a semiquantitative risk assessment can be defined as the risk (Water Resources Agency, 2015). The risk value does not represent what actually happened but only stands for the relative relationships between districts. Figure 4 depicts the 5 × 5 semiquantitative matrix used for the risk assessment. Based on the matrix, the classifications of the risk assessment can also be graded, as shown in Table 1. Different risk values can be attributed to different degrees of risk.

Hazard (H) \ Vulnerability (V)	Extreme low (1)	Low (2)	Medium (3)	High (4)	Extreme high (5)
Extreme low (1)	1	2	3	4	5
Low (2)	2	4	6	8	10
Medium (3)	3	6	9	12	15
High (4)	4	8	12	16	20
Extreme high (5)	5	10	15	20	25

Figure 4. 5 × 5 matrix for the semiquantative map used for the risk assessment.

3.9. Indices of Model Performance

To ascertain the simulated discharge values using the HEC-HMS model and the simulated water level using the FLO-2D model, three indices, flood peak error percentage (EY_p), flood peak arrival time error (ET_p), and Nash-Sutcliffe efficiency coefficient (NSE), were employed for model performance. These indices can be calculated as follows:

$$|EY_p| = \left|\frac{(Y_p)_{sim} - (Y_p)_{obs}}{(Y_p)_{obs}}\right| \times 100\% \tag{9}$$

$$|ET_p| = \left|(T_p)_{sim} - (T_p)_{obs}\right| \tag{10}$$

$$NSE = 1 - \frac{\sum\limits_{i=1}^{N}[Y_{obs}(t_i) - Y_{sim}(t_i)]^2}{\sum\limits_{i=1}^{N}[Y_{obs}(t_i) - \overline{Y}_{obs}]^2} \tag{11}$$

where N denotes the total number of observational/simulation data, $(Y_p)_{obs}$ and $(Y_p)_{sim}$ represent the observed and simulated peak discharge/water level, respectively, $(T_p)_{obs}$ and $(T_p)_{sim}$ indicate the observed and simulated peak arrival time for the discharge/water level, respectively, $Y_{obs}(t_i)$ and $Y_{sim}(t_i)$ denote the observed and simulated discharge/water level at time t_i, respectively, and $\overline{Y}_{obs}$ is the average value of the observations ($=\frac{1}{N}\sum\limits_{i=1}^{N} Y_{obs}(t_i)$).

The NSE value ranges from negative infinity to 1, and when the NSE is close to 1, it indicates that the model is of good quality and that the model has high reliability. If the NSE value is close to 0, the simulation result is close to the average level of the observed values; that is, the overall result is credible, but the error in the simulation process is large. If the NSE value is far less than 0, the model is not credible.

4. Results

To ascertain the availability of the models, two typhoon events, Typhoon Kongrey, which occurred from August 27 to 2 September 2013, and Typhoon Megi, which hit central Taiwan in the period of 25–28 September 2016, were adopted for the calibration and validation of the models, respectively. The validated models, HEC-HMS and FLO-2D, were then applied to predict inundation extents and depths, and these values were used to evaluate the hazard indices for different return periods.

4.1. HEC-HMS Model Calibration and Validation

To ensure the correctness of the rainfall runoff simulated by the HEC-HMS model, calibrating and validating the HEC-HMS model are important procedures. The observed discharges at the Wulilin Bridge flow station were compared with the simulated results.

Figure 5a displays the comparison between the observed and simulated discharges at the Wulilin Bridge flow station during Typhoon Kongrey (2013); this comparison was used for the model calibration. The figure shows that the simulated results reproduced temporal variations in the discharge. Twin observed and simulated peak discharges were selected for the analysis of statistical errors. The EY_p values for peak discharge 1 and peak discharge 2 were 0.6% and 19.3%, respectively, while the ET_p values were 1 h and 0 h, respectively. The NSE value was 0.89.

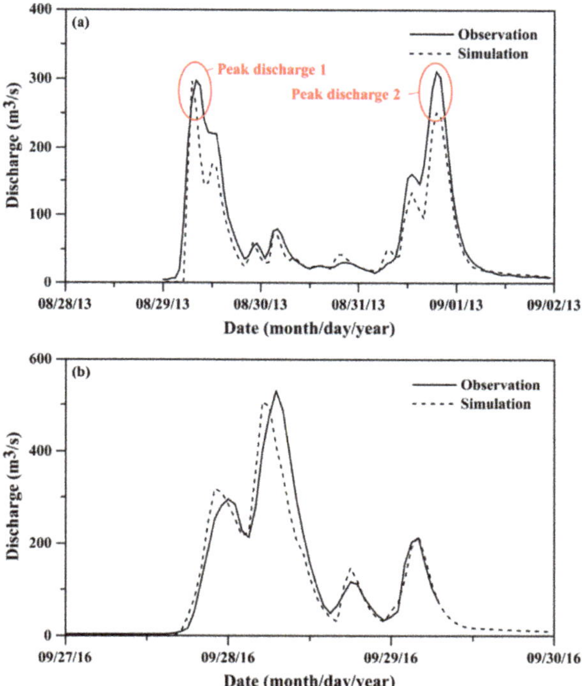

Figure 5. Comparison of the simulated and observed discharges at the Wulilin Bridge (**a**) during Typhoon Kongrey (2013), used for the model calibration, and (**b**) during Typhoon Megi (2016), used for the model validation.

A comparison of the simulated and observed discharges during Typhoon Megi (2016), used for the model validation, is shown in Figure 5b. The figure indicates that the computed water levels faithfully reproduced the observations. Based on the analysis of statistical errors, the EY_p and ET_p values were 4.6% and 2 h, respectively. The NSE value was 0.91.

4.2. FLO-2D Model Calibration and Validation

The observed water levels at the Changrun Bridge flow station during Typhoon Kongrey (2013) and Typhoon Megi (2016) were used to calibrate and validate the FLO-2D model, respectively. A comparison of the simulated and observed water levels during Typhoon Kongrey, used for the model calibration, is presented in Figure 6a. The figure shows twin peaks in the water level as a result of runoff discharge. The statistical errors represented by the EY_p values at peak water level 1 and peak water level 2 were 13.5% and

23.1%, respectively, while the ET_p values were 1 h and 4 h, respectively. The simulated water level indicated an underestimation of the observed water level at peak water level 1, while the simulation overestimated the observation at peak water level 2. The NSE value was 0.69.

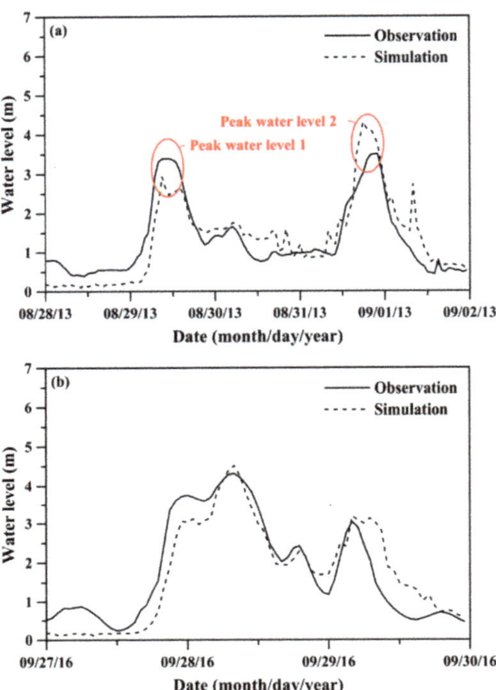

Figure 6. Comparison of the simulated and observed water levels at the Changrun Bridge (**a**) during Typhoon Kongrey (2013), used for the model calibration, and (**b**) during Typhoon Megi (2016), used for the model validation.

Figure 6b illustrates a comparison between the computed and observed water levels during Typhoon Megi, as used for the model validation. The computed water levels matched the tendencies of the observations. The EY_p and ET_p values were 4.7% and 0 h, respectively, and the NSE value was 0.79. The model performance observed during the model validation was better than that observed during the model calibration.

4.3. Inundation Disaster and Hazard Map

4.3.1. Inundation Depths and Extents under Different Return Periods

To obtain rainfall information under different return periods, this study used historical rainfall data from eight rainfall stations in the catchment area from 2012 to 2017 for frequency analysis, applied the Horner equation to establish the intensity-duration-frequency curve [43], and finally employed the alternating group method to obtain the designated rainfall of each rainfall station under different return periods. The designated rainfall was then input into the validated HEC-HMS model to obtain the runoff discharges in the upstream catchment area; then, the runoff discharges served as inputs into the validated FLO-2D model, and inundation disasters under different return periods regarded as hazards were obtained. For example, the analysis result for the rainfall hyetograph under a 200-year return period is illustrated in Figure 7.

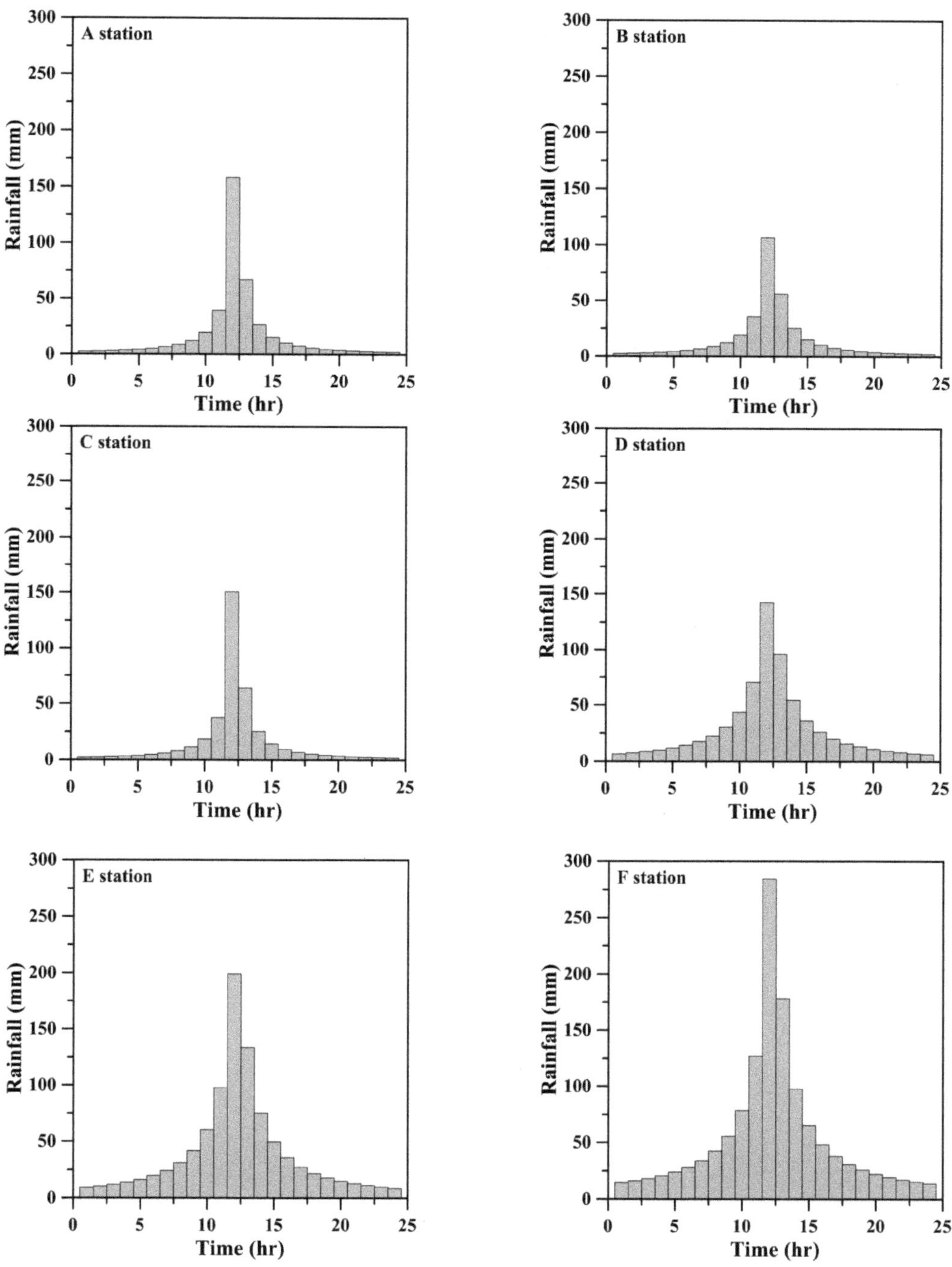

Figure 7. Rainfall hyetograph under a 200-year return period at stations (**A–F**) used as inputs for the HEC-HMS model.

The simulated results of the inundation depths and extents for 50-, 100-, and 200-year rainfall return periods are displayed in Figure 8. It can be seen from the figure that the flooding extent gradually expands from the 50-year rainfall return period to the 200-year rainfall return period, and the flooding depth gradually deepens. The maximum flooding depths under the 50-, 100-, and 200-year rainfall return periods are 5.50 m, 5.97 m, and 6.37 m, respectively. The maximum flooding site is located in the Yanchao District, resulting from the low-lying area around the riverside in the middle reach. There are 9 villages in the 50-year rainfall return period, 11 villages in the 100-year rainfall return period, and 12 villages in the 200-year rainfall return period with inundation depths greater than 3 m. Because most flooding zones are located downstream of the areas of the river where the flooding depth exceeds 0.3 m, the possible impacts on hazards are not negligible.

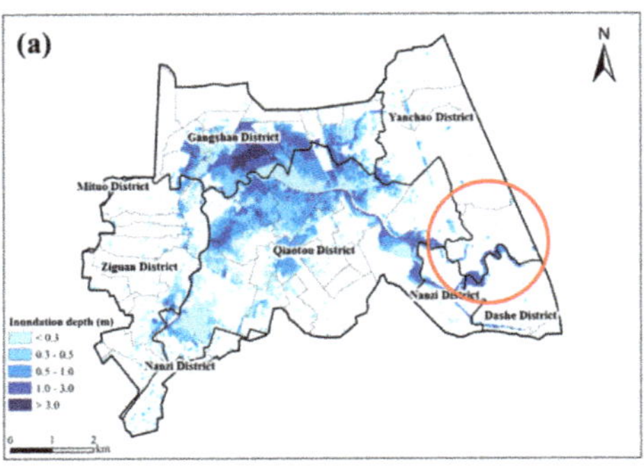

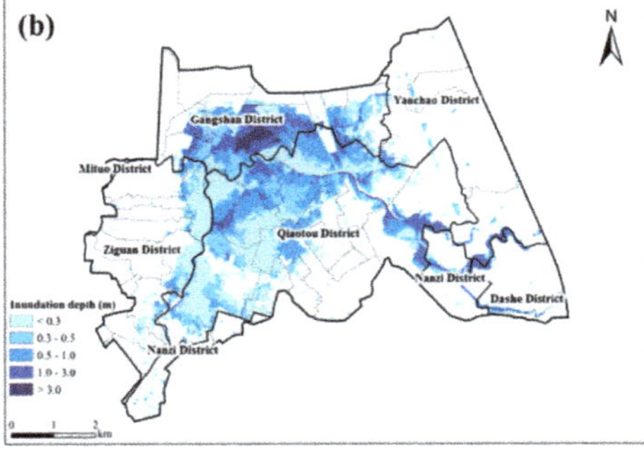

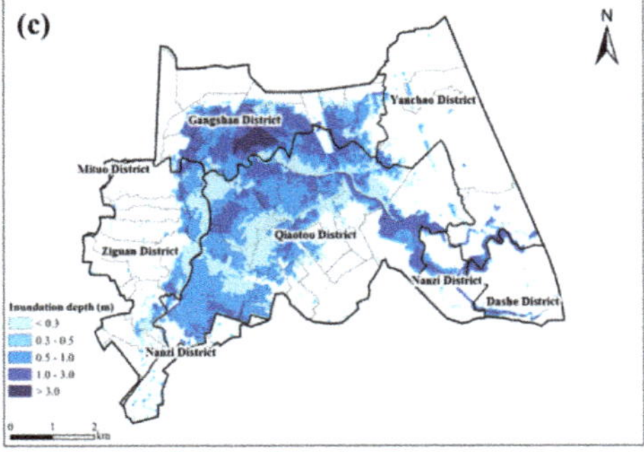

Figure 8. Inundation predictions under (**a**) 50-year, (**b**) 100-year, and (**c**) 200-year rainfall return periods. Note that the maximum flooding site is indicated with a circle.

4.3.2. Inundation Depths and Extents under Different Return Periods

The inundation depths and extents were converted to hazard maps according to 5 grades: below 0.3 m, 0.3 m to 0.5 m, 0.5 m to 1.0 m, 1.0 m to 3.0 m, and above 3.0 m (see Table 1). Figure 9 presents hazard maps under the 50-, 100-, and 200-year rainfall return periods. The figure shows that there are 7 villages in Qiaotou District, 2 villages in Gangshan District, and one village each in Dashe, Nanzi, and Yanchao Districts belonging to the high and extremely high hazard levels under different rainfall return periods. Under the 200-year rainfall return period, the hazard grades of 12 villages increase to the extremely high level.

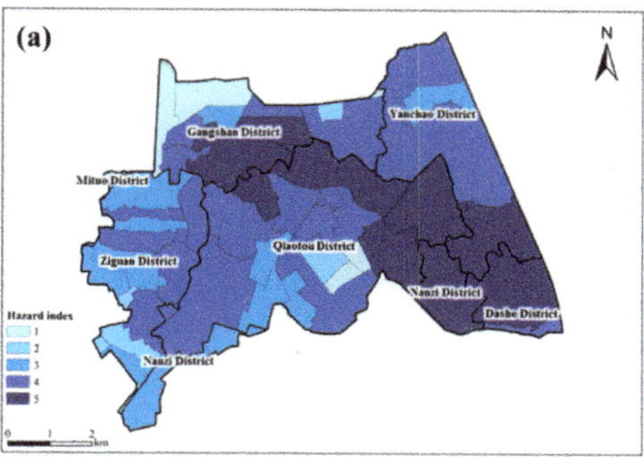

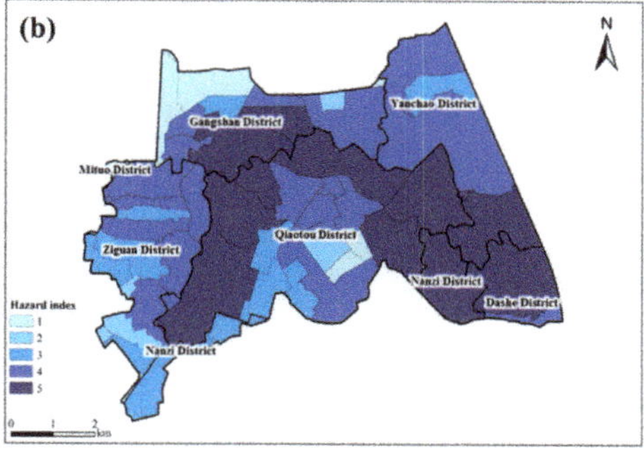

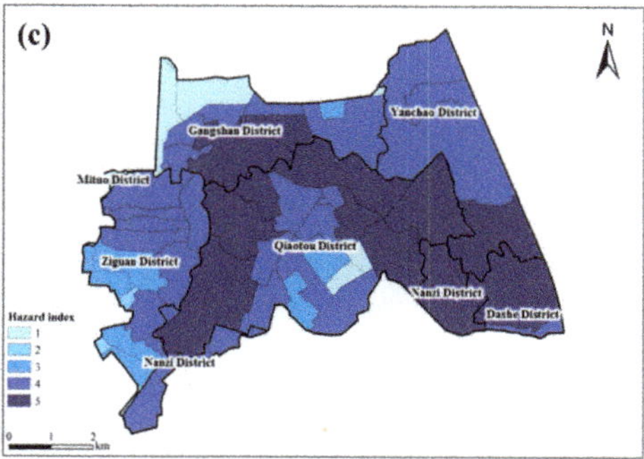

Figure 9. Hazard map under (**a**) 50-year, (**b**) 100-year, and (**c**) 200-year rainfall return periods.

4.4. Analysis of Social Vulnerability and the Social Vulnerability Map

This study used two questionnaires to calculate social vulnerability. The first was the fuzzy Delphi questionnaire, which was adopted to screen the vulnerability factors. Through this questionnaire, experts were asked to identify the main factors affecting social vulnerability in urban areas. The second was the AHP questionnaire. The main purpose of this questionnaire was to set the weight value of each selected social vulnerability factor.

Because some factors had a greater degree of influence than others, it was necessary to determine the weight value of each factor through the AHP method.

4.4.1. Process with Fuzzy Delphi (FD)

Four assessment elements, which included twenty social vulnerability factors, were selected for the expert questionnaire, as shown in Table 2. Through the 41 collected expert questionnaires and the FD screen, the social vulnerability factors with consensus values higher than 6.09 were selected. Therefore, ten main social vulnerability factors, including individuals over 65 years old, individuals under 14 years old, elderly individuals living alone, individuals with disabilities, rubber boats, mobile pumps, firefighters, refuge shelters, schools (kindergarten, elementary, and high schools), and nursing homes for elderly individuals (see Table 2), were selected for further APH analysis.

Table 2. Social vulnerability factors and consensus values according to the expert questionnaire.

Assessment Element	Social Vulnerability Factor	Consensus Value
Demographic characteristic	Over 65 years old	6.89
	Under 14 years old	7.6
	Low-income household	5.47
	Elderly living alone	6.89
	Disability	6.89
	Aboriginal	5.5
	New immigrants (foreign spouse)	4.5
	Agricultural population	5.69
	Illiterate	3.89
	University graduate	4.8
Social economy	Household income	5.09
	Total value of building	4.5
Rescue equipment	Rubber boat	7.6
	Mobile pump	7
	Firefighter	7.6
Special agency	Refuge shelter	7.6
	School (kindergarten, elementary, and high school)	6.8
	Nursing home for the elderly	6.32
	Dialysis center	6
	Homeless shelter	5.19
Mean consensus value		6.09

4.4.2. Process with the AHP

After completing the fuzzy Delphi questionnaire and selecting the main vulnerability factors, an AHP questionnaire was conducted to determine the weight of each vulnerability factor. A total of 40 questionnaires were completed and returned by experts. Table 3 shows the weights of the vulnerability factors and the cascading weights of the hierarchy. The sum of the cascading weights of the hierarchy is 1.0. The table also indicates that experts believe that being a firefighter is the most important social vulnerability factor. From the AHP method, the weight of each factor can be determined, and the SVIs of urban areas can be evaluated.

Table 3. Weights of the social vulnerability factors.

Assessment Element (A)	Social Vulnerability Factor	Weighting (B)	Cascade Weighting of Hierarchy (C = A × B)
Demographic characteristic (0.3957)	Over 65 years old	0.1812	0.0717
	Under 14 years old	0.2170	0.0859
	Elderly living alone	0.2675	0.1058
	Disability	0.3343	0.1323
Rescue equipment (0.2084)	Rubber boat	0.2276	0.0679
	Mobile pump	0.2780	0.0830
	Firefighter	0.4944	0.1476
Special agency (0.3059)	Refuge shelter	0.3316	0.1014
	School (kindergarten, elementary, and high school)	0.3168	0.0969
	Nursing home for the elderly	0.3516	0.1076

4.4.3. Social Vulnerability Map

According to Equations (4)–(6), the social vulnerability index (SVI) can be obtained. To normalize the factors according to Equation (7), the SVI value was converted to the I_{SVI} value. Since five social vulnerability levels were graded based on their I_{SVI} values (see Table 1), the social vulnerability map was built, as illustrated in Figure 10. The map shows that the social vulnerability in the Nanzi District is the highest, at level 5 (i.e., extremely high vulnerability). The main reason for this result is that there are more nursing homes for elderly individuals, individuals over 65 years old, and individuals under 14 years old in this region than in other regions, resulting in the highest social vulnerability. The social vulnerability values of the two villages in Qiaotou District are both level 3. This is due to there being more nursing homes for elderly individuals in this district than in other districts.

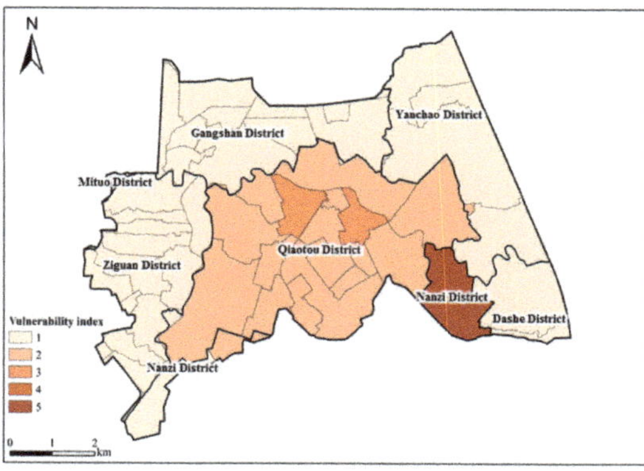

Figure 10. Social vulneriability map.

Social vulnerability values express relative, comparative relationships and do not represent actual vulnerability values. A village with a higher degree of vulnerability means that the village comprises a disadvantaged group with a denser population or fewer social resources. Therefore, when a disaster occurs, more serious damage and losses would occur in a village with a higher degree of vulnerability.

4.5. Risk Map

After completing the hazard map under different rainfall return periods and the social vulnerability map, a semiquantitative risk assessment approach was employed to construct a risk map. The map of risks under different rainfall return periods is displayed in Figure 11. It shows that six villages in Qiaotou District and one village in Nanzi District belong to medium risk and extremely high risk levels, respectively, under the 50-year rainfall return period. There are seven villages and one village in Qiaotou District that are at medium risk and high risk levels, respectively, under the 200-year rainfall return period. It can be noted that the hazard risk under the 200-year rainfall return period is more serious than that under the 50-year rainfall return period, resulting in increased risks under the 200-year rainfall return period. However, the risk levels of most villages in the districts consist of extremely low and low levels.

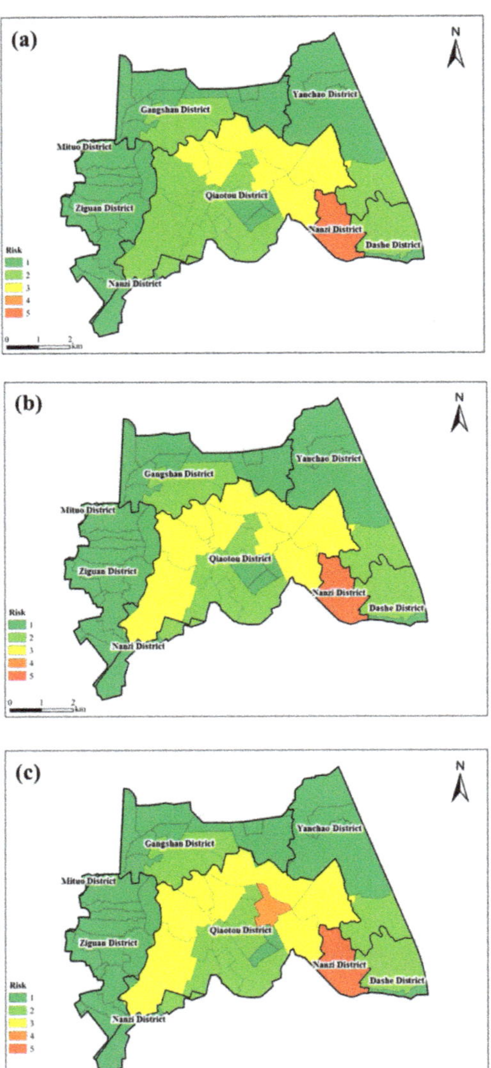

Figure 11. Risk map under (**a**) 50-year, (**b**) 100-year, and (**c**) 200-year rainfall return periods.

5. Discussion

Flood-related disasters in urban areas are subject to several factors, such as urban development, construction of transportation facilities, extreme weather and climate change, and anthropogenic changes that alter vulnerability, hazards, and risks [59]. To control the impacts of disasters caused by extreme rainfall events, an adaptation strategy must be adopted to reduce these disasters. To cope with the impacts of extreme rainfall events on urban areas, this study formulated adaptation strategies for villages with high risks levels. Generally, two measures, including nonstructural measures and structural measures, are recommended to mitigate disasters [25,60].

According to the flood risk assessment, the social vulnerability and hazards under different rainfall return periods reach extremely high levels, resulting in extremely high risk levels yielded for a village in Nanzi District. This village possesses a large population over 65 years old and under 14 years old as well as many nursing home facilities for elderly individuals. This study suggests that when planning escape routes with nonstructural measures, complex routes should be avoided, and the routes should be as simple and clear as possible to avoid dispersion in the escape process. In addition, the inundation depths are relatively high at this village, so it is recommended that a flood detention pond is added with structural measures to relieve the flooding load of the channel and reduce the flood disasters.

The social vulnerability of a village in Qiaotou District is calculated to be at level 3 (i.e., medium level), and the risk in this village for each return period is level 3 for the 50-year return period, level 3 for the 100-year return period, and level 4 for the 200-year return period. The number of mobile pumps in this area is lower than those of others area; therefore, this study proposes that mobile pumps are purchased in this area to reduce the damage caused by flooding.

The method used to reduce the social vulnerability index is a nonstructural measure, while the method employed to reduce hazards is a structural measure. To reduce the risk of flooding, it is necessary to first examine the sources of high risk from social vulnerability or hazards, as measures taken to reduce the risk must be adapted to the local conditions. In any case, the social vulnerability map, the hazard map, and the risk map can provide decision-makers with a reference that can be considered for risk management.

Most studies that have discussed hazard assessments applied a combination of hydrologic and hydrodynamic models for flood simulations [27,29,54,61]. Additionally, there are different research methods used to discuss vulnerability [23,33,62]. Further, two expert questionnaires were used to determine the main social vulnerability factors and the weights of the social vulnerability factors in the study area. This kind of research method is rarely proposed or discussed, but is specific and feasible, and the methodology can be applied to other urban areas affected by floods.

The limitation of this study is that because 20-m DTM data are available, the regional flooding simulation cannot display the flooded state of the street and because the data of the social vulnerability factor are not detailed enough, so the flood risk map can only be presented in the village.

6. Conclusions

Typhoon-induced extreme rainfall frequently results in flood damage that threatens the lives and property of people and causes economic losses in urban areas in Taiwan. Apart from structural measures that prevent flood damage, risk assessments play important roles in disaster management. To explore the flood risk map for disaster mitigation, the hypotheses of this study include the flood risk level in a village to be same, no engineering works executed to reduce hazard level, and social vulnerability for different rainfall return periods also to be same.

The flood risk assessment conducted in this study was established based on rainfall-runoff (HEC-HMS) and flooding simulation (FLO-2D) models applied for the hazard analysis and the fuzzy Delphi (FD) method and analytic hierarchy process (AHP) applied for the social vulnerability analysis. The HEC-HMS and FLO-2D models were calibrated and validated with observational data. The results indicated that the model performances during both the model calibration and validation were acceptable. The validated HEC-HMS and FLO-2D models were utilized to predict inundation depths and extents in urban areas under different rainfall return periods, and these factors were used to build hazard maps.

Furthermore, to explore social vulnerability, first, the fuzzy Delphi method was used to select the main social vulnerability factors, and then the AHP was adopted to determine the weight of each social vulnerability factor and to calculate the SVI to build a social vulnerability map.

By means of a semiquantitative risk matrix, the social vulnerability map and hazard map were combined into a risk map, and the risk map was divided into 5 levels to comprehend the risk status of urban areas in the Dianbao River watershed under different rainfall return periods.

Most studies did not clearly state that how the social vulnerability was selected for flood risk analysis. The main contribution of this study was that the hydrological-hydrodynamic model, the fuzzy Delphi, and the analytic hierarchy process were proposed to build the flood risk map in urban areas of southern Taiwan. Based on the flood risk map, the nonstructural measures and structural measures were recommended to mitigate disasters according to the regional disaster characteristics.

Due to the impacts of climate change, urban flooding caused by extreme rainfall is worsening. In future research, we will explore a risk assessment of urban flood disasters caused by climate change.

Author Contributions: Conceptualization, W.-C.L.; methodology, T.-H.H. and H.-M.L.; software, T.-H.H.; validation, W.-C.L., T.-H.H., and H.-M.L.; formal analysis, W.-C.L.; investigation, W.-C.L.; resources, W.-C.L.; data curation, T.-H.H. and H.-M.L.; writing—original draft preparation, W.-C.L.; writing—review and editing, W.-C.L. and H.-M.L.; visualization, T.-H.H.; supervision, W.-C.L.; project administration, W.-C.L.; funding acquisition, W.-C.L. All authors have read and agreed to the published version of the manuscript.

Funding: This research was funded by Ministry of Science and Technology, Taiwan, grant nos. 107-2625-M-239-002 and 108-2625-M-239-002.

Institutional Review Board Statement: The study did not require ethical approval.

Informed Consent Statement: Not applicable.

Data Availability Statement: Not applicable.

Acknowledgments: The authors would like to express their appreciation to the Water Resources Agency for providing access to their recorded data.

Conflicts of Interest: The authors declare no conflict of interest.

References

1. Intergovernmental Panel on Climate Change. *Managing the Risks of Extreme Events and Disasters to advance Climate Change Adaption*; Cambridge University Press: Cambridge, UK; New York, NY, USA, 2012.
2. Centre for Research on the Epidemiology of Disaster; United Nations Office for Disaster Risk Reduction. *The Human Cost of Weather-Related Disasters 1995–2015*; CRED: Brussels, Belgium; UNISDR: Geneva, Switzerland, 2015.
3. United Nations Office for Disaster Risk Reduction. *Sendai Framework for Disaster Risk Reduction 2015–2030*; UNISDR: Geneva, Switzerland, 2015.
4. Aerts, J.C.J.H.; Botzen, W.J.; Clarke, K.C.; Cutter, S.L.; Hall, J.W.; Merz, B.; Michel-Kerjan, E.; Mysiak, J.; Surminski, S.; Kunreuther, H. Integrating human dynamics into flood disaster risk assessment. *Nat. Clim. Chang.* **2018**, *8*, 193–199. [CrossRef]
5. Yang, T.H.; Liu, W.C. A general overview of the risk-reduction strategies for floods and droughts. *Sustainability* **2020**, *12*, 2687. [CrossRef]
6. UNISDR. *Global Assessment Report on Disaster Risk Reduction: Revealing Risk, Redefining Development. Summary and Main Findings*; United Nations: New York, NY, USA, 2011.
7. Jenkins, K.; Surminski, S.; Hall, J.; Crick, F. Assessing surface water flood risk and management strategies under future climate change. *Sci. Total Environ.* **2017**, *595*, 159–168. [CrossRef]
8. Kron, W. Flood risk = hazard·values·vulnerability. *Water Int.* **2005**, *30*, 58–68. [CrossRef]
9. European Commission. *The EU Flood Directive*; Directive 2007/60/EC of the European Parliament and of the Council on the Assessment and Management of Flood Risk; European Comission: Brussels, Belgium, 2007.

10. Apel, H.; Aronica, G.T.; Kreibich, H.; Thieken, A.H. Flood risk analyses-how detailed do we need to be? *Nat. Hazards* **2009**, *49*, 79–98. [CrossRef]
11. De Moel, H.; Aerts, J.C.J.H.; Koomen, E. Development of flood exposure in the Netherlands during the 20th and 21th century. *Glob. Environ. Chang.* **2011**, *21*, 620–627. [CrossRef]
12. Kalogrtopoulos, K.; Karalis, S.; Karymbalis, E.; Chalkias, C.; Chalkias, G.; Katsafados, P. Modeling flash floods in Vouraikos River mouth, Greece. In Proceedings of the MEDCOAST Conference, Marmaris, Turkey, 30 October–3 November 2013; pp. 1135–1146.
13. Tsanakas, K.; Gaki-Papansdtassiou, K.; Kalogeropoulos, K.; Chalkias, C.; Katsafado, P.; Karymbalis, E. Investigation of flash flood natural causes of Xirolaki Torrent, Northern Greece based on GIS modeling and geomorphological analysis. *Nat. Hazards* **2016**, *84*, 1015–1033. [CrossRef]
14. Bathrellos, G.D.; Karymbalis, E.; Skilodimou, H.D.; Gaki-Papanastassiou, K.; Baltas, E.A. Urban flood hazard assessment in the basin of Athens Metropolitan city, Greece. *Environ. Earth Sci.* **2016**, *75*, 319. [CrossRef]
15. Garrote, J.; Diez-Herrero, A.; Escudero, C.; Garcia, I. A framework proposal for regional-scale flood-risk assessment of cultural heritage sites and application to the Castile and Leon Region (Central Spain). *Water* **2020**, *12*, 329. [CrossRef]
16. Glas, H.; De Maeyer, P.; Merisier, S.; Deruyter, G. Development if a low-cost methodology for data acquisition and flood risk assessment in the floodplain of the river Moustiques in Haiti. *J. Flood Risk Manag.* **2020**, *13*, e12608. [CrossRef]
17. Grezo, H.; Mocko, M.; Izsoff, M.; Vrbicanova, G.; Petrovic, F.; Stranak, J.; Muchova, Z.; Slamova, M.; Olah, B.; Machar, I. Flood risk assessment for the long-term strategic planning considering the placement of industrial parks in Slovakia. *Sustainability* **2020**, *12*, 4144. [CrossRef]
18. Guo, E.; Zhang, J.; Ren, X.; Zhang, Q.; Sun, Z. Integrated risk assessment of flood disaster based on improved set pair analysis and the variable fuzzy set theory in central Liaoning Province, China. *Nat. Hazards* **2014**, *74*, 947–965. [CrossRef]
19. Waghwala, R.K.; Agnihotri, P.G. Flood risk assessment and resilience strategies for flood risk management: A case study of Surat City. *Int. J. Disaster Risk Red.* **2019**, *40*, 101155. [CrossRef]
20. Joo, H.; Choi, C.; Kim, J.; Kim, D.; Kim, S.; Kim, H.S. A Bayesian network-based integrated for flood risk assessment (InFRA). *Sustainability* **2019**, *11*, 3733. [CrossRef]
21. Zhang, J.; Chen, Y. Risk assessment of flood disaster induced by typhoon rainstorms in Guangdong Province, China. *Sustainability* **2019**, *11*, 2738. [CrossRef]
22. Lai, C.; Chen, X.; Wang, Z.; Yu, H.; Bai, X. Flood risk assessment and regionalization from past and future perspectives at basin scale. *Risk Anal.* **2020**, *40*, 1399–1417. [CrossRef]
23. Wang, G.; Liu, Y.; Hu, Z.; Lyu, Y.; Zhang, G.; Liu, J.; Liu, Y.; Gu, Y.; Huang, X.; Zheng, H.; et al. Flood risk assessment based fuzzy synthetic evaluation method in the Beijing-Tianjin-Hebei metropolitan area, China. *Sustainability* **2020**, *12*, 1451. [CrossRef]
24. Mani, P.; Chatterjee, C.; Kumar, R. Flood hazard assessment with multiparameter approach derived from coupled 1D and 2D hydrodynamic flow model. *Nat. Hazards* **2014**, *70*, 1553–1574. [CrossRef]
25. Van Berchum, E.C.; Mobley, W.; Jonkman, S.N.; Timmermans, J.S.; Kwakkel, J.H.; Brody, S.D. Evaluation of flood risk reduction strategies through combination of interventions. *J. Flood Risk Manag.* **2018**, *12* (Suppl. 2), e12506. [CrossRef]
26. Cai, T.; Li, X.; Ding, X.; Wang, J.; Zhan, J. Flood risk assessment based on hydrodynamic model and fuzzy comprehensive evaluation with GIS technique. *Int. J. Disaster Risk Red.* **2019**, *35*, 101077. [CrossRef]
27. Criado, M.; Martinez-Grana, A.; Roman, J.S.S.; Santos-Frances, F. Flood risk evaluation in urban spaces: The study case of Tormes River (Salamanca, Spain). *Int. J. Environ. Res. Public Health* **2019**, *16*, 5. [CrossRef] [PubMed]
28. Shen, Y.; Morsy, M.M.; Huxley, C.; Tahvildari, N.; Goodall, J.L. Flood risk assessment and increased resilience for coastal urban watersheds under the combined impact of storm tide and heavy rainfall. *J. Hydrol.* **2019**, *579*, 124159. [CrossRef]
29. Apollonio, C.; Bruno, M.F.; Iemmolo, G.; Molfetta, M.G.; Pellicani, R. Flood risk evaluation in ungauged coastal areas: The case study of Ippovampo (Southern Italy). *Water* **2020**, *12*, 1466. [CrossRef]
30. Lowe, R.; Arnbjerg-Nielsen, K. Urban pluvial flood risk assessment-data resolution and spatial scale when developing screening approaches on the microscale. *Nat. Hazards Earth Syst. Sci.* **2020**, *20*, 981–997. [CrossRef]
31. Cabrera, J.S.; Lee, H.S. Flood risk assessment for Davo Oriental in the Philippines using geographic information system-based multi-criteria analysis and maximum entropy model. *J. Flood Risk Manag.* **2020**, *13*, e12607. [CrossRef]
32. Zhang, D.; Shi, X.; Xu, H.; Jing, Q.; Pan, X.; Liu, T.; Wang, H.; Hou, H. A GIS-based spatial multi-index model for flood risk assessment in the Yangtze River Basin, China. *Environ. Impact Assess. Rev.* **2020**, *83*, 106397. [CrossRef]
33. Geng, Y.; Zheng, X.; Wang, Z.; Wang, Z. Flood risk assessment in Quzhou City (China) using a coupled hydrodynamic model and fuzzy comprehensive evaluation (FCE). *Nat. Hazards* **2020**, *100*, 133–149. [CrossRef]
34. Chang, C.Y. The Integration of HEC-RAS and FLO-2D Model for Inundation Simulation in the Dianbao River Watershed. Master's Thesis, National Cheng Kung University, Tainan, Taiwan, 2013.
35. Water Resources Agency. *The Disaster Prevention Monitoring and Model Test Base Observation in the Dianbao River and Ilan River in 2016*; Final Report; Water Resources Agency: Taichung, Taiwan, 2016.
36. US Army Corps Engineers. *Hydrologic Modeling System (HEC-HMS) Application Guide: Version 4.3*; Institute for Water Resources—Hydrologic Engineering Center: Davis, CA, USA, 2018.
37. Halwatura, D.; Najim, M.M.M. Application of the HEC-HMS model for runoff simulation in a tropical catchment. *Environ. Model. Softw.* **2013**, *46*, 155–162. [CrossRef]

38. Nourali, M.; Ghahraman, B.; Pourreza-Bilondi, M.; Davary, K. Effect of formal and informal likelihood functions on uncertainty assessment in a single event rainfall-runoff model. *J. Hydrol.* **2016**, *540*, 549–564. [CrossRef]
39. Gumindoga, W.; Rwasoka, D.T.; Nhapi, I.; Dube, T. Ungauged runoff simulation in Upper Manyame Catchment, Zimbabwe: Application of the HEC-HMS model. *Phys. Chem. Earth Parts A/B/C* **2017**, *100*, 371–382. [CrossRef]
40. Young, C.C.; Liu, W.C.; Wu, M.C. A physically based machine learning hybrid approach for accurate rainfall-runoff modeling during extreme typhoon events. *Appl. Soft Comput.* **2017**, *53*, 205–216. [CrossRef]
41. de Moraes, T.C.; dos Santos, V.J.; Calijuri, M.L.; Torres, F.T.P. Effects on runoff caused by changes in land cover in Brazilian southeast basin: Evaluation by HEC-HMS and HEC-GEOHMS. *Environ. Earth Sci.* **2018**, *77*, 250. [CrossRef]
42. Yuan, W.; Liu, M.; Wan, F. Calculation of critical rainfall for small-watershed flash flood based on the HEC-HMS hydrological model. *Water Resour. Manag.* **2019**, *33*, 2555–2575. [CrossRef]
43. Ramly, S.; Tahir, W.; Abdullah, J.; Jani, J.; Ramli, S.; Asmat, A. Flood estimation for SMART control operation using integrated radar rainfall input with the HEC-HMS model. *Water Resour. Manag.* **2020**, *34*, 3113–3127. [CrossRef]
44. Chow, V.T.; Maidment, D.R.; Mays, L.W. *Applied Hydrology*; McFraw-Hill Book Company: New York, NY, USA, 1988.
45. Straub, T.D.; Melching, C.S.; Kocher, K.E. *Equations for Estimating Clark Unit Hydrograph Parameters for Small Rural Watersheds in Illinois*; Water-Resources Investigations Report 00-4184; Illinois Department of Natural Resources—Office of Water Resources U.S. Geological Survey: Urbana, IL, USA, 2000; pp. 4–6.
46. Flo-2D Manual. Reference Manuals. 2012. Available online: http://www.flo-2d.com/download (accessed on 20 July 2019).
47. Dimitriadis, P.; Tegos, A.; Oikonomou, A.; Pagana, V.; Koukouvinos, A.; Mamassis, N.; Koutsoyiannis, D.; Efstratiadis, A. Comparative evaluation of 1D and quasi-2D hydraulic models based on benchmark and real-world applications for uncertainty assessment in flood mapping. *J. Hydrol.* **2016**, *534*, 478–492. [CrossRef]
48. Hu, M.; Sayama, T.; Zhang, X.; Tanaka, K.; Takara, K.; Yang, H. Evaluation of low impact development approach for mitigating flood inundation at a watershed scale in China. *J. Environ. Manag.* **2017**, *193*, 430–438. [CrossRef] [PubMed]
49. Dalkey, N.C.; Helmer-Hirschberg, O. An experimental application of the Delphi method to the use of experts. *Manag. Sci.* **1963**, *9*, 458–467. [CrossRef]
50. Murray, T.J.; Pipion, L.L.; van Gigch, J.P. A pilot study of fuzzy set modification of Delphi. *Hum. Syst. Manag.* **1985**, *5*, 76–80. [CrossRef]
51. Ishikawa, A.; Amagasa, M.; Shiga, T.; Tomizawa, G.; Tatsuta, R.; Mieno, H. The max-min Delphi method and fuzzy Delphi method via fuzzy integration. *Fuzzy Sets Syst.* **1993**, *55*, 241–253. [CrossRef]
52. Jeng, S.J. Fuzzy Assessment Method for Maturity of Software Organization in Improving Its Staff's Capability. Master's Thesis, National Taiwan University of Science and Technology, Taipei, Taiwan, 2001.
53. Saaty, T. *The Analytic Hierarchy Process*; McGraw-Hill: New York, NY, USA, 1980.
54. Shaw, D.G.; Li, H.C.; Yang, H.H. *Disaster Loss and Social Vulnerability Assessment*; Technical Report; National Science and Technology Center for Disaster Reduction: New Taipei City, Taiwan, 2010.
55. Lin, T.H. Flood Hazard and Risk Analysis for Kaohsiung City. Master's Thesis, National Taiwan University, Taipei, Taiwan, 2015.
56. Maskrey, A. *Disaster Mitigation: A Community Based Approach*; Oxfam: Oxford, UK, 1989.
57. Crichton, D.; Mounsey, C. How the insurance industry will use its flood research. In Proceedings of the 33rd MAFF Conference of River and Coastal Engineers. Keele University, Staffordshire, UK, 1–3 July 1998; pp. 131–134.
58. Water Resources Agency. *Research, Development and Value-Added Application of Flood Potential Risk Map Update*; Technical Report; Water Resources Agency: Taichung, Taiwan, 2015.
59. Adikari, Y.; Osti, R.; Noro, T. Flood-related disaster vulnerability: An impending crisis of megacities in Asia. *J. Flood Risk Manag.* **2010**, *3*, 185–191. [CrossRef]
60. Mai, T.; Mushtaq, S.; Reardon-Smith, K.; Webb, P.; Stone, R.; Kath, J.; An-Vo, D. Defining flood risk management strategies: A systems approach. *Int. J. Disaster Risk Reduct.* **2020**, *47*, 101550. [CrossRef]
61. Sarchani, S.; Seiradakis, K.; Coulibaly, P.; Tsanis, I. Flood inundation mapping in an ungauged basin. *Water* **2020**, *12*, 1532. [CrossRef]
62. Zou, Q.; Liao, L.; Qin, H. Fast comprehensive flood risk assessment based on game theory and cloud model under parallel computation (P-GT-CM). *Water Resour. Manag.* **2020**, *34*, 1625–1648. [CrossRef]

Article

The Spatiotemporal Distribution of Flash Floods and Analysis of Partition Driving Forces in Yunnan Province

Junnan Xiong [1], Chongchong Ye [1], Weiming Cheng [2,*], Liang Guo [3], Chenghu Zhou [2] and Xiaolei Zhang [3]

1 School of Civil Engineering and Architecture, Southwest Petroleum University, Chengdu 610500, China; neu_xjn@163.com (J.X.); yechongchong107319@163.com (C.Y.)

2 State Key Laboratory of Resources and Environmental Information System, Institute of Geographic Sciences and Natural Resources Research, CAS, Beijing 100101, China; zhouch@lreis.ac.cn

3 China Institute of Water Resources and Hydropower Research, Beijing 100038, China; guol@iwhr.com (L.G.); zhangxl@iwhr.com (X.Z.)

* Correspondence: chengwm@lreis.ac.cn; Tel.: +86-010-64889777

Received: 10 April 2019; Accepted: 20 May 2019; Published: 23 May 2019

Abstract: Flash floods are one of the most serious natural disasters, and have a significant impact on economic development. In this study, we employed the spatiotemporal analysis method to measure the spatial–temporal distribution of flash floods and examined the relationship between flash floods and driving factors in different subregions of landcover. Furthermore, we analyzed the response of flash floods on the economic development by sensitivity analysis. The results indicated that the number of flash floods occurring annually increased gradually from 1949 to 2015, and regions with a high quantity of flash floods were concentrated in Zhaotong, Qujing, Kunming, Yuxi, Chuxiong, Dali, and Baoshan. Specifically, precipitation and elevation had a more significant effect on flash floods in the settlement than in other subregions, with a high r (Pearson's correlation coefficient) value of 0.675, 0.674, 0.593, 0.519, and 0.395 for the 10 min precipitation in 20-year return period, elevation, 60 min precipitation in 20-year return period, 24 h precipitation in 20-year return period, and 6 h precipitation in 20-year return period, respectively. The sensitivity analysis showed that the Kunming had the highest sensitivity (S = 21.86) during 2000–2005. Based on the research results, we should focus on heavy precipitation events for flash flood prevention and forecasting in the short term; but human activities and ecosystem vulnerability should be controlled over the long term.

Keywords: flash flood; driving factor; sensitivity analysis; subregion of landcover; Yunnan Province

1. Introduction

Flash floods are one of the most severe natural disasters to try to prevent and deal with in the aftermath. They are responsible for loss of life and serious destruction to property and infrastructure, severely affecting a region's economic development [1–3]. According to an investigation by the World Meteorological Organization, the loss of property resulting from flash floods ranks in the top 10 among a range of natural disasters in 75% of countries [4]. As an economically developed country, flash floods in the United States ranked first in causes of death, with approximately 100 lives lost each year [5]. In addition, a total of 28,826 flash floods occurred between 2015 and 2017, and 10% of these flash floods resulted in property losses exceeding $100,000 (US dollars) per flash flood disaster [6]. In Europe, 40% of flood-related casualties during 1950–2006 were due to flash floods [7], which is already more than 80% in southern Europe [8]. China is one of the countries that experiences the most flash floods, and about 4.63 million km^2 of land tend to be impacted by flash floods, which have threatened 560 million people [3]. Yunnan is one of the most important areas of ecological value in the world [9] where flash floods are rapidly increasing, causing serious threat to people's lives and property [10].

In recent years, the vast majority of studies conducted on flash floods have indicated that flash floods are the combined result of various spatiotemporal factors [11–13]. Existing studies on flash floods have focused primarily on the following aspects: (1) the risk assessment [14–16]; (2) the occurrence, development, and influence studied from the perspective of a disaster mechanism [3,17,18]; and (3) the spatial–temporal distribution and influencing factors [19–21]. Further, risk assessment of flash floods has been used to classify a region as low risk, mid risk, or high risk and this is primarily based on the county and province scale [22]. The formation of flash floods is very complicated and has led to many such studies being carried out only in typical watersheds [23]. Regarding the spatiotemporal pattern and driving factors of flash floods, most of the research uses the global spatial autocorrelation, kernel density estimation, hot spot (Getis-Ord Gi*) analysis, standard deviational ellipse, and spatial gravity center migration to explore the spatial and temporal characteristics of flash floods [24–26]. The driving force analysis of flash floods is primarily based on geographical detectors, and the driving factors selected by the researchers have included precipitation, terrain, and human activities, et al., which were dominated by static indices [27–29]. Such research generally selects the province (or country) as the research object.

Some research has been conducted on the spatiotemporal distribution and driving factors of flash floods, although some problems still remain. First, Yunnan has a complex number of characteristics, including its mountain range, human activity, precipitation, and ecosystems. In such complicated conditions, the characteristics of flash floods there remain unknown. Second, Yunnan is one of the most important ecological areas in the world, and the species diversity has been acknowledged by the International Union and the World Wildlife Fund for the Conservation of Nature [9]. There are differences in the weather, landform, and background conditions in each subregion of landcover. As a result, the nonpartitioned approach no longer applies in such a complex ecosystem region. Additionally, the study area is divided into different subregions of landcover, which reflect well the extent of environmental damage caused by human activities. For example, the human activity in the settlement is significantly greater than in other subregions. Third, previous studies have explored how driving factors interact based on two factors that use geographical detectors [30,31], which cannot reflect the contribution rate of multiple factor interactions on the flash floods. Finally, most studies have revealed the influence of human activity on the occurrence of flash floods [32,33], but the response of flash floods on economic development is not well understood in Yunnan. Hence, it is necessary to explore the spatiotemporal variations and driving factors (in different subregions of the landcover, which include grassland, settlement, farmland, and forest) of flash floods and to conduct a sensitivity analysis of the response of flash floods on economic development in Yunnan Province.

The objectives of this study were to (1) measure the spatial–temporal variation of flash floods using the changepoint, kernel density estimation, spatial mismatch analysis, standard deviational ellipse (SDE), and spatial gravity center model; (2) analyze the driving factors for the spatial pattern of flash floods in different subregions of the landcover using the Pearson correlation coefficient, multiple linear regression, and principal component analysis; and (3) conduct a sensitivity analysis to investigate the response of flash floods on economic development. The results can provide references for the prevention of flash floods.

2. Materials and Methods

2.1. Study Area

Yunnan Province (21°–29° N, 97°–106° E) is located in southwestern China and is situated on the eastern edge of the collision zone between the Eurasian and Indian continental plates. This region is characterized by some of the most complicated and neotectonic activity and tectonic movements on the Chinese mainland (Figure 1) [34]. Yunnan follows the spatial extent of the land area for this region and covers roughly 394,100 km^2; the population was 47.7 million at the end of 2016 (http://www.yn.gov.cn/yn_yngk/gsgk/index.html). Economic activities are gradually gaining momentum, and the yearly economy has exceeded 1.63 trillion yuan (RMB; http://www.stats.gov.cn/) since 2017 in Yunnan. Although Yunnan's economy has made significant progress, the level of urbanization remains lows, and the township density is relatively sparse [35]. Yunnan is located in the Yunnan–Guizhou Plateau; and most of the areas are dominated by high mountains, deep valleys, and dense mountain streams where vertical differentiation in heat and water occurs as a result of changes in landform and altitude [36]. In addition, the precipitation in northwestern and central Yunnan has increased in recent years [37]. All of these conditions have contributed to making this study area one of the most geophysical hazard-prone regions in China.

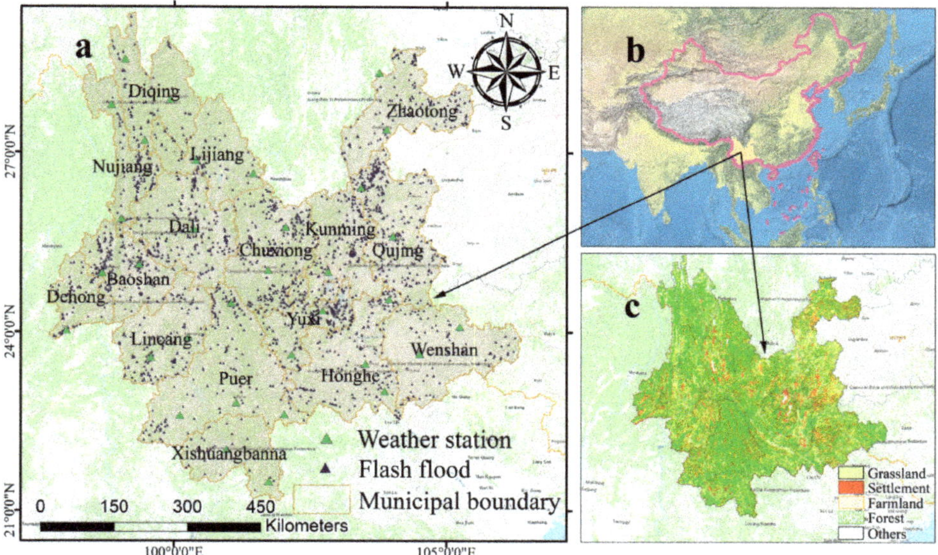

Figure 1. The study area: (**a**) the distribution of flash floods in Yunnan Province; (**b**): the geographical position of Yunnan in China; (**c**): the different subregions of landcover in Yunnan Province.

2.2. Selection and Pretreatment of Data

The flash flood data used in this research are from the National Flash Flood Investigation and Evaluation Project, which was launched in 2013 to research the historical flash floods that occurred from 1949 to 2015 on a national scale, and the disaster areas were assumed as a point for collection [29]. In Yunnan, the number of flash floods was 3166, which we derived from the above-mentioned project. Current research has shown that the occurrence of flash flood is closely related to precipitation, topography, and human influence [14,21,29]. To explore the relationship between flash floods and the driving factors of flash floods, we selected the following factors: precipitation factors (unit: mm), elevation (ELE, unit: m), topographic relief (TR, unitless), slope (SLP, unit: °), normalized difference vegetation index (NDVI, unitless), population density (PD, unit: person/km^2), and GDP (unit: yuan/km^2). The precipitation factors included the 10 min precipitation (M10_20), 60 min precipitation (M60_20), 6 h precipitation (H6_20), and 24 h precipitation (H24_20) in 20-year return period, which reflect the impact of various precipitation intensities on flash floods. In addition, the station data of annual mean precipitation used to calculate the gravity center in Yunnan Province were acquired from the National Weather Service of China (http://www.cma.gov.cn/). Table 1 shows the source and resolution of all raw data sets.

In this study, we interpolated precipitation raster data by the Inverse Distance Weighted in ArcGIS10.2 (ESRI, Inc., Redlands, CA, USA), which was based on the vector data of precipitation factors, and the final resolution was 1 km × 1 km. We calculated topographic relief using focal statistics in the ArcGIS 10.2 software, which was represented by elevation standard deviation based on the DEM data; the SLP also was generated by this data.

Table 1. Source and resolution of the raw data sets.

Factors	Source
Flash floods	China: Flash Flood Investigation and Evaluation Dataset of China, 1949–2015, 1:50,000, vector data.
Precipitation factors	China: Flash Flood Investigation and Evaluation Dataset of China, vector data.
DEM	China: Geospatial Data Cloud, 2000, 90 m × 90 m, raster data.
Population density	China: Resources and Environmental Sciences Data Center, 2000, 2005,2010, and 2015, 1 km × 1 km, raster data.
GDP	China: Resources and Environmental Sciences Data Center, 1995, 2000, 2005, 2010, and 2015, 1 km × 1 km, raster data.
Normalized difference vegetation index	USA: NASA's MODIS web site, 2001–2015, 1 km × 1 km, raster data.
Landcover	China: Global Change Research Data Publishing & Repository, 2010, 1 km × 1 km, raster data.

2.3. Methodology

2.3.1. Temporal Pattern

To explore the temporal pattern of flash floods in Yunnan, we analyzed the characteristics of year and month in this study, wherein we calculated the characteristic of year by the package "changepoint" in R software (R Core Development Team, R Foundation for Statistical Computing, Vienna, Austria), which implements various mainstream and specialized changepoint methods to find single and multiple changepoints within data [38]. The month characteristic was analyzed (from January to December) by the year characteristic.

2.3.2. Spatial Pattern

(1) Kernel density estimation

In this paper, kernel density estimation techniques were used for establishing representations for the flash floods, since the kernel function can regard this estimate as averaging the impact of a kernel function centered at the flash flood point and evaluated at each point. We calculated spatial intensity of the flash floods using a kernel density estimation:

$$\lambda_h(P) = \frac{1}{nh}\sum_{i=1}^{n} k\left(\frac{P-P_i}{h}\right) \quad (1)$$

where $\lambda_h(P)$ is the estimated spatial intensity of the flash floods, $P_1, \cdots, P_n$ are the locations of n observed flash floods, $k(.)$ is the kernel, a meristic but not necessarily positive function that unites to one, and h is the bandwidth and determines the radius of a circle centered on P that the flash flood points P_i will contribute to the intensity $\lambda_h(P)$. We calculated the spatial density of the flash floods to conduct the kernel density estimation in ArcGIS 10.2 and determined the search radius (bandwidth) based on the mean random distance (RD mean) calculations [39].

(2) Spatial mismatch analysis

Spatial mismatch analysis has a profound impact in the economy to measure misbalance between two spatially distributed factors, like unemployment and job distribution etc. In this study, the spatial distribution of number of flash floods in different time periods was taken as the input factor. We used spatial mismatch analysis to measure the imbalance between the number of flash floods in two different time periods. The formula is given as follows [40]:

$$SMI_i = \left(\frac{B_i}{\sum_{i=1}^{n} B_i} - \frac{F_i}{\sum_{i=1}^{n} F_i} \right) \times 100 \tag{2}$$

where SMI_i expresses the spatial balance of city i between the number of the flash floods in two different time periods, F_i and B_i refer to the number of flash floods during start and end time periods. To judge whether the spatial relationship of the number of flash floods in two time periods was balanced, we selected a standard value to assist us by Jenks Natural Breaks Classification, which has a minimum variance sum for each class and indicates that features in the same class are more consistent. We divided the value of SMI_i into six classes: the middle two classes were balanced, and the absolute value of the two classes was less than 1.5. If the absolute value of SMI_i was more than 1.5, it showed that the relationship between the number of flash floods in two different time periods was unbalanced in city i [41].

(3) Standard deviational ellipse

The standard deviational ellipse (SDE) has long served as a general geographic information system tool used to measure bivariate distributed characteristics. The tool typically is used to delineate the geographic distribution trend of features by summarizing both their orientation and dispersion. Since our flash flood data is presented in the form of points, this method can be used to determine the direction and trend. In ArcGIS10.2, the SDE is primarily determined by three elements: average location, dispersion (or concentration), and orientation. In addition, the SDE can be described as one, two, or three standard deviations, which can contain approximately 68%, 95%, or 99% of centroids of all input features, respectively. In this study, we selected one standard deviation [42].

(4) Spatial gravity center model

The spatial gravity center model has the advantage to research the spatial-temporal migrations of factors by analyzing the trajectory of their gravity center. We used the spatial gravity center model visually and accurately to reveal the distribution and evolution characteristics of factors in two-dimensional space [40]. To analyze the change trajectories of flash floods, precipitation, and human activities, we calculated the gravity center coordinates of the flash floods, annual mean precipitation, and population density [43]. The time interval of annual mean precipitation and population data was from 2001 to 2015, and we analyzed flash floods consistent with this interval. The gravity center coordinate of the flash floods is given as follows:

$$G_f(x) = \frac{\sum_{i=1}^{n} f_i(x)}{n}, G_f(y) = \frac{\sum_{i=1}^{n} f_i(y)}{n} \tag{3}$$

where $G_f(x)$ and $G_f(y)$ show the gravity center coordinate of annual flash floods, n is the annual number of flash floods, and $f_i(x)$ and $f_i(y)$ are the geometric coordinates of i-th flash floods ($i = 1, 2, \cdots n$) [25]. Moreover, the gravity center coordinate of the annual mean precipitation and population density are given as follows:

$$G(x) = \frac{\sum_{i=1}^{n} x_i \times k_i}{\sum_{i=1}^{n} k_i}, G(y) = \frac{\sum_{i=1}^{n} y_i \times k_i}{\sum_{i=1}^{n} k_i} \quad (4)$$

where $G(x)$ and $G(y)$ express the yearly gravity center coordinate of the factor (annual mean precipitation or population density), (x_i, y_i) is the geometric coordinate of i-th meteorological station of annual mean precipitation (or municipal administrative unit of population density), and k_i refers to the attribute value of the i-th meteorological station (or municipal administrative unit) [40].

2.3.3. Driving Force Analysis

In this study, we explored the relationship between flash floods and driving factors in subregions of the landcover. These subregions include grassland, settlement, farmland, and forest, as almost all flash floods occurred in these subregions. We generated 1000 random points in each subregion and extracted the values of the driving factors and the results of kernel density estimation for the flash floods at each random point on the spatiotemporal scale. On this basis, correlation and interaction were analyzed.

(1) Pearson correlation coefficient

The Pearson correlation coefficient is a classic mathematical method used to show the correlation between two factors x and y where the value of it ranges between (−1) and (+1). The mathematical formula is given by the following:

$$r = \frac{\sum_{i=1}^{n} x_i y_i - \frac{\sum_{i=1}^{n} x_i \sum_{i=1}^{n} y_i}{n}}{\sqrt{\left(\sum_{i=1}^{n} x_i^2 - \frac{\left(\sum_{i=1}^{n} x_i\right)^2}{n}\right)\left(\sum_{i=1}^{n} y_i^2 - \frac{\left(\sum_{i=1}^{n} y_i\right)^2}{n}\right)}} \quad (5)$$

where r is the Pearson correlation coefficient between values x and y, and n refers to the numbers of values x and y. The r values and corresponding correlation levels are shown in Table 2 [44].

Table 2. Pearson correlation coefficient and corresponding correlation levels.

r Value	Correlation Level
$r = 0$	no correlation
$0 < \lvert r\rvert < 0.2$	very weak correlation
$0.2 < \lvert r\rvert < 0.4$	weak correlation
$0.4 < \lvert r\rvert < 0.6$	intermediate correlation
$0.6 < \lvert r\rvert < 0.8$	strong correlation
$0.8 < \lvert r\rvert < 1$	very strong correlation
$\lvert r\rvert = 1$	perfect correlation

(2) Driving factors interaction on the flash floods

Different factors have different effects on the flash floods, and the same factors have different influence on the flash floods in the different subregions of the landcover. In this paper, many factors were selected to explore the driving force, there may be a correlation between the factors. To avoid the multicollinearity of driving factors and explore the interaction of driving factors on flash floods and understand which factors were dominant in the interaction, we selected principal component analysis and multiple linear regression, which enabled us to summarize and visualize the information in the driving factors.

Principal component analysis is an oldest multivariate technique and it is employed by almost all scientific disciplines. Its goal is to reflect the important information from the original data. Meanwhile, it can transform the multicollinearity of driving factors into a set of linearly independent principal components [45]. Detailed information about the principal component analysis were described by previous researches. We determined the number of principal components by the total variance, which was more than 85% [46]. In addition, we measured the contribution of the variable on the principal component by using the "FactoMineR" R package.

Multiple linear regression model is a statistical tool that can be considered as the course of fitting models to data. It can summarize data as well as investigate the relationship between input variables. In this study, the input variables were the principal components of the output of the principal component analysis. Multivariate linear regression is described in detail in reference to previously published studies [47].

2.3.4. Sensitivity analysis

To assess response of flash floods on economic development, we employed a sensitivity assessment model to calculate the flash floods' sensitivity to GDP. This method is a quantitative analysis that researches the impacts of changing GDP on the flash floods. The formula is given as follows:

$$S = \frac{[(F_{t_2} - F_{t_1})/F_{t_1}]}{[(G_{t_2} - G_{t_1})/G_{t_1}]} \tag{6}$$

where S represents the value of sensitivity; F_{t1} and F_{t2} refer to, respectively, the flash floods at the start and end of the time period; and G_{t1} and G_{t2} refer to, respectively, GDP at the start and end of the time period. We divided sensitivity into four levels according to the published studies (Table 3) [40,48].

Table 3. *S* value and corresponding sensitivity level.

S Value	Sensitivity Level	Description
$S \leq 0$	Non-sensitivity	It indicates the changes in the flash floods and economic development do not have any synchronized characteristics.
$0 < S \leq 10$	Low-sensitivity	It indicates the flash floods have low sensitivity to economic development.
$10 < S \leq 20$	Mid-sensitivity	It indicates the flash floods have medium sensitivity to economic development.
$S > 20$	High-sensitivity	It indicates the flash floods have high sensitivity to economic development.

3. Results

3.1. Temporal Pattern of the Flash Floods

Figure 2 shows the characteristics of the temporal variation of flash floods in Yunnan. The annual flash floods gradually increased from 1949 to 2015 (Figure 2b, Table 4). On the basis of the descriptions in Section 2.3.1, annual flash floods were divided into four time periods, the number of flash floods was 104, 331, 704, and 2027 during 1949–1962, 1963–1981, 1982–1995, and 1996–2015, respectively (Figure 2a). For the four time periods (1949–1962, 1963–1981, 1982–1995, and 1996–2015), a high level of monthly flash floods occurred in June (30, 53, 133, and 297 flash floods, respectively), July (31, 99, 213, and 643 flash floods, respectively), and August (34, 94, 131, and 552 flash floods, respectively), and no flash floods occurred in February (Figure 2c).

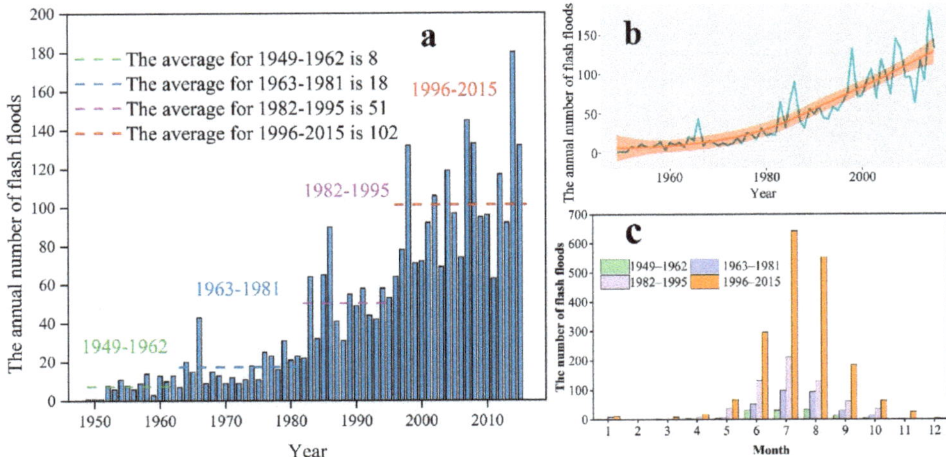

Figure 2. The characteristics of temporal variation of the flash floods: (**a**) a histogram of the number of the flash floods per year and the year interval of the flash floods based on changepoint analysis (the green dotted line is 1949–1962, the blue dotted line is 1963–1981, the purple dotted line is 1982–1995, and the red dotted line is 1996–2015); (**b**) variation trends of annual flash floods. The red line is the fitting line and the dark green line is the change line of the annual number of flash floods; and (**c**) characteristics of monthly flash floods in different time periods.

Table 4. The number of flash floods on different time scales.

Year Interval	Year	June	July	August
1949–1962	104	30	31	34
1963–1981	331	53	99	94
1982–1995	704	133	213	131
1996–2015	2027	297	643	552

3.2. Spatial Pattern of the Flash Floods

3.2.1. The Result of Kernel Density Estimation

The results of kernel density estimations of flash floods are shown in Figure 3. A relatively high number of flash floods was concentrated in Kunming (18 flash floods), Chuxiong (14 flash floods), Baoshan (10 flash floods), Yuxi (10 flash floods), and Zhaotong (10 flash floods), and the lowest level occurred in Lijiang in 1949–1962 (Figure 3a). In the 1963–1981 period, the most frequent occurrence of flash floods was in Baoshan (47 flash floods); other high-incidence areas included Kunming (32 flash floods), Dali (37 flash floods), and Yuxi (32 flash floods) (Figure 3b). Furthermore, the frequency of the flash floods has increased significantly since the 1982–1995 period. A high frequency of flash floods occurred in Qujing (113 flash floods), Dali (88 flash floods), Chuxiong (74 flash floods), and Baoshan (76 flash floods) in 1982–1995; a low frequency was concentrated in Nujiang (7 flash floods) and Xishuangbanna (6 flash floods) (Figure 3c). In general, the number of the flash floods in Yunnan reached its maximum in the 1996–2015 period. The areas with more than 200 flash floods included Qujing (275 flash floods) and Yuxi (201 flash floods), and the lowest number occurred in Xishuangbanna (30 flash floods) (Figure 3d).

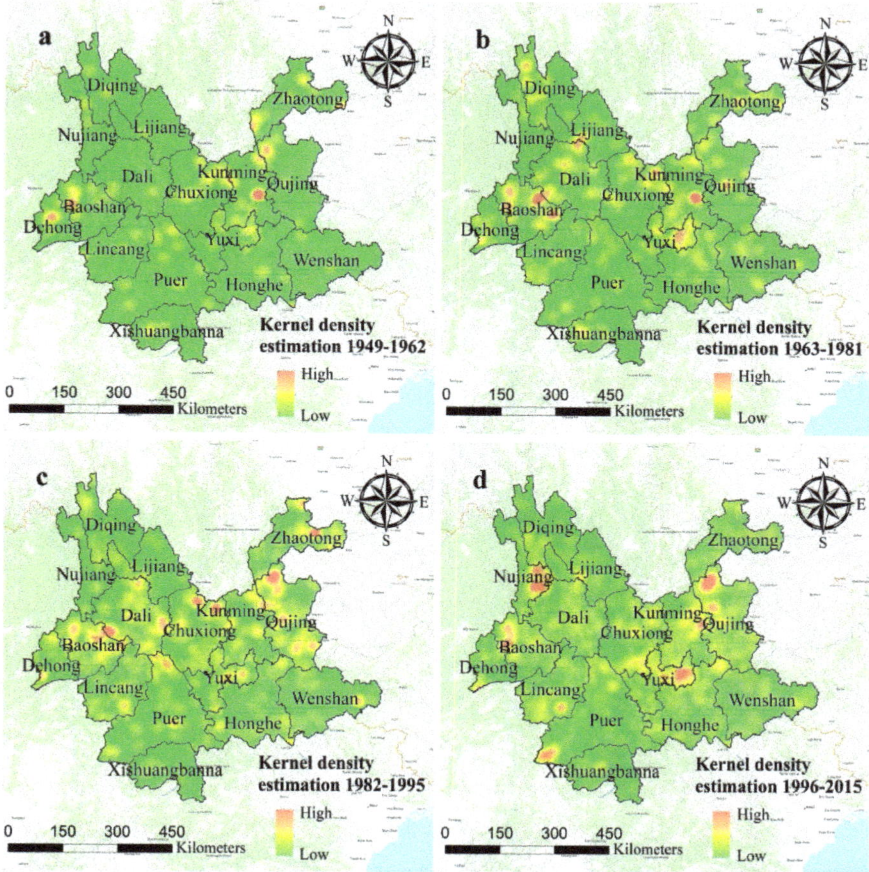

Figure 3. Kernel density estimation of the flash floods for the four time periods: (**a**) 1949–1962, (**b**) 1963–1981, (**c**) 1982–1995, and (**d**) 1996–2015.

3.2.2. Results of Spatial Mismatch Analysis and SDE

The regions with spatial unbalance were distributed mainly in Zhaotong, Qujing, Kunming, Chuxiong, Puer, Dali, Diqing, and Baoshan from the 1949–1962 to the 1963–1981 time period (Figure 4a). Similarly, standard deviational ellipses showed that the orientation and trends were located in these regions. From the 1963–1981 to the 1982–1995 time periods, the regions with an unbalanced frequency of flash floods had extended to the southern areas, which included Lincang and Xishuangbanna (Figure 4b). Additionally, the orientation and trend of standard deviational ellipses also were located in these areas. Fewer unbalanced areas, however, were evident from the 1982–1995 to the 1996–2015 time periods (Figure 4c), and the flash floods mainly occurred in these unbalanced areas (Zhaotong, Qujing, Kunming, Yuxi, Dali, Nujiang, and Baoshan). Overall, the orientation and trend of standard deviational ellipses were consistent with the unbalanced areas.

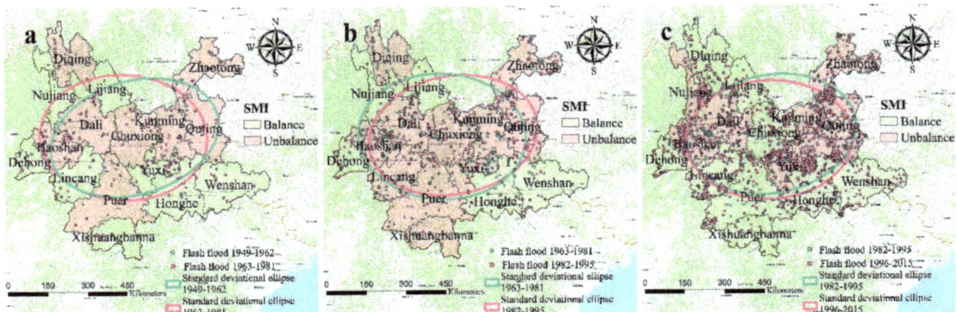

Figure 4. The spatial pattern of spatial mismatch index and standard deviational ellipse at the start and end time periods: (**a**) from the 1949–1962 to the 1963–1981 time period, (**b**) from the 1963–1981 to the 1982–1995 time period, and (**c**) from the 1982–1995 to the 1996–2015 time period. Green dots and standard deviational ellipses indicate the start time period, and red indicates the end time period.

3.2.3. Results of the Spatial Gravity Center Model

As shown in Figure 5, the gravity center of the flash floods was located mainly in Chuxiong, and its change trajectories were random. The gravity center of annual mean precipitation was focused on the border between Chuxiong and Puer, and its change trajectories were also random, but they had some similarity to that of the flash floods. The gravity center of the population moved gradually along the southwest region from 2001 to 2015.

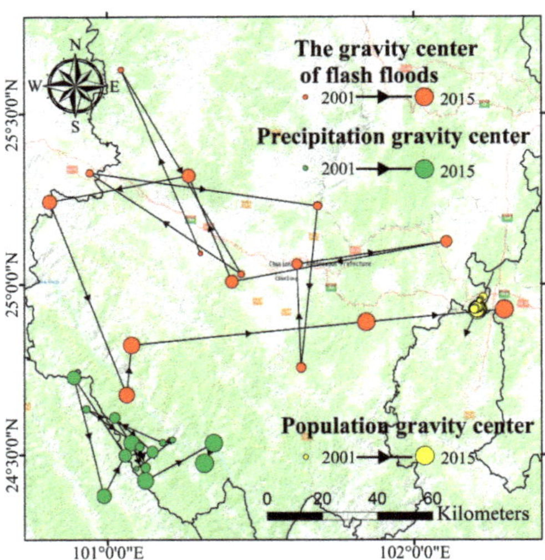

Figure 5. Change trajectories of gravity centers of flash floods, annual mean precipitation, and population from 2001 to 2015.

3.3. Driving Factors for the Spatial Distribution of the Flash Floods

We performed a correlation analysis between the flash floods and driving factors in the different subregions of the landcover, as shown in Table 5. Additionally, we analyzed the interaction of driving factors on the flash floods; the results are presented in Figure 6. In the grassland, the highest r value was the H24_20 ($r = 0.501$, $P < 0.001$). Other higher r values included M10_20 ($r = 0.487$, $P < 0.001$), H6_20 ($r = 0.449$, $P < 0.001$), ELE ($r = 0.431$, $P < 0.001$), and M60_20 ($r = 0.335$, $P = 0.001$). The highest R^2 value of multiple linear regression was 0.213, which was based on the principal component (Dim.1,

Dim.2, Dim.3, and Dim.4). Driving factors that had a greater contribution included M10_20 (17.91%), M60_20 (17.39%), H24_20 (16.99%), and H6_20 (16.93%). Regarding the settlement, the higher r values of driving factors exhibited the following ranking: M10_20 ($r = 0.675$, $P < 0.001$) > ELE ($r = 0.674$, $P < 0.001$) > M60_20 ($r = 0.593$, $P < 0.001$) > H24_20 ($r = 0.519$, $P < 0.001$) > H6_20 ($r = 0.395$, $P < 0.001$). The interaction of Dim.1 and Dim.4 contributed the highest R^2 value, which were controlled mainly by the M10_20 (17.95%), M60_20 (17.79%), H24_20 (16.84%), H6_20 (15.21%), ELE (12.73%), and PD (10.49%). Regarding the farmland, M10_20, H6_20, ELE, H24_20, and M60_20 were the most important driving factors, with r values of 0.445 ($P < 0.001$), 0.445 ($P < 0.001$), 0.402 ($P < 0.001$), 0.381 ($P < 0.001$), and 0.348 ($P = 0.001$), respectively. The highest R^2 value occurred in Dim.1, and the main contributing factors were M10_20 (23.01%), M60_20 (21.81%), H6_20 (21.07%), H24_20 (21.06%), and ELE (12.28%). The characteristics of forest were similar to those of the other three subregions. The higher r values included H6_20 ($r = 0.492$, $P < 0.001$), H24_20 ($r = 0.485$, $P < 0.001$), M10_20 ($r = 0.466$, $P < 0.001$), and M60_20 ($r = 0.326$, $P = 0.001$). In addition, the highest R^2 was determined by the interaction of Dim.1 and Dim.3, and H6_20 (17.14%), M10_20 (16.83%), M60_20 (16.19%), H24_20 (15.72%), and ELE (10.48%) were dominant.

Table 5. Pearson correlation coefficient (r) between the flash floods and driving factors in the different subregions of the landcover based on Formula (5).

Factors	Grassland		Settlement		Farmland		Forest	
	r	P	r	P	r	P	r	P
M10_20	0.487	<0.001	0.675	<0.001	0.445	<0.001	0.466	<0.001
M60_20	0.335	0.001	0.593	<0.001	0.348	0.001	0.326	0.001
H6_20	0.449	<0.001	0.395	<0.001	0.445	<0.001	0.492	<0.001
H24_20	0.501	<0.001	0.519	<0.001	0.381	<0.001	0.485	<0.001
ELE	0.431	<0.001	0.674	<0.001	0.402	<0.001	0.257	0.01
TR	−0.077	0.446	−0.146	0.172	0.219	0.032	−0.014	0.888
SLP	−0.05	0.619	−0.094	0.382	−0.176	0.085	−0.081	0.424
NDVI	−0.241	0.016	0.061	0.571	−0.222	0.029	−0.234	0.02
PD	0.082	0.42	−0.211	0.048	0.108	0.297	−0.098	0.335
GDP	0.098	0.332	0.042	0.697	0.226	0.027	0.233	0.02

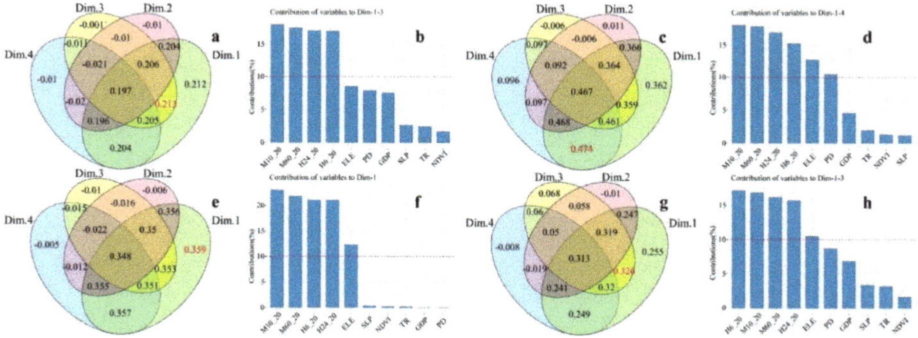

Figure 6. Driving factors interaction based on multiple linear regression and principal component analysis for flash floods, and the contribution of driving factors to the principal components (Dim.1, Dim.2, Dim.3, and Dim.4) in the different subregions of the landcover: (**a**) R^2 value of multiple linear regression of principal components to the flash floods in grassland; (**b**) the contribution of driving factors to Dim.1 and Dim.3 in grassland; (**c**) R^2 value of multiple linear regression of principal components to the flash floods in settlement; (**d**) the contribution of driving factors to Dim.1 and Dim.4 in settlement; (**e**) R^2 value of multiple linear regression of principal components to the flash floods in farmland; (**f**) the contribution of driving factors to Dim.1 in farmland; (**g**) R^2 value of multiple linear regression of principal components to the flash floods in forest; and (**h**) the contribution of driving factors to Dim.1 and Dim.3 in forest. The red number is the highest R^2 value. The contribution of driving factors to the principal components greater than the red dotted line indicates that they are dominant.

3.4. Sensitivity Analysis between the Flash Floods and Economy Development

By combining the data of the flash floods and GDP and using Formula (6), we calculated the value of sensitivity (S) of the flash floods to the economic development of each city in Yunnan Province, as shown in Figure 7. From an overall perspective, fewer regions experienced medium and high sensitivity during the study period, and the quantity of regions without sensitivity first decreased and then remained stable. During 1995–2000, all regions were non-sensitive. In comparison, the quantity of low-sensitivity regions during 2000–2005 and 2005–2010 increased to six (Nujiang, Baoshan, Lijiang, Yuxi, Honghe, and Wenshan) and nine (Zhaotong, Qujing, Kunming, Honghe, Xishuangbanna, Puer, Lincang, Baoshan, and Diqing), respectively. Simultaneously, Kunming was the highest sensitivity region (S = 21.86) and low-sensitivity (S = 0.98) during these two periods. In the 2010–2015 time period, the flash floods in half of the regions were not sensitive to GDP in Yunnan Province, whereas Diqing was the mid-sensitivity region (S = 10.67).

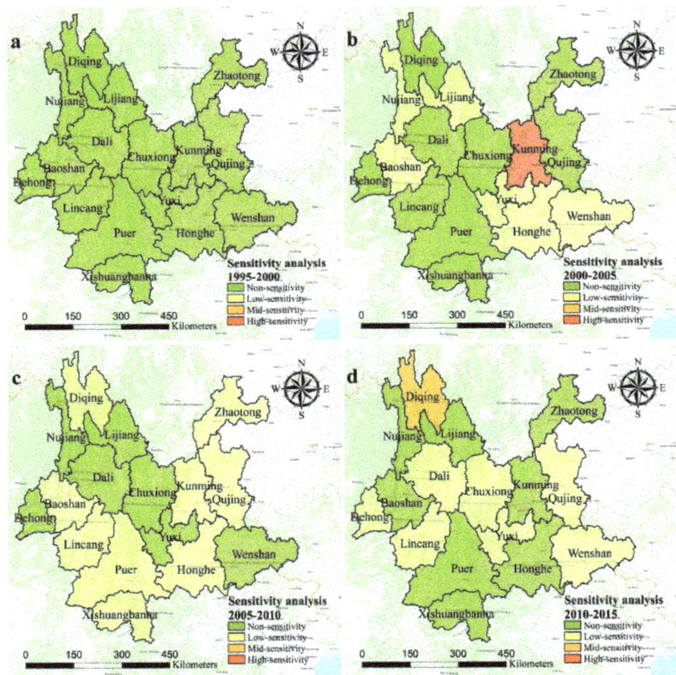

Figure 7. The spatial–temporal variation of sensitivity from 1995 to 2015: (**a**) during 1995–2000, (**b**) during 2000–2005, (**c**) during 2005–2010, and (**d**) during 2010–2015.

4. Discussion

4.1. Temporal and Spatial Distribution of the Flash Floods

In this study, a significant variation occurred in the annual and monthly quantity of the flash floods on a provincial scale (Figure 2). The annual quantity gradually increased from 1949 to 2015. The probable causes included human activity and the heavy precipitation that has become more frequent over these years [49,50]. However, the possible reason why flash floods were rare in the 1950s is that many flash floods were not recorded in the early years of the founding of the People's Republic of China. In addition, our finding revealed that the most frequent quantity of flash floods occurred during 1996–2015, and the average quantity of flash floods was 102 per year. This result showed that, in general, a similar trend occurred in China, which has been supported by existing studies [25,29]. In the meantime, Yangtze River Basin floods have triggered multiple flash floods, which can be explained by the fact that economic development achieved through reform and opening-up has damaged the environment [21]. Regarding monthly variations, flash floods have been concentrated in June, July, and August. The finding was further evidence of the effects of precipitation on flash floods, because heavy precipitation mainly occurred in these three months [51].

The kernel density estimation, unbalanced regions, standard deviational ellipse, and gravity centers showed that flash floods were mainly concentrated in Baoshan, Dali, Chuxiong, Kunming, Yuxi, Qujing, and Zhaotong (Figures 3 and 4). This high concentration of flash floods may have been caused by the elevated levels of precipitation in the central and western regions of Yunnan Province, which have played an important role as external factors [52]. The formation of flash floods was affected by precipitation, environmental background conditions, and human activities, etc. Of the above-mentioned regions, environmental background condition and human activities were other important influencing factors, with the exception of precipitation. In Yunnan, most of the regions are predominately mountainous landscapes [53], which provided ideal geographical conditions for the formation of flash floods. Additionally, the urban agglomerations and major economic area are located in the central regions of Yunnan [54], which leads to the proliferation of environmental destruction, which was another important reason for the formation of flash floods in these regions. Furthermore, previous researchers have investigated the species distribution in Yunnan, their study have revealed low-species areas that were consistent with the high-incidence regions of flash floods [55]; these areas have a high density of human activity, which leads to the destruction of hydrological processes, thus providing conditions for the occurrence of flash floods. According to the results of the change trajectories of gravity centers (Figure 5), the trajectories of flash floods were somewhat similar to those for precipitation but not for population. This finding can be explained by the existing research. Liu et al., for example, found that the influence of precipitation on the flash floods was greater than that of human activities over the short term (i.e., during the 15-year period from 2001 to 2015) [25].

4.2. Driving Factors Influencing the Flash Floods in Yunnan Province

Our findings showed that precipitation and ELE were the main factors affecting flash floods from 2001 to 2015 (Table 5 and Figure 6). This result indicated that flash floods might be easily induced in high-altitude areas by heavy precipitation, and vegetation cover was negatively correlated with flash floods in grassland ($r = -0.241$, $P = 0.016$), farmland ($r = -0.222$, $P = 0.029$), and forest ($r = -0.234$, $P = 0.02$), which is consistent with existing studies [5,21,29]. Regarding the different subregions of the landcover, the driving factors revealed different effects on flash floods. In the grassland, the H24_20 had the maximum correlation with the flash floods, but the ELE did not exceed the average contribution rate for the interaction of driving factors. This is because the topographic relief of the grassland was relatively small. Additionally, the economic activities were not obvious [56], and the short periods of heavy precipitation did not produce strong scour. In this surrounding, the TR, SLP, PD, and GDP had less influence on flash floods (Table 5). In comparison, the contribution rate of ELE and PD exceeded the average value for the interaction of driving factors in settlement. As the population increased, the economic activities expanded to the high altitude area and the ecological system was damaged, which determined the elevation and human activities had a certain boost on flash floods [57]. In contrast, the TR, SLP, and NDVI showed a pretty low correlation to flash floods, which likely occurred because the topographic relief, slope, and vegetation coverage held to a relatively low level in the settlement [58]. This subregion was dominated by small basins and presented the most relevant relationship between flash floods and M10_20. Such a relationship was consistent with the short duration precipitation, which could cause flooding in small basins that have a low time of concentration. For the farmland, the TR and SLP had a greater effect on flash floods than on other subregions (Table 5). This topographic relief and slope mainly included the lower hilly areas of the farmland, which had a higher population density and greater influence on the formation of the flash floods [29,59]. Regarding the forest, GDP had a higher correlation to flash floods than other subregions (Table 5). A possible reason for this correlation was that economic development led to the destruction of forests [60]; we can draw a similar conclusion from the effects of NDVI on flash floods.

4.3. Sensitivity Analysis of the Flash Floods to Economic Development

The sensitivity analysis of the flash floods to GDP is important for sustainable socioeconomic development. The phenomenon of sensitivity rising can influence many development issues, including an imbalance in regional structures and tourism damage [61,62]. The results of our sensitivity analysis showed that all regions were non-sensitive during 1995–2000 (Figure 7a). The early stage of the reform and opening-up of China's western region was relatively hysteretic for economic development; the intensity of human activity was maintained at a low level and the ecosystem structure was relatively complete [63], which determined this non-sensitivity pattern. During 2000–2005, however, the regions with low-sensitivity increased significantly, and Kunming was the most sensitive area (Figure 7b). In this period, the Development of the West Region, a significant plan for western development launched in 2000, saw economic development in the western city, which was promoted through a massive investment of national capital [64]. Consequently, the quantity of regions with low-sensitivity increased gradually. As the economic and cultural center in Yunnan, Kunming had the strongest human activity [65], which led to the highest sensitivity occurring in this region. With the management of mineral resources in Yunnan Province and regulations to restore farmland to forest, the sensitivity of flash floods to economic development obviously decreased in Kunming during 2005–2010. The quantity of regions with low-sensitivity increased gradually in this period, which can be explained primarily by the macroscopic development of the area [66]. Subsequently, more attention was paid to ecological protection and harmonious economic development, including primary functional area planning (2011) and national new urbanization planning (2014) [60], which determined that the sensitivity of flash floods to economic development decreased gradually during 2010–2015.

To prevent and control flash floods, we need to not only control human activities, such as excessive deforestation and mining, but also implement harmonious economic development and protect the integrity of ecosystem. Regarding the high-incidence areas of flash floods, the rational use of land and flexible barriers should be utilized in these regions. In addition, the government should strengthen its ability to predict flash floods during heavy periods of precipitation.

Our study shows that precipitation and elevation in settlement have a significant impact on flash floods than in other subregions. This result reveals the potential impact of human activities on the environment. Similarly, partition research should be considered when studying the relationship between flash floods and driving forces in other parts of the world, as the flash floods differ in background conditions in different subregions of landcover. However, the study has some drawbacks, including, for example, the time series of driving factors did not begin in 1949. In the future, a longer time series of data can be performed. Precipitation data interpolation should consider mountain terrain in order to reflect the change of climate with elevation. In addition, the spatial analysis methods used in this paper included the kernel density estimation, spatial mismatch analysis, standard deviational ellipse, and spatial gravity center model, which can highlight the spatial distribution characteristics of flash floods on a large regional scale, but have a weak application effect on small regions (e.g., flood protection works, excessive human exposure within the floodplain, basin shape, local geomorphology, and others). As to the principal component analysis, the method to determine the number of principal components has not been unified, which may lead to the loss of local information.

5. Conclusions

In this study, we examined the temporal and spatial distribution of flash floods; the driving factors for the spatial distribution of the flash floods in different subregions of the landcover; and the sensitivity (S) of flash floods to economic development using the changepoint, kernel density estimation, spatial mismatch analysis, standard deviational ellipse, spatial gravity center model, Pearson's correlation coefficient, multiple linear regression, principal component analysis, and sensitivity analysis. Yunnan

Province, in southwestern China, was selected as the study region because of the high incidence area of flash floods and its representative major mountainous landform. The temporal variation of the flash floods showed that the annual quantity of flash floods gradually increased from 1949 to 2015, and the months with the greatest quantity of flash floods were June, July, and August. As to spatial patterns, the regions with a high quantity of flash floods included Zhaotong, Qujing, Kunming, Yuxi, Chuxiong, Dali, and Baoshan. Furthermore, an analysis of the driving factors for flash floods in different subregions of the landcover showed that precipitation and ELE were the main factors from 2001 to 2015. In the settlement, precipitation and ELE had a greater effect on flash floods than in other subregions. Finally, the spatial–temporal variation of sensitivity revealed that the quantity of regions with non-sensitivity first decreased and then remained stable (from 1995 to 2015). During 2000–2005, Kunming had the highest sensitivity; then, this sensitivity gradually began to weaken.

On the basis of the results of this study, the most effective measure is early warning preferential to heavy precipitation events in the short term. In the long term, we should control human activities and reduce the ecosystem vulnerability caused by damage to species and to vegetation and by population growth to effectively prevent flash floods. According to the sensitivity of flash floods to economic development, Yunnan Province should take note of the previous lessons of imbalance and continue to introduce a balanced development policy.

Author Contributions: Conceptualization, J.X. and C.Y.; formal analysis, C.Y.; data and resources, W.C. and C.Z.; writing—original draft preparation, J.X. and C.Y.; writing—review and editing, L.G. and X.Z.; supervision, W.C.; funding acquisition, W.C. and C.Z.

Funding: This research was supported by the Strategic Priority Research Program of Chinese Academy of Sciences (XDA20030302), China Geological Survey Project (DD20190637), Open Subject of Big Data Institute of Digital Natural Disaster Monitoring in Fujian (NDMBD2018003), Southwest Petroleum University of Science and Technology Innovation Team Projects (2017CXTD09), and National Flash Flood Investigation and Evaluation Project (SHZH-IWHR-57).

Conflicts of Interest: The authors declare no conflict of interest.

References

1. Gascón, E.; Laviola, S.; Merino, A.; Miglietta, M.M. Analysis of a localized flash-flood event over the central Mediterranean. *Atmos. Res.* **2016**, *182*, 256–268. [CrossRef]
2. Rahman, M.T.; Aldosary, A.S.; Nahiduzzaman, K.M.; Reza, I. Vulnerability of flash flooding in Riyadh, Saudi Arabia. *Nat. Hazards* **2016**, *84*, 1807–1830. [CrossRef]
3. Lian, J.; Yang, W.; Xu, K.; Ma, C. Flash flood vulnerability assessment for small catchments with a material flow approach. *Nat. Hazards* **2017**, *88*, 699–719. [CrossRef]
4. Ashley, S.T.; Ashley, W.S. Flood Fatalities in the United States. *J. Appl. Meteorol. Clim.* **2008**, *47*, 805–818. [CrossRef]
5. He, B.; Huang, X.; Ma, M.; Chang, Q.; Tu, Y.; Li, Q.; Zhang, K.; Hong, Y. Analysis of flash flood disaster characteristics in China from 2011 to 2015. *Nat. Hazards* **2017**, *90*, 407–420. [CrossRef]
6. Gourley, J.J.; Flamig, Z.L.; Vergara, H.; Kirstetter, P.-E.; Clark, R.A.; Argyle, E.; Arthur, A.; Martinaitis, S.; Terti, G.; Erlingis, J.M.; et al. The FLASH project: Improving the tools for flash flood monitoring and prediction across the United States. *Bull. Am. Meteorol. Soc.* **2017**, *98*, 361–372. [CrossRef]
7. Barredo, J.I. Major flood disasters in Europe: 1950–2005. *Nat. Hazards* **2006**, *42*, 125–148. [CrossRef]
8. Pereira, S.; Diakakis, M.; Deligiannakis, G.; Zêzere, J.L. Comparing flood mortality in Portugal and Greece (Western and Eastern Mediterranean). *Int. J. Disaster Risk Reduct.* **2017**, *22*, 147–157. [CrossRef]
9. Xu, J.; Wilkes, A. Biodiversity impact analysis in northwest Yunnan, southwest China. *Biodivers. Conserv.* **2004**, *13*, 959–983. [CrossRef]
10. Zeng, Z.; Tang, G.; Long, D.; Zeng, C.; Ma, M.; Hong, Y.; Xu, H.; Xu, J. A cascading flash flood guidance system: Development and application in Yunnan Province, China. *Nat. Hazards* **2016**, *84*, 2071–2093. [CrossRef]
11. Bajabaa, S.; Masoud, M.; Al-Amri, N. Flash flood hazard mapping based on quantitative hydrology, geomorphology and GIS techniques (case study of Wadi Al Lith, Saudi Arabia). *Arab. J. Geosci.* **2013**, *7*, 2469–2481. [CrossRef]

12. Shi, P.; Lili, L.; Wang, M.; Wang, J.; Chen, W. Disaster system: Disaster cluster, disaster chain and disaster compound. *J. Nat. Disasters* **2014**, *23*, 1–12.
13. Zhong, D.; Xie, H.; Wei, F.; Liu, H.; Tang, J. Discussion on Mountain Hazards Chain. *J. Mt. Sci.* **2013**, *31*, 314–326.
14. Xiong, J.; Wei, F.; Liu, Z. Hazard assessment of debris flow in Sichuan Province. *J. Geo-Inf. Sci.* **2017**, *19*, 1604–1612.
15. Zeleňáková, M.; Gaňová, L.; Purcz, P.; Satrapa, L. Methodology of flood risk assessment from flash floods based on hazard and vulnerability of the river basin. *Nat. Hazards* **2015**, *79*, 2055–2071. [CrossRef]
16. Guo, E.; Zhang, J.; Ren, X.; Zhang, Q.; Sun, Z. Integrated risk assessment of flood disaster based on improved set pair analysis and the variable fuzzy set theory in central Liaoning Province, China. *Nat. Hazards* **2014**, *74*, 947–965. [CrossRef]
17. Sangati, M.; Borga, M. Influence of rainfall spatial resolution on flash flood modelling. *Nat. Hazards Earth Syst. Sci.* **2009**, *9*, 575–584. [CrossRef]
18. Lin, Z.; Nimaji; Huang, Z. Hydrological dynamics simulation and critical rainfall for flash flood in southeastern Tibet. *Bull. Soil Water Conserv.* **2017**, *37*, 183–187.
19. Jun, D.U.; Wen-Feng, D.; Hong-Yu, R. Relationships between different types of flash flood disasters and their main impact factors in the Sichuan Province. *Resour. Environ. Yangtze Basin* **2015**, *24*, 1977–1983.
20. Wan, S.; Zhao, N.; Wei, D. Correlation and multi-timescale characteristics of strong precipitations and landslide debris flows in Yunnan Province. *J. Catastrophol.* **2015**, *30*, 45–50.
21. Xiong, J.; Zhao, Y.; Cheng, W.; Guoliang; Wang, N.; Li, W. Temporal and spatial variation rule and influencing ractors of flash floods in Sichuan Province. *J. Geo-Inf. Sci.* **2018**, *20*, 1443–1456.
22. Chen, C.-C.; Tseng, C.-Y.; Dong, J.-J. New entropy-based method for variables selection and its application to the debris-flow hazard assessment. *Eng. Geol.* **2007**, *94*, 19–26. [CrossRef]
23. Tang, C.; van Asch, T.W.; Chang, M.; Chen, G.Q.; Zhao, X.H.; Huang, X.C. Catastrophic debris flows on 13 August 2010 in the Qingping area, southwestern China: The combined effects of a strong earthquake and subsequent rainstorms. *Geomorphology* **2012**, *139*, 559–576. [CrossRef]
24. Vennari, C.; Parise, M.; Santangelo, N.; Santo, A. A database on flash flood events in Campania, southern Italy, with an evaluation of their spatial and temporal distribution. *Nat. Hazards Earth Syst. Sci.* **2016**, *16*, 2485–2500. [CrossRef]
25. Liu, Y.; Yang, Z.; Huang, Y.; Liu, C. Spatiotemporal evolution and driving factors of China's flash flood disasters since 1949. *Sci. China Earth Sci.* **2018**, *61*, 1804–1817. [CrossRef]
26. Jun, D.U.; Ren, H.; Zhang, P.; Zhang, C. Comparative study of the hazard assessment of mountain torrent disasters in macro scale. *J. Catastrophol.* **2016**, *31*, 66–72.
27. Rozalis, S.; Morin, E.; Yair, Y.; Price, C. Flash flood prediction using an uncalibrated hydrological model and radar rainfall data in a Mediterranean watershed under changing hydrological conditions. *J. Hydrol.* **2010**, *394*, 245–255. [CrossRef]
28. Doocy, S.; Daniels, A.; Murray, S.; Kirsch, T.D. The human impact of floods: A historical review of events 1980-2009 and systematic literature review. *PLoS Curr. Disasters* **2013**, *5*, 1808–1815. [CrossRef] [PubMed]
29. Liu, Y.; Yuan, X.; Guo, L.; Huang, Y.; Zhang, X. Driving force analysis of the temporal and spatial distribution of flash floods in Sichuan Province. *Sustainability* **2017**, *9*, 1527. [CrossRef]
30. Wang, J.; Xu, C. Geodetector: Principle and prospective. *Acta Geogr. Sin.* **2017**, *72*, 116–134.
31. Zhou, L.; Zhou, C.; Yang, F.; Wang, B.; Sun, D. Spatio-temporal evolution and the influencing factors of PM2.5 in China between 2000 and 2011. *Acta Ecol. Sin.* **2017**, *72*, 161–174.
32. Shi, J.; Li, B.; Li, P.; Huang, J.; Sun, F.; Liu, B. Analisis of characteristics and formation mechanism for the 9·17 giant debris flow in Yuanmou Country, Yunnan Province. *Geol. Rev.* **2018**, *64*, 665–673.
33. Xu, S.; Yuan, Z.; Yang, Z. Based on EKC analysis of landslide and debris flow disasters. *Soil Water Conserv. China* **2015**, 54–56.
34. Zhan, W.; Li, F.; Hao, W.; Yan, J. Regional characteristics and influnencing factors of seasonal vertical crustal motions in Yunnan, China. *Geophys. J. Int.* **2017**, *210*, 1295–1304. [CrossRef]
35. Yang, Y.; Jia, W. Types and geographical environment elements of township placenames in Yunnan Province. *Geogr. Res.* **2016**, *4*, 60.

36. Cao, X.; Pan, W.; Zhou, S.; Han, Z.; Han, T.; Shui, Z. Seasonal variability of oxygen and hydrogen isotopes in a wetland system of the Yunnan-Guizhou Plateau, southwest China: A quantitative assessment of groundwater inflow fluxes. *Hydrogeol. J.* **2017**, *26*, 1–17. [CrossRef]
37. Liu, Y.; Zhao, E.; Huang, W.; Zhou, J.; Ju, J. Characteristic analysis of precipitation and temperature change trend in Yunnan in recent 46 years. *J. Catastrophol.* **2010**, *25*, 39–44.
38. Killick, R.; Fearnhead, P.; Eckley, I.A. Optimal Detection of Changepoints With a Linear Computational Cost. *J. Am. Stat. Assoc.* **2012**, *107*, 1590–1598. [CrossRef]
39. Guo, F.; Innes, J.L.; Wang, G.; Ma, X.; Sun, L.; Hu, H.; Su, Z. Historic distribution and driving factors of human-caused fires in the Chinese boreal forest between 1972 and 2005. *J. Plant Ecol.* **2015**, *8*, 480–490. [CrossRef]
40. Chai, J.; Wang, Z.; Yang, J.; Zhang, L. Analysis for spatial-temporal changes of grain production and farmland resource: Evidence from Hubei Province, central China. *J. Clean. Prod.* **2019**, *207*, 474–482. [CrossRef]
41. Li, T.; Long, H.; Zhang, Y.; Tu, S.; Ge, D.; Li, Y.; Hu, B. Analysis of the spatial mismatch of grain production and farmland resources in China based on the potential crop rotation system. *Land Use Policy* **2017**, *60*, 26–36. [CrossRef]
42. Wang, B.; Shi, W.; Miao, Z. Confidence analysis of standard deviational ellipse and its extension into higher dimensional euclidean space. *PLoS ONE* **2015**, *10*, e0118537. [CrossRef]
43. Li, M.; Ren, X.; Zhou, L.; Zhang, F. Spatial mismatch between pollutant emission and environmental quality in China—A case study of NOx. *Atmos. Pollut. Res.* **2016**, *7*, 294–302. [CrossRef]
44. Irhoumah, M.; Pusca, R.; Lefèvre, E.; Mercier, D.; Romary, R. Diagnosis of induction machines using external magnetic field and correlation coefficient 2017. In Proceedings of the 2017 IEEE 11th International Symposium on Diagnostics for Electrical Machines, Power Electronics and Drives (SDEMPED), Tinos, Greece, 29 August–1 September 2017; pp. 531–536.
45. Wang, H.; Zhang, B.; Liu, Y.; Liu, Y.; Xu, S.; Deng, Y.; Zhao, Y.; Chen, Y.; Hong, S. Multi-dimensional analysis of urban expansion patterns and their driving forces based on the center of gravity-GTWR model: Acase study of the Beijing-Tianjin-Hebei urban agglomeration. *Acta Geogr. Sin.* **2018**, *73*, 1076–1092.
46. Abdi, H.; Williams, L.J. Principal component analysis. *Wiley Interdiscip. Rev. Comput. Stat.* **2010**, *2*, 433–459. [CrossRef]
47. Khademi, F.; Akbari, M.; Jamal, S.M.; Nikoo, M. Multiple linear regression, artificial neural network, and fuzzy logic prediction of 28 days compressive strength of concrete. *Front. Struct. Civ. Eng.* **2017**, *11*, 90–99. [CrossRef]
48. Han, Z.; Song, W.; Deng, X. Responses of ecosystem service to land use change in Qinghai Province. *Energies* **2016**, *9*, 303. [CrossRef]
49. Yu, W.; Shao, M.; Ren, M.; Zhou, H.; Jiang, Z.; Li, D. Analysis on spatial and temporal characteristics drought of Yunnan Province. *Acta Ecol. Sin.* **2013**, *33*, 317–324. [CrossRef]
50. Liu, L.; Xu, Z.X. Regionalization of precipitation and the spatiotemporal distribution of extreme precipitation in southwestern China. *Nat. Hazards* **2015**, *80*, 1195–1211. [CrossRef]
51. Liu, J.; Li, L.; Li, J.; Wang, Z.; Chen, S.; Zhang, K. Characteristics of precipitation variation and potential drought-flood regional responses in Yunnan Province from 1954 to 2014. *J. Geo-Inf. Sci.* **2016**, *18*, 1077–1086.
52. Tao, Y.; He, Q. The temporal and spatial distribution of precipitation over Yunnan province and its response to global warming. *J. Yunnan Univ.* **2008**, *30*, 587–595.
53. Luo, J.; Zhan, J.; Lin, Y.; Zhao, C. An equilibrium analysis of the land use structure in the Yunnan Province, China. *Front. Earth Sci.* **2014**, *8*, 393–404. [CrossRef]
54. Liu, X.; Salmeron, D.; Luo, Q.; Rana, S.; Liu, Z. Research on optimal development pattern of Yunnan central economic region. *China City Plan. Rev.* **2012**, *21*, 38–45.
55. Zhang, M.; Zhou, Z.; Chen, W.; Slik, J.W.F.; Cannon, C.H.; Raes, N. Using species distribution modeling to improve conservation and land use planning of Yunnan, China. *Boil. Conserv.* **2012**, *153*, 257–264. [CrossRef]
56. Han, Z.; Song, W.; Deng, X.; Xu, X. Grassland ecosystem responses to climate change and human activities within the Three-River Headwaters region of China. *Sci. Rep.* **2018**, *8*, 9079. [CrossRef] [PubMed]
57. Špitalar, M.; Gourley, J.J.; Lutoff, C.; Kirstetter, P.-E.; Brilly, M.; Carr, N. Analysis of flash flood parameters and human impacts in the US from 2006 to 2012. *J. Hydrol.* **2014**, *519*, 863–870. [CrossRef]

58. Jenerette, G.D.; Harlan, S.L.; Brazel, A.; Jones, N.; Larsen, L.; Stefanov, W.L. Regional relationships between surface temperature, vegetation, and human settlement in a rapidly urbanizing ecosystem. *Landsc. Ecol.* **2006**, *22*, 353–365. [CrossRef]
59. Luo, H. Dynamic of vegetation carbon storage of farmland ecosystem in hilly area of central Sichuan Basin during the Last 55 years—A case study of Yanting County, Sichuan Province. *J. Nat. Resour.* **2009**, *24*, 251–258.
60. Peng, L.; Deng, W.; Zhang, H.; Sun, J.; Xiong, J. Focus on economy or ecology? A three-dimensional trade-off based on ecological carrying capacity in southwest China. *Nat. Resour. Model.* **2018**, *32*, e12201. [CrossRef]
61. Kourgialas, N.N.; Karatzas, G.P.; Nikolaidis, N.P. Development of a thresholds approach for real-time flash flood prediction in complex geomorphological river basins. *Hydrol. Process.* **2012**, *26*, 1478–1494. [CrossRef]
62. Chou, J.; Xian, T.; Dong, W.; Xu, Y. Regional temporal and spatial trends in drought and flood disasters in China and assessment of economic losses in recent years. *Sustainability* **2018**, *11*, 55. [CrossRef]
63. Tang, W.; Zhou, T.; Sun, J.; Li, Y.; Li, W. Accelerated urban expansion in Lhasa City and the implications for sustainable development in a Plateau city. *Sustainability* **2017**, *9*, 1499. [CrossRef]
64. Yi, S.; Yang, J.; Hu, X. How economic globalization affects urban expansion: An empirical analysis of 30 Chinese provinces for 2000–2010. *Qual. Quant.* **2016**, *50*, 1117–1133.
65. Hu, S.; Yan, X.; Qu, D.; Li, X. Approach to the legislative problem of the recycling economy in Kunming. *Ecol. Econ.* **2005**, *1*, 92–96.
66. Gao, J.; Wei, Y.; Chen, W.; Yenneti, K. Urban land expansion and structural change in the Yangtze River Delta, China. *Sustainability* **2015**, *7*, 10281–10307. [CrossRef]

Article

Flood Hazard Assessment Mapping in Burned and Urban Areas

Hariklia D. Skilodimou [1], George D. Bathrellos [2,*] and Dimitrios E. Alexakis [3]

1 Department of Geography and Climatology, Faculty of Geology and Geoenvironment, National and Kapodistrian University of Athens, University Campus, 15784 Athens, Greece; hskilodimou@geol.uoa.gr
2 Department of Geology, Division of General Marine Geology and Geodynamics, University of Patras, 26504 Rio Patras, Greece
3 Laboratory of Geoenvironmental Science and Environmental Quality Assurance, Department of Civil Engineering, University of West Attica, 12244 Athens, Greece; d.alexakis@uniwa.gr
* Correspondence: gbathrellos@upatras.gr

Abstract: This study proposes a simple method to produce a flood hazard assessment map in burned and urban areas, where primary data are scarce. The study area is a municipal unit of Nea Makri, a coastal part of the eastern Attica peninsula (central Greece), which has been strongly urbanized and suffered damage from urban fires in 2018. Six factors were considered as the parameters most controlling runoff when it overdraws the drainage system's capacity. The analytical hierarchy process (AHP) method and a geographical information system (GIS) were utilized to create the flood hazard assessment map. The outcome revealed that the areas with highest flood hazard are distributed in the eastern and southern parts of the study area, as a result of the combination of lowlands with gentle slopes, torrential behavior of the streams, streams covered by construction, increasing urbanization and burned areas. The uncertainty and the verification analyses demonstrate a robust behavior for the model predictions, as well as reliability and accuracy of the map. Comparing the existing urban fabric and road network to the potential flood hazard areas showed that 80% of the urban areas and 50% of the road network were situated within areas prone to flood. The method may be applied to land use planning projects, flood hazard mitigation and post-fire management.

Keywords: Athens metropolitan area; urban fires; analytical hierarchy process; GIS; flooding

Citation: Skilodimou, H.D.; Bathrellos, G.D.; Alexakis, D.E. Flood Hazard Assessment Mapping in Burned and Urban Areas. *Sustainability* **2021**, *13*, 4455. https://doi.org/10.3390/su13084455

Academic Editor: Andrzej Wałęga

Received: 21 March 2021
Accepted: 14 April 2021
Published: 16 April 2021

1. Introduction

The Earth landscape is a result of the interaction of several parameters, such as the geomorphic process, climate and human activity. Natural hazards are physical events, able to influence the natural and human environment significantly. In this context, the morphological changes of landforms due to natural disasters can restrict human interaction with the ecosystem [1–12].

Floods have affected the natural environment, before humans' appearance on Earth. Nowadays, they are considered natural hazards and a significant global problem threatening human life [13,14].

Floods have a significant impact on the development of a region. Historically, floods have large consequences for individuals and communities which include loss of human life, damage to buildings and infrastructure, destruction of crops, disruption to social affairs and loss of livestock [15,16].

In Europe, floods are among the most expensive types of natural disasters with costs about €4 billion per year over the period 2000–2020. Flooding is less significant in inner continental countries compared to Mediterranean countries [17].

On the other hand, floods after a fire are typically more extensive than before wildfires. Worldwide, fires burn more than 300 Mha h^{-1} per year, and in the Euro-Mediterranean region, wildfires burn about 450,000 ha per year [18–20]. Greece in particular experienced unusually high burned areas in 2007 and 2018. In the aftermath of wildfires, all surrounding and impacted areas are at risk of flooding for several years. Damaged areas are left without the vegetation that can help absorb rainfall. Wildfires dramatically change the landscape and the hydrologic response of a watershed and thus, a simple storm can cause extreme flash floods [21,22]. Another invisible threat is the deposited ash layer which contains potentially toxic elements [23]. However, much work should be performed on the topic of the impact of wildfire ash on human health and ecological terrestrial receptors [23–25].

Burned areas, along with human activities, influence flooding. Proactive planning is among the powerful tools to minimize losses and reduce the economic impact that accompanies floods' occurrence. Moreover, accurate spatial data related to the distribution of floods are a powerful tool for planers and engineers to decide whether or not the land of an area is safe for urban development [3,13,20–22,26,27].

Flood peaks, depths, volumes and mapping of flood hazard areas may be estimated by applying hydrologic and hydraulic–hydrodynamic models. Their produced flood hazard maps linked to the probability of occurrence of a flooding phenomenon in a certain period of time [28–30]. On the other hand, these models involve data that in many cases are often unavailable. Alternatively, several researchers have used deterministic methods to define flood-prone areas [27,28,31,32]. A deterministic method, proposed by [33] is the analytical hierarchy process (AHP) method, which is a multi-criteria decision-making technique. It is a weight evaluation procedure which uses qualitative and quantitative criteria to evaluate alternative solutions for a particular problem, among which the most suitable solution is selected [34,35]. Besides, the AHP can easy couple with a geographical information system (GIS) to estimate natural hazard maps. However, the AHP method cannot identify the uncertainty associated with spatial outputs [36,37].

This study's main scope is to propose a method to produce a flood hazard map in burned and urban areas. Hence, a coastal part of the northeastern Attica peninsula in central Greece has been selected as the case study. The study area is a part of the Athens metropolitan area and has been affected by extreme summer fires in the recent past.

To achieve this goal, geomorphologic, geologic and land use data affecting flood events were considered. These factors were evaluated by using the AHP method and GIS capabilities and their uncertainty was assessed by using a sensitivity analysis. Previous floods which occurred in the study area were utilized for the validation of the flood hazard map. As a last step, the urban activities were evaluated with the produced flood hazard map.

2. Materials and Methods

The case study area is the municipal unit of Nea Makri and is a popular holiday resort. It is located on the coast of Marathon Bay, a bay of the South Euboean Gulf, in the northeastern part of Attica Prefecture (Figure 1A). It is bounded on its western—southwestern part by Mount Pentelikon, while the hill Agriliki dominates its northern part. The area is a popular tourist destination with holiday houses, hotels and yacht marinas.

The study area occupies approximately 34 km^2, and it is a low altitude area, with elevation varying from 0 to 782 m above sea level. Gentle slopes characterize its morphology, while only the western and northwestern parts of the basin are mountainous (Figure 1B).

As regards climatic conditions, the study area has a Mediterranean climate type. The mean annual temperature reaches 18 °C and the yearly precipitation is 400 mm.

Several streams drain the study area. They have a limited length and their drainage basins have small areas. The length of their mainstream channel varies from 0.2 to 5.5 km. The steams debouch northeastwards from the eastern flank of Mount Pentelikon to the South Euboean Gulf (Figure 1B). Their flow is ephemeral, and they have torrential behavior. Although such fluvial systems in semiarid areas are frequently dry, extreme flash floods happen.

The study area was severely hit by fire on 23 July 2018. The fire started on the Pentelikon Mountain, spread very quickly to the seaside towns of Mati and Neos Voutzas (Figure 1B) and grew into one of the largest urban wildfires in recent European history, which left 102 people dead and 164 injured. Seventy percent of Mati was utterly burnt from the wildfires, while the remaining 30% was either partially burned or contains burnt areas. Houses and vehicles were totally burned, leaving Mati and nearby towns completely destroyed [38,39].

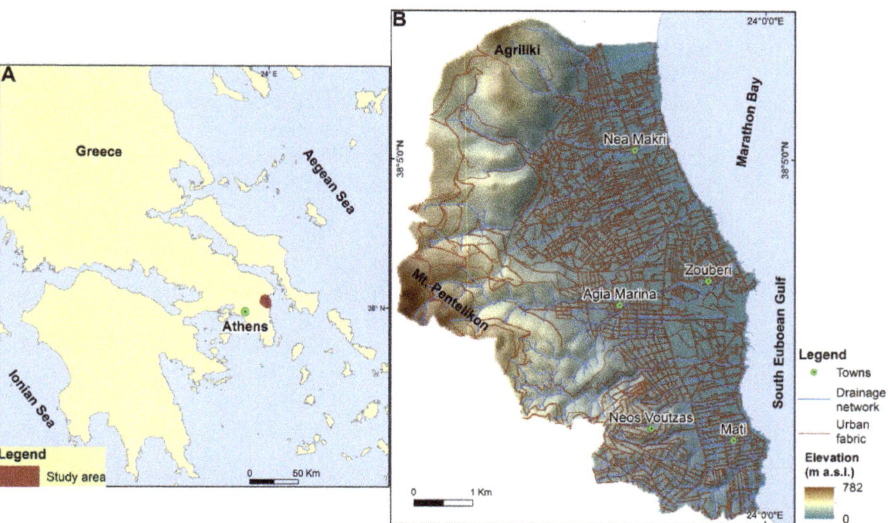

Figure 1. (**A**) Map showing the location of the study area. (**B**) Map illustrating the variation of elevations, the spatial extent of the drainage network and the roads, along with the main towns, in the study area.

Since the early 21st century, the region has strongly urbanized and its population has greatly increased [40]. Unchecked constructions block parts of the drainage network. The lowland areas of the towns Nea Makri, Zouberi and Mati have repeatedly been affected by flood events during extreme storm events [41].

3. Methodology

3.1. Data

The used data and their sources in the present work were:

- topographic maps at different scales (1:50,000, 1:25,000) along with topographic diagrams at scale 1:5000 from the Hellenic Military Geographical Service;
- the geological map sheet Kifisia (scale 1:50,000) from the Institute of Geology and Mineral Exploration of Greece, published in 2002;
- historical flood data [38];
- The Land Cover map of CORINE 2012 and Urban Atlas, 2018 [42,43];
- field data involving observations and mapping of the drainage networks of the study area.

3.2. Applied Method

Flood hazard mapping can be achieved by using hydraulic–hydrodynamic methods. An advantage of these models is that the indicated hazard maps are usually connected to certain return periods according to the European flood directive, even though they need data which are often not able to be obtained at a scale of the drainage basin [26–32,44,45].

In the present study, the lack of data related to precipitation and especially discharge gauges resulted in limitations to the hydrologic analysis. Thus, the link of the flood hazard assessment to return periods was not feasible. Otherwise, the AHP method was applied to produce the flood hazard assessment model. The method has been commonly used to delineate flood-prone areas since this approach warrants accurate and reliable predictions. For instance, Rahmati, et al. [27] used the AHP to map the flood-susceptible areas in Yasooj region, Iran, while Fernández and Lutz [28] applied the same method to assess flood hazard zoning in Tucumán Province, Argentina. In Greece, several researchers used the AHP method to assess potential flood-prone areas; i.e., Stefanidis and Stathis [29] in Kassandra Peninsula (northern Greece), Papaioannou et al. [45] in Thessaly (central Greece), Bathellos et al. [46] in northeastern

Peloponnesus (southern Greece) and Skilodimou et al. [47] in the drainage basin of Pinios River (western Greece). Additionally, Bathrellos et al. [32] applied this method to produce an urban flood hazard map in the metropolitan urban area of Athens, capital of Greece.

Saaty [33–35] proposed the AHP method, which implements a linear correlation between the factors involved. The application of the method in flood hazard assessment, including the pair-wise comparison, the creation of the table-matrix, the calculating of the weighting coefficients, as well as the computation of the consistency ratio, was analyzed in detail in the published work [32].

3.3. Involved Factors

The applied model is based on the factors which control the route of water when the capacity of the drainage network is exceeded by an extreme flood event. Previous published works [12,13,26,27,32,45–49] as well as data availability were utilized for the selection of the appropriate factors. Six physical and anthropogenic factors were taken into account. The physical factors were slope, elevation, distance from open channel streams, and hydrolithology, while the anthropogenic factors were distance from totally covered streams, along with land cover.

3.3.1. Slope

Since steep slopes contribute to flood occurrences and gentle slopes are more susceptible to flooding, slope is an essential factor for flood hazard assessment. Since a high-resolution digital terrain model (DEM) of the study area was not available, contours (4 m interval) and elevation points of the topographic diagrams were used to produce a DEM. The data were interpolated by using the topo to raster method, which is based on the ANUDEM program developed by Hutchinson [50]. Although DEMs derived from contour lines and elevation points often have bias, this method has a minor bias in the interpolation algorithm that may result in misleading results when calculating the profile curvature of the output surface, but is otherwise not noticeable [50]. According to Arun [51], the ANUDEM interpolation method is preferable for streams and adapted itself to terrain variations. Therefore, this method was considered to be the most appropriate to meet the requirements of the present work. The DEM was used to create the slope layer which was separated into five classes (Figure 2A), which are (i) <2°, (ii) 2–6°, (iii) 6–12°, (iv) 12–20° and (v) >20°.

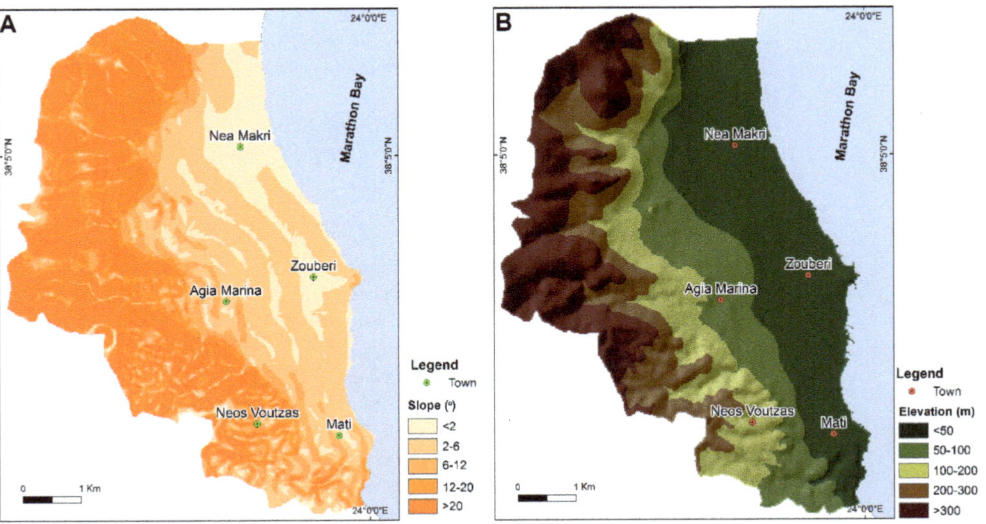

Figure 2. Thematic maps showing: (**A**) the classes of slopes; (**B**) the classes of elevations.

3.3.2. Elevation

Lowlands are flood-prone areas, and thus elevation is a vital factor in flood hazard estimation. As in the case of slope, the DEM was used to create the elevation layer. It was classified into the following, i.e., (i) <50 m a.s.l., (ii) 50–100 m a.s.l., (iii) 100–200 m a.s.l., (iv) 200–300 m a.s.l. and (v) >300 m a.s.l. (Figure 2B).

3.3.3. Distance from Open Channel Streams

Streams are a significant parameter for flood hazard assessment. The drainage networks were obtained from the topographic maps (scale 1:5000) and were digitized as a line layer. The Strahler's stream order system was used to define the streams. Third and higher order streams were studied, as they can cause flood events. The drainage networks of the study area consist of open channel streams and streams covered by urban fabric (Figure 3). The open channel streams were mapped through extensive fieldwork.

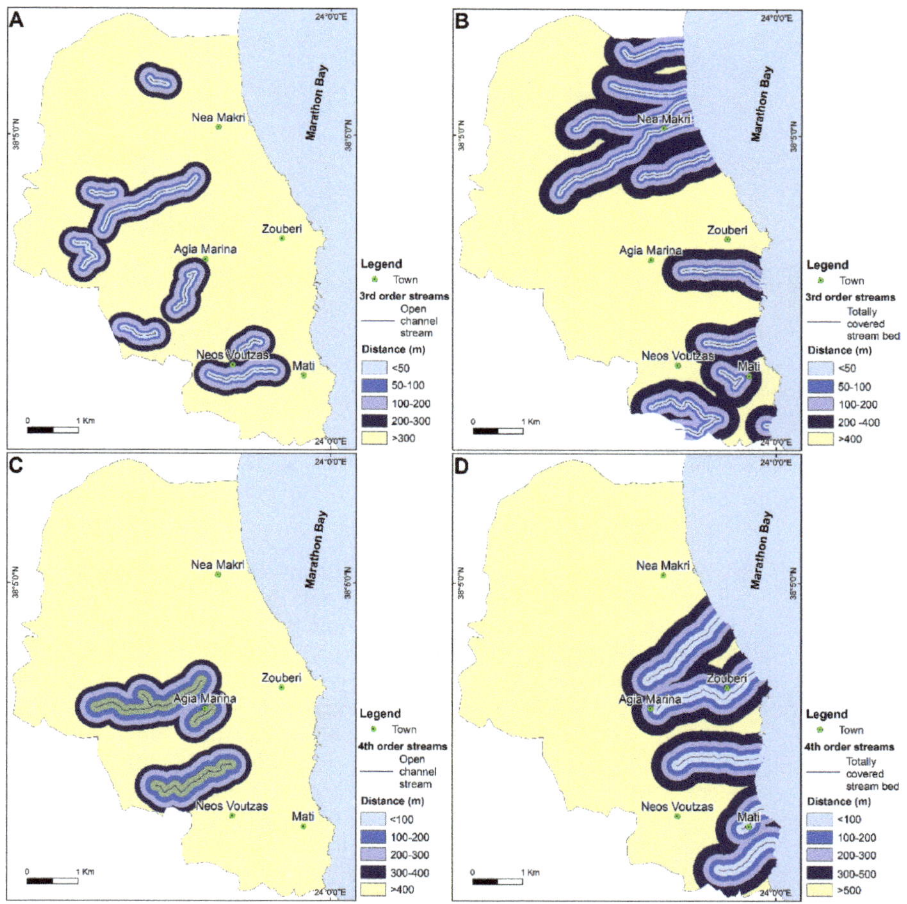

Figure 3. Maps of streams depicting distance from: (**A**) open channel streams and third-order streams; (**B**) totally covered streams and third-order streams; (**C**) open channel streams and fourth-order streams; (**D**) totally covered streams and fourth-order streams.

Buffer zones were created around third-order streams at distances of 50, 100, 200 and 300 m (Figure 3A), and fourth-order streams at distances of 100, 200, 300 and 400 m (Figure 3C).

3.3.4. Distance from Totally Covered Streams

As in the case of open channel streams, the totally covered streams were identified and mapped in detail through extensive fieldwork. Buffer zones were generated around the totally covered streams of the study area, i.e., third-order streams at distances of 50, 100, 200 and 400 m (Figure 3B) and fourth-order streams at distances of 100, 200, 300 and 500 m (Figure 3D).

3.3.5. Hydrolithology

The geologic conditions influence the surface runoff and may also contribute to flooding occurrences and flood hazard. The geological formations derived from the geological map of the study area were categorized into three classes according to their hydrolithological status [52]. They are (i) permeable, (ii) semi-permeable and (iii) impermeable formations (Figure 4A).

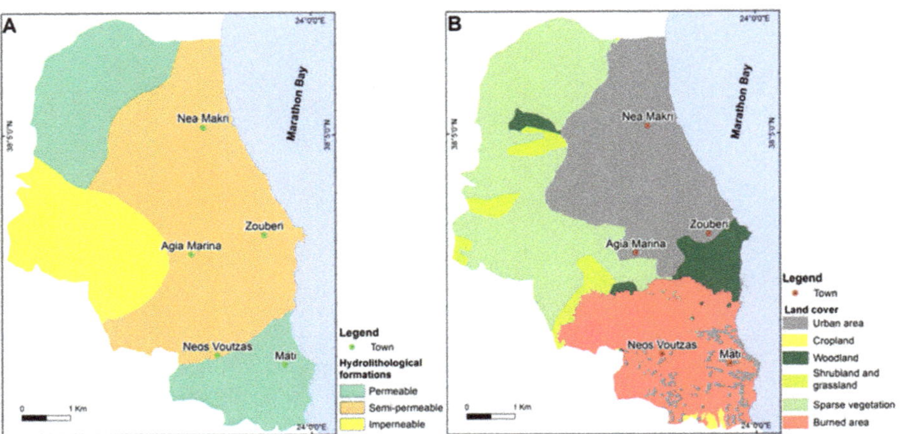

Figure 4. Thematic maps showing: (**A**) the spatial distribution of hydrolithological formations and (**B**) land cover in the study area.

3.3.6. Land Cover

Since the vegetation can help absorb rainfall, land cover influences the occurrence of floods. Since high-resolution land use data of the study area were not available, the land cover of the study area was taken from the CORINE 2012 Land Cover (CLC) map Copernicus Program, and the Urban Atlas, 2018. The CORINE 2012 program contains land cover data for Europe including land cover classes published by the European Commission. The European Urban Atlas provides reliable, high-resolution land use maps for more than 300 large urban zones [39,40]. For the necessities of this study, the Urban Atlas, 2018 was used to map the urban and burned areas while the CORINE 2012 was utilized to identify the rest of the land uses. The land uses, which reflect the vegetation covering, were classified into six categories as follows: (i) urban area, (ii) cropland, (iii) woodland, (iv) shrubland and grassland, (v) sparse vegetation area and (vi) burned area (Figure 4B). The urban and burned areas covered 33% and 22% respectively of the whole study area, while the remaining area consisted of 1% cropland, 5% woodland, 5% shrubland and grassland and 34% sparse vegetation.

3.4. The Rating of Every Class and the Weighting of the Factors

Each class of the involved factors had to be normalized to a standard rating scale. The determination of the rating of the classes was based on published work [32,46,47] and personal experience from previous studies of flood activity.

A five-grade scale was used for the normalization of the classes. Positive, whole numbers, ranging from 0 to 4, were given to every class. Each class expresses a different hazard level. Thus, the class rated as 0 expresses a low flood hazard, while the class rated as 4 expresses a high flood hazard. Table 1 shows the factors, their classes and their ratings.

Table 1. The rate values for each class of the involved factors.

Factors	Classes	Rating
Slope (°)	<2	4
	2–6	3
	6–12	2
	12–20	1
	>20	0
Elevation (m a.s.l.)	<50	4
	50–100	3
	100–200	2
	200–300	1
	>300	0
Distance from open channel streams (m)	third-order stream	
	0–50	4
	50–100	3
	100–200	2
	200–300	1
	>300	0
	fourth-order stream	
	0–100	4
	100–200	3
	200–300	2
	300–400	1
	>400	0
Distance from totally covered streams (m)	third-order stream	
	0–50	4
	50–100	3
	100–200	2
	200–400	1
	>400	0
	fourth-order stream	
	0–100	4
	100–200	3
	200–300	2
	300–500	1
	>500	0
Hydrolithology	permeable	1
	semipermeable	2
	impermeable	3
Land cover	urban area	4
	cropland	2
	woodland	0
	shrubland and grassland	1
	sparse vegetation area	3
	burned area	4

Table 2 shows the weights of every involved factor in flood hazard assessment. The CR was calculated to check the table—matrix and the valuation of the factors. It was found to be 0.04, so the decisions which are represented in Table 2 were well estimated.

Table 2. The weights of the factors and the consistency ratio (CR), where F1 = slope, F2 = elevation, F3 = distance from open channel streams, F4 = distance from totally covered streams, F5= hydrolithology and F6 = land cover.

	F1	F2	F3	F4	F5	F6	Weights, W_i
F1	1	4	1/2	3	1/2	2	0.207
F2		1	1/3	1/2	1/4	1/4	0.058
F3			1	3	1	1/3	0.088
F4				1	1/3	1	0.260
F5					1	1/3	0.104
F6							0.283

CR = 0.04.

3.5. *The Flood Hazard Index Values and the Uncertainty*

The flood hazard index values (FI) were commutated by using Equation (1):

$$FI = \sum_{i=1}^{n} R_i W_i, \tag{1}$$

where n is the number of factors, R_i is the rating of factor i and W_i is the weight of factor i. Thus, the basic flood hazard assessment map was produced.

Uncertainty has an important role in natural hazard evaluation [49,53]. Uncertainties in the weights of the adopted factors can bias the outcome of every hazard assessment. Published works [32,36,37,46,47], where geoenvironmental factors were evaluated through the AHP method, were used to calculate the uncertainties, ΔW_i, of the weighting coefficient. The values of the weighting coefficient which are presented in Table 2 were changed by 20%. The new weight values (ΔW_i) of every factor are shown in Table 3.

Table 3. The uncertainties (ΔW_i) of the weighting coefficient for each factor.

Factors	ΔW_i
Slope	0.041
Elevation	0.012
Open streams	0.018
Covered streams	0.052
Hydrolithology	0.021
Land cover	0.057

Equation (2) was used to calculate the error (ΔS):

$$\Delta S = \sqrt{\sum_{i=1}^{n} (\Delta W_i X_i)^2} \tag{2}$$

The values of the error map were multiplied by 1.96 to calculate the 95% confidence level of the flood hazard values and in the next step, were added and subtracted from the basic flood hazard values. Thus, two other maps characterizing the maximum and minimum flood hazard values were produced.

3.6. *Verification and Correlation of the Flood Hazard Evaluation*

The accuracy of the actual hazard level was examined by using past flood events. They were included in published works [32], and the historical flood database of Greece [41]. A flood inventory map was generated which includes nine flood occurrences that were observed in the municipality unit of Nea Makri for the period 2000–2020 (Figure 5).

Figure 5. Map showing the spatial distribution of flood events in the study area.

The verification was based on the statistical analysis of the frequency ratio (FR) method. It expresses the association between the spatial distribution of floods and hazard zones. In the published works [11,32,54,55], the methodology of the calculation of the FR area is presented.

Finally, the zones of the basic flood hazard map were compared with the active urban fabric and road network of the study area.

4. Results

The implementation of the previously described flood hazard assessment method led to the production of three flood hazard maps. The standard deviation method was used to categorize the maps. The areas of the maps were categorized into five classes, which were very high, high, moderate, low and very low flood hazard.

Figure 6 illustrates the basic flood hazard assessment map (FHb). In this map, the areas of very high and high flood hazard are located primarily in lowlands and particularly in eastern and southern parts of the study area. The moderate hazard zone is distributed mainly in the central part and the eastern and southern parts of the map. In the northern and western parts of the study area, many locations are situated in low and very low hazard areas.

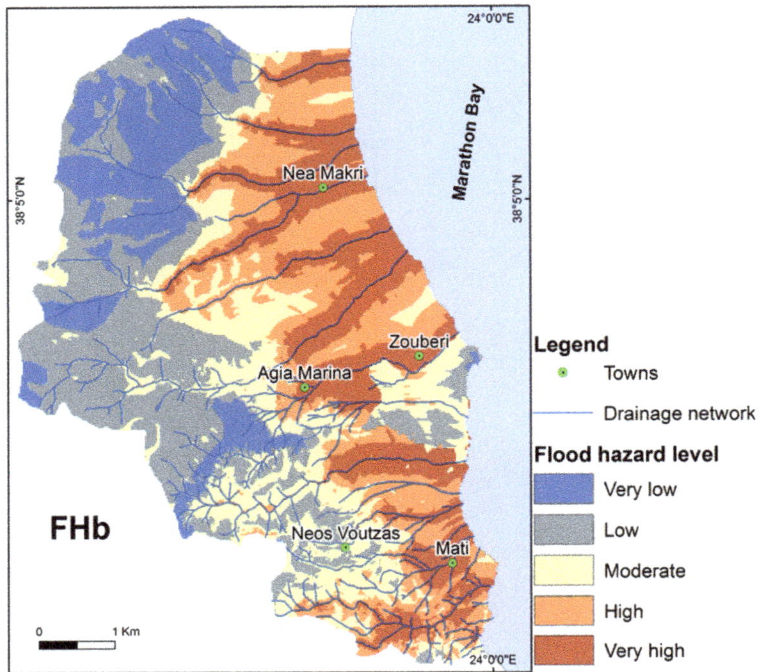

Figure 6. The basic flood hazard (FHb) assessment map.

Figure 7A,B demonstrates the maximum and minimum value of each pixel's flood hazard assessment, respectively, after considering the uncertainty in the weighting coefficients. According to the flood hazard analysis, the northeastern eastern and southeastern parts of the study area host the most flood-prone regions.

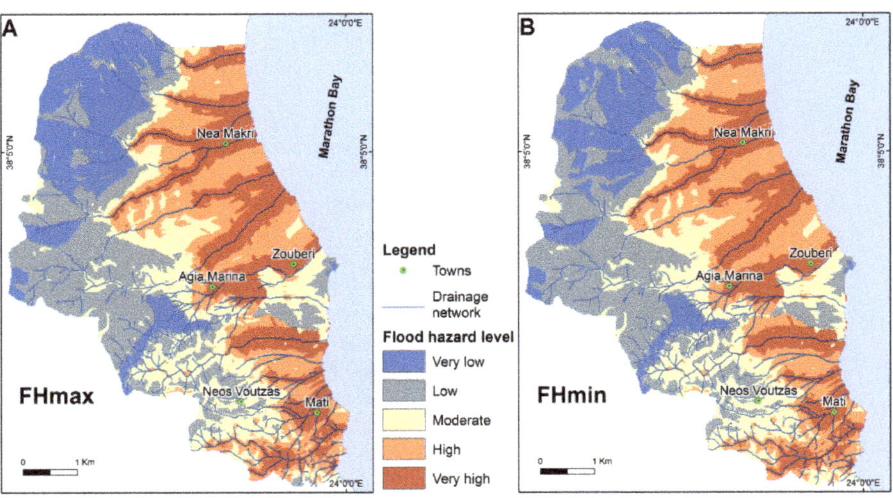

Figure 7. Maps illustrating (**A**) the higher (FHmax) and (**B**) the lower (FHmin) values of the flood hazard assessment.

The percentages corresponding to the area of each hazard zone for the three maps of Figures 6 and 7 are tabulated in Table 4. In the areas of the high and very high zones of the map with the upper flood hazard (FHmax), values slightly increased compared to the corresponding zones of the basic map (FH). The similar correlation proves that the spatial extents of moderate and low hazard zones decreased, while the very low hazard zone increase. Comparing the zones of the map with the lower flood hazard values (FHmin) to those of the basic map (FH), the areas of very high and high hazard zones showed a little increase, the moderate and low hazard areas slightly decreased, while the area of very low hazard zone remained unchanged.

Table 4. Percentages represent the area of each flood hazard zone (areas are shown in Figures 6 and 7).

Flood Hazard Zones	Percentage % of Total Study Area		
	FH	FHmax	FHmin
Very low	14.4	16.5	14.4
Low	25.6	23.5	25.3
Moderate	20.1	19.6	19.0
High	20.1	20.6	21.4
Very high	19.8	20.1	20.1

Table 5 tabulates the estimated frequency ratio of the flood events for each flood hazard zone. According to [54–56], when the ratio values are more than 1, then there is a strong relationship between events and the given natural hazard zone.

Table 5. The frequency ratio (FR) values of floods in flood hazard zones; A = area of each flood hazard zone in Km^2, AC = the area ratio for each flood hazard zone to the total area in a percentage form, Fn = number of floods in each zone, FF = the frequency of floods expressed in percentage.

Flood Hazard Zones	A	AC	FN	FF	FR
Very low	4.7	14.4	0	0.0	0.0
Low	8.4	25.6	0	0.0	0.0
Moderate	6.6	20.1	1	11.1	0.6
High	6.6	20.1	3	33.3	1.7
Very high	6.5	19.8	5	55.6	2.8

Table 5 shows that flood events were significantly correlated to very high and high hazard zones. In the very high hazard area, the frequency ratio was higher than 2, demonstrating a very high probability of flooding. Similarly, the ratio value was observed to be 1.7, in the high hazard zone, showing a high likelihood of flood occurrences. Because the ratio value was lower than 1 in the moderate hazard zone, the probability of floods is low. Since the ratio value was found to be equal to 0 in the low and the very low hazard zones, the likelihood of flooding is minimal.

Moreover, the urban area and the road network were superimposed on the basic flood hazard map. In this context, the area of the urban fabric and the length of the road network that overlaps into each flood hazard zone were estimated (Table 6).

Table 6. Urban fabric area (UF) and length of the road network (RN) that fall into each flood hazard zone.

Flood Hazard Zones	UF (Km^2)	%	RN (Km)	%
Very low	0.0	0	22.5	6
Low	0.5	4	79.2	22
Moderate	2.7	19	82.0	22
High	5.4	37	83.4	23
Very high	5.7	40	99.4	27
Total	14.3	100	366.5	100

The results demonstrate that the vast majority of the urban areas are located within very high and high flood hazard zones. Half of the road network's total length is situated within very high and high flood hazard zones. In contrast, small parts of the urban fabric and the other half of the road network are located within very low, moderate and low landslide hazard zones.

5. Discussion

The urban area constitutes about 33% of the entire study area. The study area has suffered from urban fires in 2018, resulting in about 25% of the area being burned. According to Fox [21], Euro-Mediterranean countries have common post-fire risk characteristics. They have relatively high population densities, and much of the coastal lowland area is developed. Hence, even small fires can be hazardous, and after fire these coastal areas are vulnerable to flooding. So since the study area is a coastal area and more than half of it is occupied by urban and burned areas, is at high risk of flooding.

In the present study, a method is presented to assess the flood hazard in burned areas and it can be applied even outside the localities affected by the fire. It provides an estimation of flood-prone areas where data are not available, which are typically used more often in connection with floods. Six physical and anthropogenic factors were considered as the most relevant parameters affecting the watercourse when intense storm events cause high runoff that exceeds drainage system capacity. The physical factors, related to flood activity, were slope, elevation, distance from open channel streams, and hydrolithology. The anthropogenic factors associated with floods were distance from totally covered streams, along with burnt and urban areas derived from land cover, as well as supplementary land uses.

A flood hazard assessment map was created via the AHP method and a GIS. According to Hervás and Bobrowsky [57], flood hazard refers to the probability of flood occurrence within a reference period of time. In the present study, the link of the flood hazard assessment to return periods was not possible. So, a flood susceptibility map was produced which refers to the spatial likelihood or probability for a flood to occur in the future. This map represents the potential flood hazard in the study area, and was further categorized by using the standard deviation method with five classes (Figure 6) of hazard level: very low, low, moderate, high and very high.

The areas of flood hazard varying between very high and high were recorded mostly in the eastern and southern parts of the Nea Makri area. The area of these two hazard zones covers almost 40% of the whole area (Table 4). In Greece, extreme flash flood events of low frequency and high magnitude occur in many fluvial systems with a relatively small extent drained by ephemeral streams [58]. Thus, the small drainage basins, along with the torrential behavior of the streams, influence the flood manifestations in the study area. Moreover, the lowland morphology, gentle slope, the covered streams and intense urbanization in the eastern part of the municipality unit of Nea Makri increase the runoff and produce encouraging locations for flood occurrences. According to Shakesby [59], in Euro-Mediterranean regions wildfires in forests leave the terrain bare and vulnerable to storms which can cause severe runoff and soil erosion. The runoff increases often in the upper parts of watersheds, and thus poses flooding risks in urban zones located downstream. Thus, the burned area in the southern part of the study area creates favorable conditions for flooding.

In contrast, low and very low flood hazard areas are situated in the mountainous northern, northwestern and western parts of the municipality unit of Nea Makri. These areas are not urbanized, contain open stream channels and do not include burned areas. Therefore, the hazard of flooding in these areas is very low. The spatial distribution of the moderate hazard zone is observed primarily in the central, eastern and southern parts of the municipality unit of Nea Makri. These three hazard zones cover almost 60% of the whole area (Table 4).

The AHP method is a powerful and valuable tool for flood hazard analysis [32,36,46,47], although it shows several constraints in assessing the impact of uncertainties [60,61]. This fact was evaluated by calculating two additional scenarios that present the upper and lower flood hazard values after accounting for the uncertainty in the weighting coefficients (Table 3). The findings of the uncertainty analysis showed that the two additional maps (FHmin and FHmax) illustrate a low variation of the presence of flood hazard zones (Table 5) concerning the basic flood hazard map (FHb). Consequently, the analysis proved slight differences in the spatial and quantitative distribution of the flood hazard zones. This fact indicates robust behavior for the predictions of the applied method.

The combination of several different maps can create a map that does not contain an area's actual hazard level [62]. Thus, the basic flood hazard map was verified by means of frequency ratio and flood events which affected the study area over the past 20 years. The findings established that the vast majority of the floods, almost 90% (Table 5), were located within the limits of the high and very high flood hazard zones. Moreover, the frequency ratio was higher than 1 in the high and very high hazard zones, indicating that these hazard zones are strongly correlated with flood occurrences (Table 5). The flood hazard intensity rises as the frequency ratio values increase.

The applied method brings many advantages; first of all, that of offering completeness of the analysis of the territory and methodological homogeneity. It is simple and easy to use where primary data are scarce. The results provide reasonable values, which could lead to its further application. The used approach is, therefore, useful in synergy with methods linked to the attribution of a probability.

Comparing the existing urban fabric and road network to the proposed hazard areas of the basic flood hazard map, it is recorded that the vast majority (77%) of the urban fabric and 50% of the road network's length are situated within the boundaries of high and very high flood hazard zones (Table 6). This result demonstrated that they have been constructed in areas prone to flood. In these areas, proactive planning and selection of the proper construction rules are essential for the prevention and mitigation of the consequences of flood hazards.

Detailed works relating to the condition of events and evolution of natural hazards and the vulnerability of the exposed elements at risk (i.e., buildings) will help propose the proper protection measures [46]. The application of the procedure facilitates identifying those sites from the already existing urban fabric and road network situated in non-safe areas. Thus, the appropriate mitigation or hazard reduction approaches can be more effectively designed and applied. Furthermore, the awareness of the non-safe regions related to flood hazard might be helpful in emergency preparedness planning.

Therefore, scientists, engineers, stakeholders, planners and decision makers may utilize the proposed approach in forthcoming spatial planning projects [63–68]. Additionally, the local authorities may use the produced map to guide the adoption of measures and strategies aiming towards flood hazard mitigation and post-fire management.

6. Conclusions

In the present work, a burned and urban area of the Attica peninsula in central Greece was selected to produce a flood hazard assessment map. The proposed areas of very high and high flood hazard are located mainly in the study area's eastern and southern parts as a consequence of the synthesis of lowlands with gentle slope, ephemeral streams with torrential behavior, totally covered streams and urban areas. The burned area in southern part of the study area will increase surface runoff and create favorable conditions for flooding. The uncertainty analysis showed insignificant variations of the spatial distribution of the flood hazard zones. The flood hazard map presents a significant correlation between the flood hazard zones and the spatial distribution of flood phenomena. This verification demonstrated reliable results and high accuracy of the created map. The spatial distribution of the existent urban fabric and road network was analyzed to identify the parts situated in unsafe areas to design and apply flood hazard mitigation measures.

About 80% of the urban areas and 50% of the road network's length were located within areas prone to flood. The proposed method offers completeness of the analysis of the territory, is simple to apply where primary data are scarce and provides reasonable results, which could lead to its further application. Therefore, scientists, stakeholders, engineers and decision makers may apply the proposed approach in forthcoming land use planning projects. Flood hazard mitigation techniques can be more effectively designed, while the proposed approach may be useful also in emergency preparedness planning and post-fire management.

Author Contributions: Conceptualization, G.D.B. and H.D.S.; methodology, G.D.B. and H.D.S.; software, G.D.B. and H.D.S.; validation, G.D.B. and H.D.S.; formal analysis, G.D.B. and H.D.S.; investigation, G.D.B., H.D.S. and D.E.A.; resources, G.D.B., H.D.S. and D.E.A.; data curation, G.D.B. and H.D.S.; writing—original draft preparation, G.D.B., H.D.S. and D.E.A.; writing—review and editing, G.D.B., H.D.S. and D.E.A.; visualization, G.D.B. and H.D.S.; supervision, G.D.B. and H.D.S. All authors have read and agreed to the published version of the manuscript.

Funding: This research received no external funding.

Institutional Review Board Statement: Not applicable.

Informed Consent Statement: Not applicable.

Data Availability Statement: Data that support the findings of this study are available from the corresponding author upon reasonable request.

Acknowledgments: The constructive and thorough reviews of anonymous reviewers are warmly acknowledged.

Conflicts of Interest: The authors declare no conflict of interest.

References

1. Cerdà, A. Effect of Climate on Surface Flow Along a Climatological Gradient in Israel: A Field Rainfall Simulation Approach. *J. Arid Environ.* **1998**, *38*, 145–159. [CrossRef]
2. Bathrellos, G.D.; Skilodimou, H.D.; Maroukian, H. The spatial distribution of Middle and Late Pleistocene cirques in Greece. *Geogr. Ann.* **2014**, *96*, 323–338. [CrossRef]
3. Skilodimou, H.D.; Bathrellos, G.D.; Maroukian, H.; Gaki-Papanastassiou, K. Late Quaternary evolution of the lower reaches of Ziliana stream in south Mt. Olympus (Greece). *Geogr. Fis. Din. Quat.* **2014**, *37*, 43–50.
4. Bathrellos, G.D.; Skilodimou, H.D.; Maroukian, H.; Gaki-Papanastassiou, K.; Kouli, K.; Tsourou, T.; Tsaparas, N. Pleistocene glacial and lacustrine activity in the southern part of Mount Olympus (central Greece). *Area* **2017**, *49*, 137–147. [CrossRef]
5. Kamberis, E.; Bathrellos, G.; Kokinou, E.; Skilodimou, H. Correlation between the structural pattern and the development of the hydrographic network in a portion of the Western Thessaly basin (Greece). *Cent. Eur. J. Geosci.* **2012**, *4*, 416–424. [CrossRef]
6. Kokinou, E.; Skilodimou, H.D.; Bathrellos, G.D.; Antonarakou, A.; Kamberis, E. Morphotectonic analysis, structural evolution/pattern of a contractional ridge: Giouchtas Mt., Central Crete, Greece. *J. Earth Syst. Sci.* **2015**, *124*, 587–602.
7. Chousianitis, K.; Del Gaudio, V.; Sabatakakis, N.; Kavoura, K.; Drakatos, G.; Bathrellos, G.D.; Skilodimou, H.D. Assessment of Earthquake-Induced Landslide Hazard in Greece: From Arias Intensity to Spatial Distribution of Slope Resistance Demand. *Bull. Seismol. Soc. Am.* **2016**, *106*, 174–188. [CrossRef]
8. Bathrellos, G.D.; Skilodimou, H.D.; Maroukian, H. The significance of tectonism in the glaciations of Greece. *Geol. Soc.* **2017**, *433*, 237–250. [CrossRef]
9. Papadopoulou-Vrynioti, K.; Bathrellos, G.D.; Skilodimou, H.D.; Kaviris, G.; Makropoulos, K. Karst collapse susceptibility mapping considering peak ground acceleration in a rapidly growing urban area. *Eng. Geol.* **2013**, *158*, 77–88. [CrossRef]
10. Tsolaki-Fiaka, S.; Bathrellos, G.D.; Skilodimou, H.D. Multi-Criteria Decision Analysis for an Abandoned Quarry in the Evros Region (NE Greece). *Land* **2018**, *7*, 43. [CrossRef]
11. Skilodimou, H.D.; Bathrellos, G.D.; Koskeridou, E.; Soukis, K.; Rozos, D. Physical and anthropogenic factors related to landslide activity in the Northern Peloponnese, Greece. *Land* **2018**, *7*, 85. [CrossRef]
12. Bathrellos, G.D.; Kalivas, D.P.; Skilodimou, H.D. GIS-based landslide susceptibility mapping models applied to natural and urban planning in Trikala, Central Greece. *Estud. Geol.* **2009**, *65*, 49–65. [CrossRef]
13. Bathrellos, G.D.; Skilodimou, H.D.; Soukis, K.; Koskeridou, E. Temporal and Spatial Analysis of Flood Occurrences in the Drainage Basin of Pinios River (Thessaly, Central Greece). *Land* **2018**, *7*, 106. [CrossRef]
14. Skilodimou, H.D.; Livaditis, G.; Bathrellos, G.D.; Verikiou-Papaspiridakou, E. Investigating the flooding events of the urban regions of Glyfada and Voula, Attica, Greece: A contribution to Urban Geomorphology. *Geogr. Ann.* **2003**, *85*, 197–204. [CrossRef]

15. Merz, B.; Kreibich, H.; Schwarze, R.; Thieken, A. Assessment of economic flood damage. *Nat. Hazards Earth Syst. Sci.* **2010**, *10*, 1679–1724. [CrossRef]
16. European Environment Agency (EEA). *Mapping the Impacts of Natural Hazards and Technological Accidents in Europe: An Overview of the Last Decade*; Report 13/2010; Office for Official Publications of the European Communities, European Environment Agency (EEA): Luxembourg, 2010.
17. Gaume, E.; Bain, V.; Bernardara, P.; Newinger, O.; Barbuc, M. A compilation of data on European flash floods. *J. Hydrol.* **2009**, *367*, 70–78. [CrossRef]
18. Chuvieco, E.; Mouillot, F.; Pereira, J.M.C.; Oom, D. A new global burned area product for climate assessment of fire impacts. Glob. *Ecol. Biogeogr.* **2016**, *5*, 619–629. [CrossRef]
19. Bassi, S.; Kettunen, M. *Forest Fires: Causes and Contributing Factors in Europe*; IP/A/ ENVI/ST/2007-15,PE401.003; European Parliament, Policy Department, Economic and Scientific Policy: Brussels, Belgium, 2008.
20. Dennis, M.S.; Joseph, E.G.; Jason, W.K. Objective Definition of Rainfall Intensity-Duration Thresholds for Post-fire Flash Floods and Debris Flows in the Area Burned by the Waldo Canyon Fire, Colorado, USA. In *Engineering Geology for Society and Territory*; Lollino, G., Ed.; Springer: Cham, Switzerland, 2015; Volume 2.
21. Fox, D.M.; Laaroussi, Y.; Malkinson, L.D.; Maselli, F.; Andrieu, J.; Bottai, L.; Wittenberg, L. POSTFIRE: A model to map forest fire burn scar and estimate runoff and soil erosion risks. *Remote Sens. Appl. Soc. Environ.* **2016**, *4*, 83–91. [CrossRef]
22. Bathrellos, G.D.; Gaki-Papanastassiou, K.; Skilodimou, H.D.; Papanastassiou, D.; Chousianitis, K.G. Potential suitability for urban planning and industry development using natural hazard maps and geological–geomorphological parameters. *Environ. Earth Sci.* **2012**, *66*, 537–548. [CrossRef]
23. Alexakis, D. Suburban areas in flames: Dispersion of potentially toxic elements from burned vegetation and buildings. Estimation of the associated ecological and human health risk. *Environ. Res.* **2020**, *183*, 109153. [CrossRef] [PubMed]
24. Alexakis, D.; Kokmotos, I.; Gamvroula, D.; Varelidis, G. Wildfire effects on soil quality: Application on a suburban area of West Attica (Greece). *Geosci. J.* **2021**, *25*, 243–253. [CrossRef]
25. Alexakis, D. Contaminated land by wildfire effect on ultramafic soil and associated human health and ecological risk. *Land* **2020**, *9*, 409. [CrossRef]
26. Rahmati, O.; Zeinivand, H.; Besharat, M. Flood hazard zoning in Yasooj region, Iran, using GIS and multi-criteria decision analysis. *Geomat. Nat. Hazards Risk* **2016**, *7*, 1000–1017. [CrossRef]
27. Fernández, D.S.; Lutz, M.A. Urban flood hazard zoning in Tucumán Province, Argentina, using GIS and multicriteria decision analysis. *Eng. Geol.* **2010**, *111*, 90–98. [CrossRef]
28. Merwade, V.; Cook, A.; Coonrod, J. GIS techniques for creating river terrain models for hydrodynamic modeling and flood inundation mapping. *Environ. Model. Softw.* **2008**, *23*, 1300–1311. [CrossRef]
29. Tate, E.C.; Maidment, D.R.; Olivera, F.; Anderson, D.J. Creating a terrain model for floodplain mapping. *J. Hydrol. Eng.* **2002**, *7*, 100–108. [CrossRef]
30. De Moel, H.; van Alphen, J.; Aerts, J.C.J.H. Flood maps in Europe—Methods, availability and use. *Nat. Hazards Earth Syst. Sci.* **2009**, *9*, 289–301. [CrossRef]
31. Stefanidis, S.; Stathis, D. Assessment of flood hazard based on natural and anthropogenic factors using analytic hierarchy process (AHP). *Nat. Hazards* **2013**, *68*, 569–585. [CrossRef]
32. Bathrellos, G.D.; Karymbalis, E.; Skilodimou, H.D.; Gaki-Papanastassiou, K.; Baltas, E.A. Urban flood hazard assessment in the basin of Athens Metropolitan city, Greece. *Environ. Earth Sci.* **2016**, *75*, 319. [CrossRef]
33. Saaty, T.L. A scaling method for priorities in hierarchical structures. *J. Math. Psychol.* **1977**, *15*, 234–281. [CrossRef]
34. Saaty, T.L. How to make a decision: The Analytic Hierarchy Process. *Eur. J. Oper. Res.* **1990**, *48*, 9–26. [CrossRef]
35. Saaty, T.L. Decision making—The Analytic Hierarchy and Network Processes (AHP/ANP). *J. Syst. Sci. Syst. Eng.* **2004**, *13*, 1–35. [CrossRef]
36. Bathrellos, G.D.; Gaki-Papanastassiou, K.; Skilodimou, H.D.; Skianis, G.A.; Chousianitis, K.G. Assessment of rural community and agricultural development using geomorphological-geological factors and GIS in the Trikala prefecture (Central Greece). *Stoch. Environ. Res. Risk Assess.* **2013**, *27*, 573–588. [CrossRef]
37. Bathrellos, G.D.; Kalivas, D.P.; Skilodimou, H.D. Landslide Susceptibility Assessment Mapping: A Case Study in Central Greece. In *Remote Sensing of Hydrometeorological Hazards*; Petropoulos, G.P., Islam, T., Eds.; CRC Press, Taylor & Francis Group: London, UK, 2017; pp. 493–512. ISBN 13978-1498777582.
38. Georgiopoulos, G.; Kambas, M. 'Armageddon' Fire in Greece Kills at Least 80, Many Missing. In Reuters. 2018. Available online: https://www.reuters.com/article (accessed on 25 July 2018).
39. Mitsopoulos, I.; Mallinis, G.; Dimitrakopoulos, A.; Xanthopoulos, G.; Eftychidis, G.; Goldammer, J.G. Vulnerability of peri-urban and residential areas to landscape fires in Greece: Evidence by wildland-urban interface data. *Data Brief* **2020**, *31*, 106025. [CrossRef]
40. Authority, H.S. 2011 Population-Housing Census. 2011. Available online: www.statistics.gr/en/2011-census-pop-house (accessed on 27 December 2018).
41. Ministry of Environment and Energy. Floods, Historic Floods. 2019. Available online: http://www.ypeka.gr/Default.aspx?tabid=252&language=el-GR (accessed on 21 February 2021).

42. Copernicus. Copernicus Land Monitoring Service. 2016. Available online: http://land.copernicus.eu (accessed on 21 February 2021).
43. Copernicus. Urban Atlas 2018. Copernicus Land Monitoring Service. 2021. Available online: https://land.copernicus.eu/local/urban-atlas/urban-atlas-2018 (accessed on 21 February 2021).
44. European Council. *EU Directive of the European Parliament and of the European Council on the Assessment and Management of Flood Risks (2007/60/EU)*; European Council: Brussels, Belgium, 2007.
45. Papaioannou, G.; Vasiliades, L.; Loukas, A. Multi-criteria analysis framework for potential flood prone areas mapping. *Water Resour. Manag.* **2015**, *29*, 399–418. [CrossRef]
46. Bathrellos, G.D.; Skilodimou, H.D.; Chousianitis, K.; Youssef, A.M.; Pradhan, B. Suitability estimation for urban development using multi-hazard assessment map. *Sci. Total Environ.* **2017**, *575*, 119–134. [CrossRef] [PubMed]
47. Skilodimou, H.D.; Bathrellos, G.D.; Chousianitis, K.; Youssef, A.M.; Pradhan, B. Multi-hazard assessment modeling via multi-criteria analysis and GIS: A case study. *Environ. Earth Sci.* **2019**, *78*, 47. [CrossRef]
48. Migiros, G.; Bathrellos, G.D.; Skilodimou, H.D.; Karamousalis, T. Pinios (Peneus) River (Central Greece): Hydrological—Geomorphological elements and changes during the Quaternary. *Cent. Eur. J. Geosci.* **2011**, *3*, 215–228. [CrossRef]
49. Youssef, A.M.; Sefry, S.A.; Pradhan, B.; Al Fadail, E.A. Analysis on causes of flash flood in Jeddah city (Kingdom of Saudi Arabia) of 2009 and 2011 using multi-sensor remote sensing data and GIS. *Geomat. Nat. Haz. Risk* **2016**, *7*, 1018–1042. [CrossRef]
50. Hutchinson, M.F.; Xu, T.; Stein, J.A. Recent progress in the ANUDEM elevation gridding procedure. *Geomorphometry* **2011**, *2011*, 19–22.
51. Arun, P.V. A comparative analysis of different DEM interpolation methods. *Egypt. J. Remote Sens. Space Sci.* **2013**, *16*, 133–139.
52. Bathrellos, G.D.; Skilodimou, H.D.; Kelepertsis, A.; Alexakis, D.; Chrisanthaki, I.; Archonti, D. Environmental research of groundwater in the urban and suburban areas of Attica region, Greece. *Environ. Geol.* **2008**, *56*, 11–18. [CrossRef]
53. Van Westen, C.; Kappes, M.S.; Luna, B.Q.; Frigerio, S.; Glade, T.; Malet, J.P. Medium-scale multi-hazard risk assessment of gravitational processes. In *Mountain Risks: From Prediction to Management and Governance*; Springer: Dordrecht, The Netherlands, 2014; pp. 201–231.
54. Lee, S.; Pradhan, B. Probabilistic landslide hazards and risk mapping on Penang Island, Malaysia. *J. Earth Syst. Sci.* **2006**, *115*, 661–672. [CrossRef]
55. Rozos, D.; Skilodimou, H.D.; Loupasakis, C.; Bathrellos, G.D. Application of the revised universal soil loss equation model on landslide prevention. An example from N. Euboea (Evia) Island, Greece. *Environ. Earth Sci.* **2013**, *70*, 3255–3266. [CrossRef]
56. Pradhan, B.; Lee, S. Delineation of landslide hazard areas on Penang Island, Malaysia, by using frequency ratio, logistic regression, and artificial neural network models. *Environ. Earth Sci.* **2010**, *60*, 1037–1054. [CrossRef]
57. Hervás, J.; Bobrowsky, P. Mapping: Inventories, susceptibility, hazard and risk. In *Landslides—Disaster Risk Reduction*; Springer: Berlin/Heidelberg, Germany, 2009; pp. 321–349.
58. Karymbalis, E.; Katsafados, P.; Chalkias, C.; Gaki-Papanastassiou, K. An integrated study for the evaluation of natural and anthropogenic causes of flooding in small catchments based on geomorphological and meteorological data and modeling techniques: The case of the Xerias torrent (Corinth, Greece). *Z. Geomorphol.* **2012**, *56*, 45–67. [CrossRef]
59. Shakesby, R.A. Post-wildfire soil erosion in the Mediterranean: Review and future research directions. *Earth Sci. Rev.* **2011**, *105*, 71–100. [CrossRef]
60. Rozos, D.; Bathrellos, G.D.; Skilodimou, H.D. Comparison of the implementation of rock engineering system and analytic hierarchy process methods, upon landslide susceptibility mapping, using GIS: A case study from the Eastern Achaia County of Peloponnesus, Greece. *Environ. Earth Sci.* **2011**, *63*, 49–63. [CrossRef]
61. Nefeslioglu, H.A.; Sezer, E.A.; Gokceoglu, C.; Ayas, Z. A modified analytical hierarchy process (M-AHP) approach for decision support systems in natural hazard assessments. *Comput. Geosci.* **2013**, *59*, 1–8. [CrossRef]
62. Kappes, M.S.; Keiler, M.; von Elverfeldt, K.; Glade, T. Challenges of analyzing multi-hazard risk: A review. *Nat. Hazards* **2012**, *64*, 1925–1958. [CrossRef]
63. Bathrellos, G.D.; Skilodimou, H.D. Land Use Planning for Natural Hazards. *Land* **2019**, *8*, 128. [CrossRef]
64. Makri, P.; Stathopoulou, E.; Hermides, D.; Kontakiotis, G.; Zarkogiannis, S.D.; Skilodimou, H.D.; Bathrellos, G.D.; Antonarakou, A.; Scoullos, M. The Environmental Impact of a Complex Hydrogeological System on Hydrocarbon-Pollutants' Natural Attenuation: The Case of the Coastal Aquifers in Eleusis, West Attica, Greece. *J. Mar. Sci. Eng.* **2020**, *8*, 1018. [CrossRef]
65. Panagopoulos, G.P.; Bathrellos, G.D.; Skilodimou, H.D.; Martsouka, F.A. Mapping Urban Water Demands Using Multi-Criteria Analysis and GIS. *Water Resour. Manag.* **2012**, *26*, 1347–1363. [CrossRef]
66. Papadopoulou-Vrynioti, K.; Alexakis, D.; Bathrellos, G.D.; Skilodimou, H.D.; Vryniotis, D.; Vasiliades, E.; Gamvroula, D. Distribution of trace elements in stream sediments of Arta plain (western Hellas): The influence of geomorphological parameters. *J. Geochem. Explor.* **2013**, *134*, 17–26. [CrossRef]
67. Papadopoulou-Vrynioti, K.; Alexakis, D.; Bathrellos, G.D.; Skilodimou, H.D.; Vryniotis, D.; Vasiliades, E. Environmental research and evaluation of agricultural soil of the Arta plain, western Hellas. *J. Geochem. Explor.* **2014**, *136*, 84–92. [CrossRef]
68. Skilodimou, H.; Stefouli, M.; Bathrellos, G. Spatio-temporal analysis of the coastline of Faliro Bay, Attica, Greece. *Estud. Geol.-Madrid* **2002**, *58*, 87–93.

Article

A Bayesian Network-Based Integrated for Flood Risk Assessment (InFRA)

Hongjun Joo [1], Changhyun Choi [1], Jungwook Kim [1], Deokhwan Kim [2], Soojun Kim [1,*] and Hung Soo Kim [1]

1 Department of Civil Engineering, Inha University, Incheon 22212, Korea
2 Department of Land, Water and Environment Research, Korea Institute of Civil Engineering and Building Technology, Ilsan 10223, Korea
* Correspondence: sk325@inha.ac.kr; Tel.: +82-32-860-7563

Received: 28 May 2019; Accepted: 2 July 2019; Published: 9 July 2019

Abstract: Floods are natural disasters that should be considered a top priority in disaster management, and various methods have been developed to evaluate the risks. However, each method has different results and may confuse decision-makers in disaster management. In this study, a flood risk assessment method is proposed to integrate various methods to overcome these problems. Using factor analysis and principal component analysis (PCA), the leading indicators that affect flood damage were selected and weighted using three methods: the analytic hierarchy process (AHP), constant sum scale (CSS), and entropy. However, each method has flaws due to inconsistent weights. Therefore, a Bayesian network was used to present the integrated weights that reflect the characteristics of each method. Moreover, a relationship is proposed between the elements and the indicators based on the weights called the Integrated Index for Flood Risk Assessment (InFRA). InFRA and other assessment methods were compared by receiver operating characteristics (ROC)-area under curve (AUC) analysis. As a result, InFRA showed better applicability since InFRA was 0.67 and other methods were less than 0.5.

Keywords: flood risk; Bayesian networks; integrated index for flood risk assessment

---

1. Introduction

Recent major disasters highlight the importance of disaster preparedness around the world and emphasize the concept of disaster risk across communities. Floods are major natural disasters and have been studied with great interest worldwide [1,2]. Federal Emergency Management Agency (FEMA) [3] and National Oceanic and Atmospheric Administration (NOAA) [4] developed a risk assessment program to estimate the extent of damage related to disasters. Munich Re Group [5] classified disasters into four categories after evaluating disaster scenarios using four factors (natural, technological, socio-political, and economic factors) and direct/indirect damages. The Tyndall Centre [6] classified flood vulnerability into social and biological categories and proposed a basic framework to improve vulnerability-specific adaptability.

Rygel et al. [7] suggested that the most important issue in a vulnerability assessment is selecting appropriate indicators. They proposed a method to evaluate flood risk using the Pareto ranking process after selecting and collecting vulnerabilities into two categories (exposure and sociological). Chang and Huang [8] selected potential impact indicators (PIs) for urban areas in Taiwan and estimated the flood risk index by combining PIs with adaptive capacity indicators (AIs). Kablan et al. [9] estimated a flood vulnerability index based on the concept of climate change vulnerability assessment through proxy variables that are relevant to disaster risk management and adaptation to climate changes using three flooding indices: (1) an exposure index (EI), (2) sensitivity index (SI), and (3) adaptive capacity index (AI). In addition, some studies have examined the impact of forest areas on flood risk [10] and the relationship between intensity-duration-frequency (IDF) curves and flood risk [11].

Many research institutes around the world have also assessed flood risk, including the Environment Agency (EA), Jeollabuk-do Total Human Institute National of Korea (JTHINK), Korea Institute of Civil Engineering and Building Technology(KICT), Korea Research Institute for Human Settlements (KRIHS), Korea Environment Institute(KEI), Ministry of Land, Transport and Maritime Affairs (MOLTMA), National Disaster Management Research Institute (NDMI), National Institute for Land and Infrastructure Management (NILIM), and Seoul Institute (SI) [12–21]. The above studies commonly selected indicators that are expected to have an effect on flood risk. Then, the conclusive flood risk is derived from the weight between each indicator (mainly expert questionnaire or subjective judgment). Their goal is to address flood management through flood risk assessment.

To assess flood risks, it is essential to select indicators that affect floods and assign them reasonable weights. However, as mentioned earlier, most studies lack a basis for the selection of indicators, and they have typically selected indicators based on the frequency of their use in other studies or a subjective view on the importance of such indicators. In addition, most flood risk assessment methods are not differentiated because they use similar estimation methods. These problems have not been validated for the methods to be actually applied.

The purpose of this study is to develop a flood risk assessment method that can address the problems of undifferentiated and accurate estimates of previous methods. To this end, a methodology is proposed to derive the integrated weights of components and indicators using Bayesian networks (BNs) as an integrated decision model after selecting representative indicators among the existing flood risk indicators through factor analysis and principal component analysis (PCA). Section 2 explains the basic theories behind the methodologies that are material in this study, and Section 3 discusses the result of the methodology used on the target areas. Finally, Section 4 presents the conclusions.

2. Materials and Theories

2.1. Existing Flood Risk Assessment Indices

In general, flood risk is computed by multiplying three factors of vulnerability related to flood occurrence: (1) hazard; (2) asset or human exposure; and (3) lack of flood protection [22]. Based on these definitions, many flood risk assessment indicators have been developed for flood risk management. The indicators mainly used in Korea are the potential flood damage (PFD) [17], excess flood vulnerability index (EFVI) [18], flood disaster risk reduction index (FDRRI) [19], flood vulnerability assessment (FVA) [13], flood damage index (FDI) [15], and regional safety assessment (RSA) [21]. These six methods have been used mainly in Korea because they are well known to be applicable with general indicators for assessing flood risk. Each assessment method is estimated by the general procedure with almost similar methods presented in introduction.

In this study, the factors of these indicators were reviewed and classified into four components: (1) hydro-geology; (2) socio-economics; (3) protection; and (4) climate. Furthermore, 28 indicators were also used for the components, as shown in Table 1. The flood risk index (FRI) increases as the H (hydro-geology), S (socio-economics), and C (climate) components increase. The index decreases as the P (protection) component increases. With these indicators, the FRI can be expressed as follows:

$$\text{FRI} = (\text{H} \times \text{S} \times \text{C})/\text{P} \quad (1)$$

Table 1. Classification of components and indicators using six assessment methods.

Classification		MOLTMA		NDMI	JTHINK	KRIHS	SI
Components	**Indicators**	**EFVI**	**PFD**	**FDRRI**	**FVA**	**FDI**	**RSA**
A. Hydro-geology	(1) Flood hazard area	○					○
	(2) Flood damage cost for public facilities	○	○	○	○		○
	(3) Imperviousness	○		○		○	○
	(4) Urban rate		○				
	(5) Curve number (CN)				○		
	(6) Basin slope	○		○	○		
	(7) Lowland area rate				○		○
	(8) Stream density					○	
B. Socio-economy	(9) Population density	○	○	○		○	
	(10) Asset (reference land price)		○	○			○
	(11) Financial independence rate	○		○	○	○	
	(12) Infrastructure		○			○	○
	(13) Dependence population	○		○			
	(14) Manufacturing output in value					○	
	(15) Total number of houses					○	
C. Climate	(16) Frequency of hourly rainfall (P ≥ 50 mm)	○		○			
	(17) Frequency of intensive rainfall per day (P ≥ 150 mm)	○		○			
	(18) Maximum hourly rainfall						○
	(19) Annual precipitation					○	
	(20) Probability rainfall		○				
D. Flood Protection	(21) Levee maintenance	○		○			
	(22) Levee length	○		○			
	(23) Pump station (number)						○
	(24) Pump station (capacity)					○	
	(25) Dam and reservoir		○				
	(26) Drainage capacity						
	(27) Number of public servants per resident					○	
	(28) Index of damage reduction ability			○			○

2.2. Methodology for Selecting Representative Indicators

Factor analysis is used to reduce the complexity of data by grouping measurement variables into common factors and determining whether the measured variables measure the desired data in the same construct [23]. Factor analysis is advantageous because it is relatively free of constraints of multicollinearity, it can classify variables by factors in the development process of a measurement scale, and it can analyze them based on the correlation of variables. principal component analysis (PCA) in factor analysis is a technique for creating a small number of new variables by combining many highly correlated variables. It is a method of forming a group of components that represent many variables by reducing the dimensionality of data [24]. In addition, it can easily identify the component that has the highest explaining power in the group because the component of each variable can be quantified.

Figure 1 illustrates the method of selecting representative indicators using factor analysis and PCA. Assuming that there are six indicators A–F, they can be grouped by factor analysis, and the ones that have high explanatory power can be selected using PCA. In this way, the complexity of the FRI can be effectively reduced.

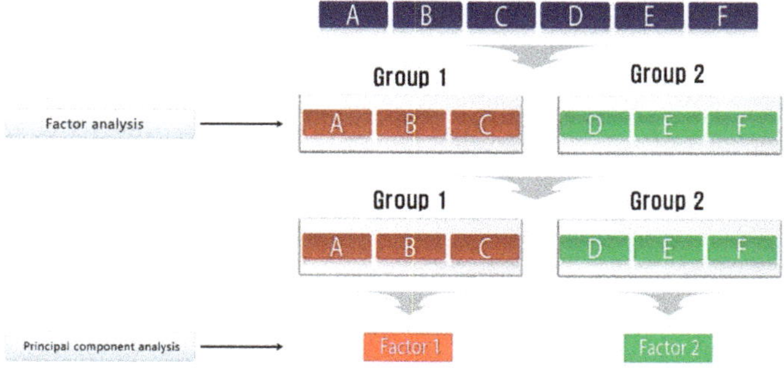

Figure 1. Overview of factor analysis and principal component analysis (PCA).

2.3. Methodology for Assigning Weights

Weights should be assigned according to the importance of each indicator that affects the flood in assessing flood risk. There are many theories for weight assignment methods, and it is difficult to say that these weighting techniques differ in explanatory power or merit. Only the classification methods differ according to the purpose of a study or a subjective view about importance. The weighting techniques originate from differences in assumptions that lead to values or preferences [25], and they are categorized into direct and indirect methods, including surveys [26]. In this study, the analytic hierarchy process (AHP), constant sum scale (CSS), and entropy weight were selected for weighting the flood risk indicators identified to derive the FRI.

AHP is a typical method of multicriteria decision-making (MCDM). It forms a hierarchical structure by assessment items and evaluates alternatives through pairwise comparison. Quantitative and qualitative data can be processed on a ratio scale, which makes the method useful for verifying objectivity through the process of secondary processing of data [27,28]. CSS provides a consistent fixed total score to respondents and divides the score according to the relative importance of the attributes within the total score [29]. The score used for the total fixed scale is usually 10 or 100 based on the number of factors and indicators.

The entropy weight technique is based on the theory that information about the signal can be measured indirectly with a degree of reduction of uncertainty. In this sense, information and uncertainty are dual terms and sometimes used interchangeably [30], and weights between the indicators can be determined using the characteristics of these entropies. The estimation procedure consists of (1) constructing a matrix for each item; (2) normalizing the attribute information for each constructed indicator; (3) calculating the entropy for each attribute; (4) considering the degree of diversity between the indicators; and (5) determining the final weights (see Equations (2)–(6)).

(1) Matrix construction

$$\begin{matrix} x_{11}\cdots & x_{1j}\cdots & x_{1n} \\ \vdots & \vdots & \vdots \\ x_{i1} & x_{ij} & x_{1n} \\ \vdots & \vdots & \vdots \\ x_{m1} & x_{mj} & x_{mn} \end{matrix} \tag{2}$$

(2) Normalization of assessment items

$$p_{ij} = \frac{x_{ij}}{\sum_{i=1}^{m} x_{ij}} (i = 1,2,\cdots m;\ j = 1,2,\cdots n) \tag{3}$$

(3) Calculation of entropy of each attribute

$$E_j = -k\sum_{i=1}^{m} p_{ij}logp_{ij}\left(Here,\ k = \frac{1}{logm}; i = 1,2,\cdots m;\ j = 1,2,\cdots n\right) \tag{4}$$

(4) Weight assignment between assessments

$$d_j = 1 - E_j \tag{5}$$

$$w_j = \frac{d_j}{\sum_{j=1}^{n} d_j} (j = 1,2,\cdots n) \tag{6}$$

2.4. Bayesian Networks (BNs)

A BN is a stochastic graphical model that can represent the relationship between variables even when there is uncertainty between them. It consists of a directed acyclic graph (DAG) model of nodes and links and has the advantage of integrating variables of sources and types into a single structure. The relationship between nodes is described by conditional probability distribution (CPD), which considers dependencies between variables [31–33].

For example, the child nodes $(x_1,\ x_2)$ in Figure 2 are determined by the conditional probability of the parent nodes $(x_2,\ x_3)$ if there is a graph that has the nodes $(x_1, x_2,\ x_3)$ that follow the CPD. Any unconnected node is ignored. The joint distribution with n number of variables $\mathrm{p}(x_1,\ x_2,\ \cdots, x_n)$ is expressed in Equation (7):

$$p(x_1,\ x_2,\ \cdots, x_n) = \prod_{i=1}^{n} p(x_i \mid a_i) \tag{7}$$

where a_i denotes the set of parent nodes of x_i, and $\mathrm{p}(x_1,\ x_2,\ \cdots, x_n)$ is normalized constantly as the predistribution has been normalized.

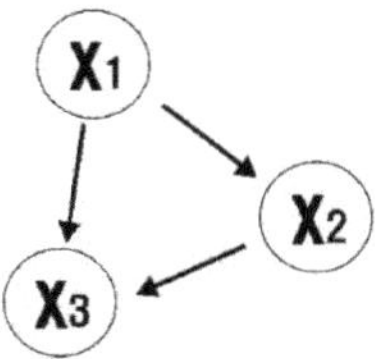

Figure 2. Example of a Bayesian network.

2.5. Integrated Index for Flood Risk Assessment (InFRA)

Equation (8) is proposed as a formula to construct the Integrated Index for Flood Risk Assessment (InFRA) based on the weight of each factor:

$$\mathrm{InFRA} = H^{a_1} \times S^{a_2} \times C^{a_3} \times (1-P)^{a_4} \tag{8}$$

where:

H = hydro-geology
S = socio-economics
C = climate
P = flood protection and
α_i = weight of each indicator

As flood protection is inversely proportional to InFRA, it is necessary to consider it in descending order when estimating flood protection. If flood protection is 0 and the other indicators are 1, the value of InFRA will be 1, and the closer the value is to 1, the higher the flood risk will be. The key components that make up an indicator can be determined by multiplying them by the weights and summing them (Equations (9)–(12)):

$$H = \beta_1 h_1 + \beta_2 h_2 + \cdots + \beta_n h_n \tag{9}$$

$$S = \gamma_1 s_1 + \gamma_2 s_2 + \cdots + \gamma_n s_n \tag{10}$$

$$C = \delta_1 c_1 + \delta_2 c_2 + \cdots + \delta_n c_n \tag{11}$$

$$P = \varepsilon_1 p_1 + \varepsilon_2 p_2 + \cdots + \varepsilon_n p_n \tag{12}$$

where:

h_i, s_i, c_i, p_i= each indicator

β_i, γ_i, δ_i, ε_i= the weights of each indicator

3. Application and Results

3.1. Selection of Target Areas and Data Collection

This study targeted the Midwest region of the Republic of Korea, including Daejeon Metropolitan City, Sejong Special Self-Governing City, Chungnam Province, and Chungbuk Province. The area consists of 28 cities and gun (districts or counties in Korea), including two municipalities, 11 cities, and 15 gun (see Figure 3). The advantage of the area is that it provides various geographical environments to select flood risks because it encompasses large cities, small and medium cities, coastal cities, mountainous regions, and rural areas.

Figure 3. Target area (Chungnam and Chungbuk Provinces).

The elements needed to construct an indicator database are summarized as statistical data and geographic information system (GIS)-based data. Demographic, social, and economic data were collected from the Korean Statistical Information Service (KOSIS), the statistical yearbooks of local governments, and the Statistical Yearbook of Natural Disasters. Meteorological data were collected from the Korea Meteorological Administration. All data were collected based on GIS for spatial analysis. The base year of the data is 2016, and statistical data for the ten years prior (2007–2016) were used. As shown in Table 1, data were collected for all 28 indicators of four components. All indicators were normalized to values between 0 and 1 using the average values estimated for each region and the standard deviations. Thus, the larger the value of an indicator for a region is, the closer it is to 1. The indicators of the H, S, and C components are positively correlated with FRI, whereas the indicators of P are negatively correlated.

3.2. Selection of Indicators Using Factor Analysis and Principal Component Analysis

Factor analysis and PCA were performed on each of the four components (hydro-geology, socio-economics, flood protection, and climate), which each consisted of several indicators. First, the indicators were grouped by factor using factor analysis, and the indicators with the highest component point for each group were selected using PCA. The procedure can prevent the duplication of meaning of the indicators and reduce dimension of the indicators in each group.

Table 2 shows the results of the factor analysis. The hydro-geology, socio-economic, and protection components were classified into three groups, whereas climate was classified into two groups. Kaiser–Meyer–Olkin (KMO) [34] and Barlett's test of sphericity [35] were used to determine the appropriateness of the analysis. The Kaiser–Harris measurement [34] was used to select principal components that have an eigenvalue of 1 or higher (see Table 2). The result of each component was determined to be significant because KMO remained at 0.5 or higher, and the probability value (p) remained below 0.05.

Table 2. Grouping and selection of representative indicators by PCA and factor analysis.

Classification		Component Points (Selected: ⊙)			Eigenvalue			KMO	Barlett's Test of Sphericity	
Comp-Onents	Indic-Ators	Group			Factor				Chi-Square	df(p)
		1	2	3	1	2	3			
1. Hydro-geology	(1)	0.420	−0.168	0.751	4.43	1.96	1.34	0.75	128.3	36(0)
	(2)	−0.479	−0.518	0.951 ○						
	(3)	0.351	0.674	0.055						
	(4)	−0.229	0.895 ○	−0.092						
	(5)	0.658	0.096	−0.080						
	(6)	−0.237	−0.172	0.378						
	(7)	0.850 ○	−0.096	−0.042						
	(8)	−0.333	0.124	0.627						
2. Socio-economy	(9)	0.764	0.286	0.198	4.35	1.54	1.01	0.67	164.7	21(0)
	(10)	0.694	0.088	0.001						
	(11)	0.147	0.957 ○	0.237						
	(12)	0.595	0.216	0.150						
	(13)	−0.412	−0.326	0.523 ○						
	(14)	0.144	0.351	0.010						
	(15)	0.902 ○	0.354	0.158						
3. Climate	(16)	0.827 ○	−0.018	-	2.30	1.18	-	0.65	27.2	10(0)
	(17)	−0.572	0.580							
	(18)	0.628	−0.500							
	(19)	0.108	0.903 ○							
	(20)	−0.342	0.776							
4. Flood protection	(21)	0.802	−0.030	−0.32	2.87	1.46	1.25	0.54	78.7	28(0)
	(22)	0.610	0.562	−0.045						
	(23)	0.928 ○	0.096	−0.119						
	(24)	0.522	−0.042	0.024						
	(25)	0.372	0.398	0.526						
	(26)	0.199	0.862 ○	−0.071						
	(27)	−0.216	−0.349	0.667 ○						
	(28)	0.103	0.624	0.073						

PCA was used to select indicators that have the most significant contribution for each group. Among 28 indicators, 17 were eliminated and 11 were chosen: (1) three indicators for hydro-geology (damage cost, urban rate, and lowland area rate); (2) three for socio-economics (total number of houses, financial independence rate, and dependence population); (3) three for flood protection (number of pump stations, drainage capacity, and number of public servants per resident); and (4) two for climate (frequency of intensive rainfall and probability rainfall), as shown in Table 2.

3.3. *Weight Assignment by Method and Calculation of Integrated Weights*

3.3.1. Weight Assignment by Method

The factors selected are expressed as normalized values between 0 and 1, and each indicator is estimated using the assigned weights. The flood risk index can be quantified using the weight of each indicator. To this end, weights can be assigned by various methods, and three weight assignment methods were applied, as described in Section 2.3. For the AHP, a survey was conducted with 30 respondents from academia and research. The survey was constructed in such a way that the importance of each indicator was compared in pairs. The terms of each indicator were defined and presented in the questionnaire to improve the accessibility for the respondents. The first-level hierarchy consists of four upper-level assessment components (hydro-geology, socio-economics, flood protection, and climate), and the second-level hierarchy consists of 11 lower-level assessment indicators.

For the CSS, a survey was conducted with 21 experts who have experience in work related to flood or wind damage and did not participate in the AHP survey. The questionnaire was structured in such a way that the sum of the four components presented and the sum of indicators for each component was 10. A sufficient explanation of the survey method was provided to supplement the questionnaire so that the respondents would not be confused.

For entropy weighting, Equations (1)–(6) from Section 2.3 were used based on the data collected for each component. The weights are shown in Table 3 and Figure 4. In summary, the weights for the socio-economic component were low in the survey methods, while they were high in the entropy weight method. In particular, the weights were evenly distributed among the rest of the components other than the socio-economic component and thus gave a relatively identical position concerning importance.

Table 3. Weight assignment for the representative indicators.

Components	Weight 1 (AHP)	Weight 2 (CSS)	Weight 3 (Entropy)	Indicators	Weight 1 (AHP)	Weight 2 (CSS)	Weight 3 (Entropy)
A. Hydro-geology	0.25	0.31	0.25	(2) Flood damage cost for public facilities	0.31	0.37	0.19
				(4) Urban rate	0.29	0.33	0.31
				(7) Lowland area rate	0.40	0.30	0.50
B. Socio-economy	0.13	0.15	0.28	(11) Financial independence rate	0.30	0.36	0.28
				(13) Dependent population	0.31	0.30	0.01
				(15) Total number of houses	0.39	0.34	0.71
C. Climate	0.31	0.28	0.20	(16) Frequency of hourly rainfall (P ≥ 50 mm)	0.77	0.72	0.99
				(19) Annual precipitation	0.23	0.28	0.01
D. Flood Protection	0.31	0.26	0.27	(23) Pump station (number)	0.37	0.38	0.56
				(26) Drainage capacity	0.50	0.45	0.35
				(27) Number of public servants per resident	0.13	0.17	0.08

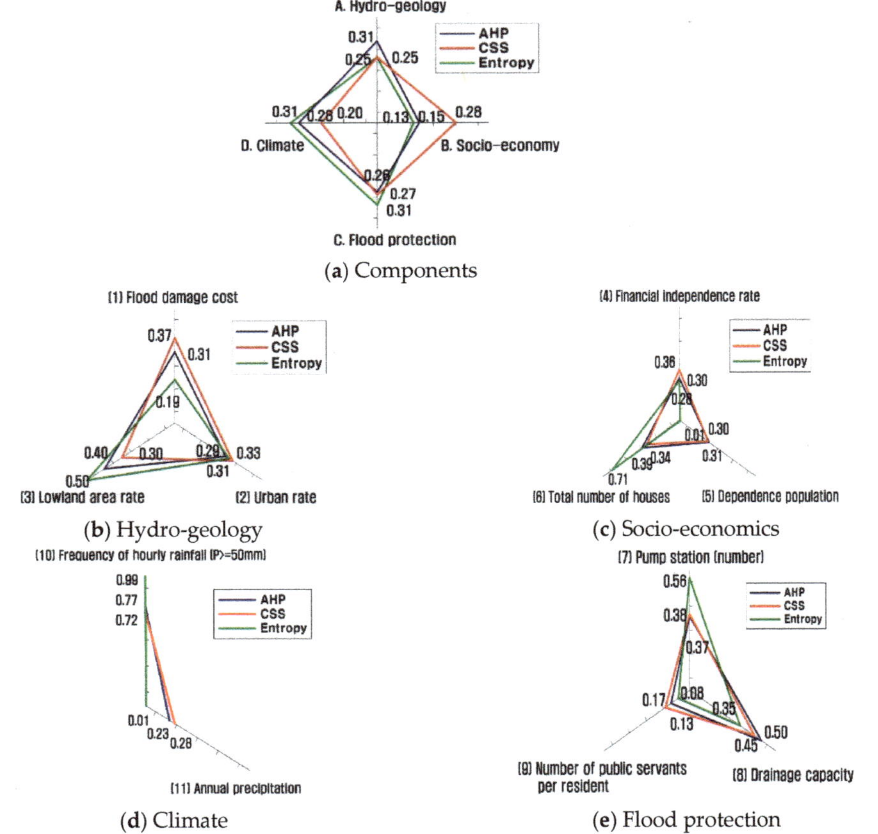

Figure 4. Weight distribution.

The indicators within a component showed a different aspect in the survey and entropy methods: the entropy method showed higher weight values for specific indicators, while the surveys showed relatively similar weight values for the indicators. This was due to the unique characteristics of entropy, which increases when deviations between alternatives are low. Moreover, the deviations between the normalized values of the indicators were small. However, the number of public servants per resident indicator of the flood protection component and the annual precipitation indicator of the climate component were low, and divisions in indicators between regions were high, which resulted in low entropy weights.

3.3.2. Integrated Weight Assignment Using Bayesian Networks (BNs)

The estimated weights affect the outcome of the estimated flood risk. It is not easy to determine the weights for each component and indicator, particularly when the entropy weight is higher than the other weights, such as for the total number of houses in Table 4. Thus, the BN method was used to estimate the combined weights while considering causal relationships between weights obtained from the AHP, CSS, and entropy techniques. First, a BN with 20 nodes and 19 links was constructed with AgenaRisk 10, as shown in Figure 5. The BN was constructed in consideration of the relationships between the components and indicators. The pre-probability assigned to each higher node can be inferred directly from the conditional probability, and the deviation of the probability determines the post-probability of the lower nodes. That is, the post-probability (the integrated weights) can be derived from pre-weights (the current weights), the conditional probability of each component (hydro-geology, socio-econometrics, flooding protection, and climate), and its indicators. As each component was weighted separately, it will not affect its indicators and can be expressed as dotted-line links that have indirect influences. Table 4 shows all of the probabilities (weights) of each component and indicator obtained from the configuration in Figure 5 and the post-probabilities (integrated weights).

Table 4. Resulting integrated weights.

Components	Weights Using Bayesian Networks (AHP, CSS, and Entropy)	Indicators	Weights Using Bayesian Networks (AHP, CSS, and Entropy)
A. Hydro-geology	0.26	(2) Flood damage cost	0.32
		(4) Urban rate	0.28
		(7) Lowland area rate	0.40
B. Socio-economy	0.20	(11) Financial independence rate	0.31
		(13) Dependent population	0.21
		(15) Total number of houses	0.48
C. Climate	0.26	(16) Frequency of hourly rainfall ($P \geq 50$ mm)	0.84
		(19) Annual precipitation	0.16
D. Flood protection	0.28	(23) Pump station (number)	0.51
		(26) Drainage capacity	0.36
		(27) Number of public servants per resident	0.13

The estimated weight of each component was relatively uniform in the range of 0.20–0.28. The weight of the socio-economic component was low in the survey method but increased significantly in the entropy method. This indicates that the entropy weight contributed to conditional probabilities as a prior probability. Similarly, the other indicators within each component were adjusted adequately by prior and conditional probabilities. For example, the drainage capacity indicator of the flood protection component was weighted as 0.50 and 0.45 in two surveys, respectively, while it was weighted as 0.35 in the entropy method, and its integrated weight became 0.36.

Moreover, the annual precipitation of the climate component was weighted with a small value of 0.01 in the entropy method, but the integrated weight was 0.06 because it was weighted with 0.23 and 0.28 in the two surveys. The BN model has an effective and optimal decision-making capability to integrate different knowledge and data [36,37]. Thus, BNs are expected to be a new alternative in assigning weights between indicators.

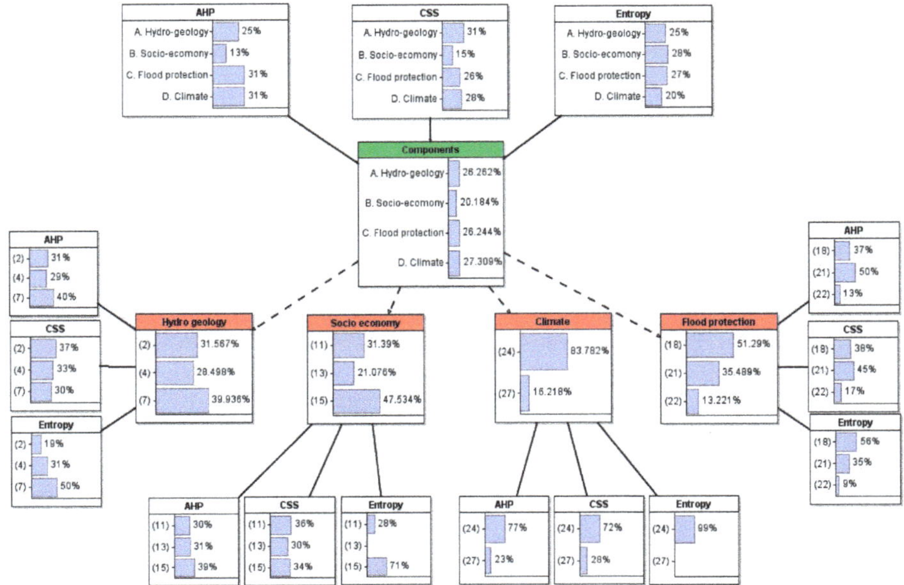

Figure 5. BN configuration for integrated weight assignment.

3.4. Results of Calculation with InFRA

The final InFRA was estimated for hydro-geology, socio-economics, flood protection, and climate components in 28 cities in the Chungcheong Province using several formulas (see Figure 4). As a result, InFRA did not show a significant gap between regions except for some areas and showed a flood risk of 0.3–0.5 in most places. The resulting values for Seosan (17), Dangjin (20), and Taean (27) were close to 0.7, despite their low risk from the socio-economic component. This occurred because the risk from the other components was high. Some village areas including Jeungpyeong (8) and Jincheon (9) showed a low InFRA level because they had a low level of flood protection and other components. The risk related to the hydro-geology component was high in the countryside because these areas are more influenced by flood damage, and there are more lowland areas than urban areas. In the socio-economic component, the indicators of the total number of houses and financial independence rate showed a high risk in large cities, followed by some villages that have a high dependent population.

In the flood protection component, large and medium cities showed a high level of protection, whereas villages showed a low level of protection because they lack flood protection systems. The flood protection component shown in Figure 6 is expressed in the concept of "1-flood protection," so it is interpreted accordingly. In the climate component, the basins were clustered in a continuous pattern and showed a constant flood risk, particularly in coastal areas according to the consistent measurement of the measurement stations in the Thiessen network along the coastal cities. This is attributed to the high frequency of intense rainfall in coastal areas, and thus, the frequency of intensive rainfall indicator is weighted higher than the annual precipitation indicator.

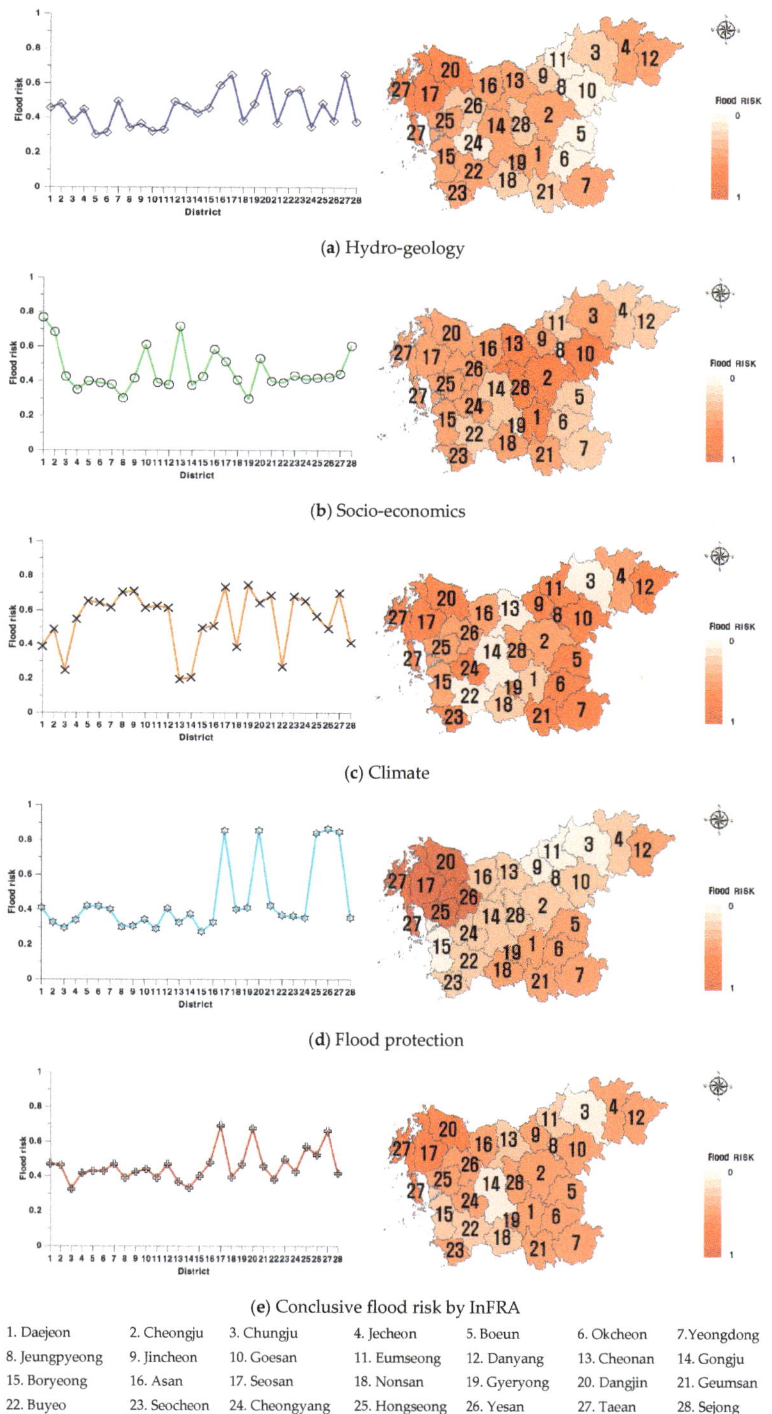

Figure 6. Integrated Index for Flood Risk Assessment (InFRA) estimation.

3.5. Comparison with Other Methods and Discussion

The proposed method was compared with other three methods used to assess flood risk: PFD, FDI, and RSA [15,17,21]. The three methods are the most popular in a practical field because they are easy to collect data and simple to apply. These methods are briefly explained in Table 5. However, the same assessment criteria must be used to compare the two sets of methods. The other assessment methods use various criteria with grades (1–5) or groups (A–D), and thus, comparing the methods using the same set of criteria is not appropriate. Therefore, they were compared in an alternative way using risk values between 0 and 1 instead of using grades or groups for a consistent comparison. In all the assessment methods, the risk of flood increased when the risk value was closer to 1, indicating that appropriate measures need to be taken for flood mitigation.

Table 5. Basic information on other flood risk assessment methods.

Other Methods	Method for Selecting Indicators	Method for Assigning Weights	Formula
PFD	Selected based on the subjective judgment of the researchers	Assigned based on the subjective judgment of the researchers	$\text{PFD} = \textit{Potential}^{a_1} \times \textit{Risk}^{a_2}$
FDI	Selected based on the subjective judgment of the researchers	CSS	$\text{FDI} = \sum_{j=1}^{n} (W_j \times Z_{ij})$
RSA	Selected based on the subjective judgment of the researchers	CSS	$\text{RSA} = \alpha \times \textit{Risk} - \times \textit{Reduction}$

The other assessment methods generally showed high flood risk in cities and low risk in villages. In particular, flood risks were high in Daejeon, Cheongju, Chungju, and Cheonan but low in Yeongdong, Jincheon, Goesan, and Geumsan. The other assessment methods were polarized in urban and rural areas and showed large regional variations compared to InFRA (see Figure 7). It seems that the duplicated meaning in the construction of indicators and the insufficient level of flood protection in cities are major reasons for such results. Nevertheless, indicators such as population, financial independence rate, and infrastructure are typically high in urban areas. Thus, the other risk assessment methods are considered to have produced somewhat overestimated values because they use a system that would inevitably estimate large flood risk in large cities.

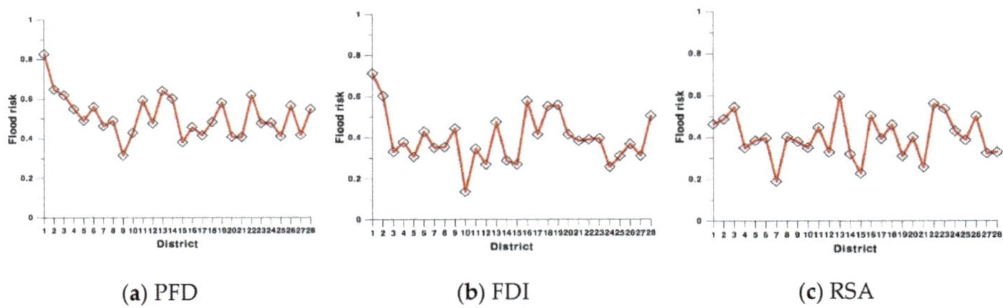

Figure 7. Results of previous flood risk assessment methods.

This comparison was qualitative, and a quantitative comparison is necessary. Therefore, the methods were validated by analyzing the area under the curve (AUC) of the receiver operating characteristic (ROC) curve. The ROC curve is a graph where the x-axis shows the specificity, which indicates the probability of the estimated value being false. The y-axis shows the sensitivity, which indicates the probability of the estimated value being true. That is, the evaluation method is better if the risk assessment is more likely to be correct and has a lower false probability rate. A higher

AUC indicates higher accuracy of the prediction, and the accuracy of the results increases as AUC approaches 1.

The limit of this validation is that the factors to be compared must be considered to evaluate the accuracy of each method, but the assessment factors are already included as indicators, and there is no suitable criterion to apply. As the best alternative, data from [38] were used, and total flood damage costs were derived, including injuries and flooding of farmland and cities, for the flood damage cost for public facilities (see Table 6). Then, the integrated sums were normalized. If the value is 0.5 or higher, the corresponding region is considered to have high damage cost. The ROC analysis was then conducted.

Table 6. Estimated total flood damage cost by district.

District	Period: 2007–2016 (Units: Thousand KRW)		District	Period: 2007–2016 (Units: Thousand KRW)	
	Flood Damage Cost for Public Facilities	Total Flood Damage Cost		Flood Damage Cost for Public Facilities	Total Flood Damage Cost
Daejeon	28,904,388	47,212,788	Boryeong	32,401,710	83,801,710
Cheongju	91,359,488	92,472,688	Asan	48,656,598	73,920,798
Chungju	104,137,429	131,245,429	Seosan	114,830,315	190,237,515
Jecheon	186,090,075	246,660,075	Nonsan	47,680,083	96,512,883
Boeun	71,415,564	91,802,764	Gyeryong	51,490,991	61,412,991
Okcheon	52,452,809	76,211,609	Dangjin	90,928,932	102,690,132
Yeongdong	538,481,881	559,584,281	Geumsan	116,518,336	248,099,536
Jeungpyeong	47,680,083	68,518,883	Buyeo	84,734,079	103,852,079
Jincheon	110,242,303	135,001,103	Seocheon	27,608,991	88,375,991
Goesan	93,072,362	130,146,762	Cheongyang	34,453,315	90,744,515
Eumseong	80,755,775	85,305,375	Hongseong	46,091,101	59,817,901
Danyang	310,987,567	332,961,167	Yesan	39,018,182	51,444,182
Cheonan	47,680,083	55,908,083	Taean	106,713,297	140,302,897
Gongju	90,773,723	128,253,723	Sejong	18,198,894	34,950,094

According to the AUC of the ROC, the accuracy of InFRA was 0.67, while that of PFD, FDI, and RSA was less than 0.5 (0.296, 0.417 and 0.174, respectively). Thus, they were withdrawn from the assessment of flood damage cost (see Figure 8). In other words, the other risk assessment methods were revealed to be inappropriate for assessing flood damage costs. The evaluation showed that InFRA is better than the classic methods for assessing flood risk and could thus be applicable in the field.

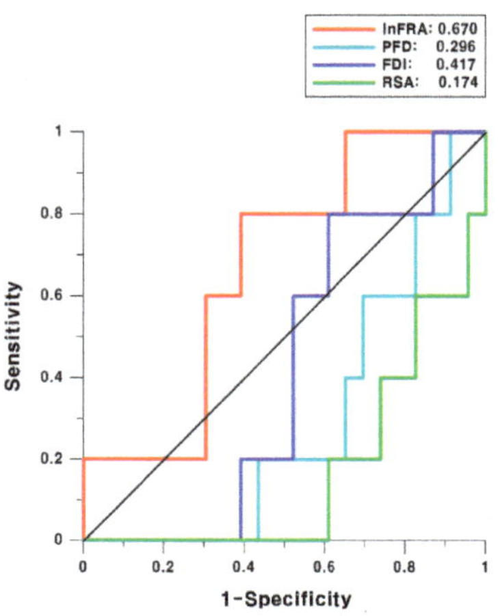

Figure 8. Validation of flood risk assessment methods using receiver operating characteristic (ROC)-area under curve (AUC) analysis.

4. Conclusions

Numerous methods have been developed to assess flood risk. In this study, a methodology was proposed for use in decision-making by integrating existing methods rather than developing another flood risk assessment method. First, previous flood risk assessment methods were evaluated (PFD, EFVI, FDRRI, FVA, FDI, and RSA), and four components and 28 indicators were extracted. Factor analysis and PCA were carried out, and 11 of the 28 indicators were selected as a result. Then, the weights of each component and indicator were estimated by the AHP, CSS, and entropy methods, and the results of each method differed from one another. Therefore, BNs that integrate the conventional weight assignment methods were structured to estimate integrated weights.

The BN-based InFRA was applied to target regions to estimate the flood risk of each region. The result of both qualitative and quantitative comparisons between InFRA and the conventional methods demonstrated the excellent applicability of InFRA. The InFRA methodology can integrate various other flood risk assessment methods and it could be used as a useful tool in decision-making for flood risk management.

Author Contributions: Conceptualization and methodology, H.J. and S.K.; statistical analysis, C.C., J.K. and D.K.; writing-original draft preparation, H.J.; Final review, H.S.K.; supervision, S.K. and H.S.K.

Funding: This research was supported by a grant [MOIS-DP-2015-05] through the Disaster and Safety Management Institute funded by Ministry of the Interior and Safety of Korean government.

Conflicts of Interest: The authors declare no conflicts of interest.

References

1. Luo, P.; Mu, D.; Xue, H.; Duc, T.N.; Dinh, K.D.; Takara, K.; Nover, D.; Schladow, S.G. Flood inundation assessment for the Hanoi Central Area, Vietnam under historical and extreme rainfall conditions. *Sci. Rep. (Nat.)* **2018**, *8*, 12623. [CrossRef] [PubMed]
2. Luo, P.; He, B.; Takara, K.; Xiong, Y.E.; Nover, D.; Duan, W.; Fukushi, K. Historical Assessment of Chinese and Japanese Flood Management Policies and Implications for Managing Future Floods. *Environ. Sci. Policy* **2015**, *48*, 265–277. [CrossRef]
3. Federal Emergency Management Agency. *Flood Information Tool User Manual (Rev. 7)*; FEMA: Washington, DC, USA, 2003.
4. National Oceanic and Atmospheric Administration. *Risk Vulnerability Assessment Tool*; NOAA: Washington, DC, USA, 2007.
5. *Annual Report 2004: Advancing Innovation*; Munich Re Group: München, Germany, 2004.
6. Tyndall Centre. *Vulnerability, Risk and Adaptation: A Conceptual Framework, Tyndall Centre for Climate Change Research*; Tyndall Centre: Norwich, Norfolk, UK, 2003.
7. Rygel, L.; O'Sullivan, D.; Yarnal, B. A method for constructing a social vulnerability index: An application to hurricane storm surges in a developed country. *Mitig. Adapt. Strateg. Glob. Chang.* **2006**, *11*, 741–764. [CrossRef]
8. Chang, L.F.; Huang, S.L. Assessing urban flooding vulnerability with an emergy approach. *Landsc. Urban Plan* **2015**, *143*, 11–24. [CrossRef]
9. Kablan, M.K.A.; Dongo, K.; Coulibaly, M. Assessment of social vulnerability to flood in urban Côte d'Ivoire using the MOVE Framework. *Water* **2017**, *9*, 292. [CrossRef]
10. Luo, P.; Zhou, M.; Deng, H.; Lyu, J.; Cao, W.; Takara, K.; Nover, D.; Schladow, S.G. Impact of forest maintenance on water shortages: Hydrologic modeling and effects of climate change. *Sci. Total Environ.* **2018**, *615*, 1355–1363. [CrossRef] [PubMed]
11. Luo, P.; He, B.; Duan, W.; Takara, K.; Nover, D. Impact assessment of rainfall scenarios and land-use change on hydrologic response using synthetic Area IDF curves. *J. Flood Risk Manag.* **2015**, *11*, S84–S97. [CrossRef]
12. Environment Agency. *Catchment Flood Management Plans: CFMP Guidelines Volume 1 Consultant Report*; EA: Almondsbury, Bristol, UK, 2004.
13. Jeollabuk-do Total Human Institute National of Korea. *Improv. Flood Plans Against Clim. Chang. Jeollabuk-do* **2010**, *107*, 1415–1419.

14. Korea Institute of Civil Engineering and Building Technology. *Strengthen Facility Standards Against Excess Climate*; KICT: Ilsan, Gyeonggi, Korea, 2009.
15. Korea Research Institute for Human Settlements. *Analysis of Flood Damage Characteristics and Development of Flood Damage Index*; KRIHS: Sejong, Korea, 2005.
16. Korea Environment Institute. *Development and Introduction of Indicators to Assess Vulnerability of Climate Change*; KEI: Sejong, Korea, 2008.
17. Ministry of Land, Transport and Maritime Affairs. *National Water Resource Plan*; MOLTMA: Sejong, Korea, 2001.
18. Ministry of Land, Transport and Maritime Affairs. *National Water Resource Plan (2010–2020)*; MOLTMA: Sejong, Korea, 2010.
19. National Disaster Management Research Institute. *Development of Assessment System for Flood Vulnerability Index*; NDMI: Ulsan, Korea, 2012.
20. National Institute for Land and Infrastructure Management. *Flood Vulnerability Index*; NIMIL: Tsukuba, Asahi, Japan, 2009.
21. Seoul Institute. *Development of the Regional Safety Assessment Model in Seoul–Focusing on Flood*; SI: Seoul, Korea, 2006.
22. World Meteorological Organization (WMO). *Guide to Hydrological Practices*, 5th ed.; WMO: Geneva, Switzerland, 2008.
23. Rummel, R.J. *Applied Factor Analysis*, 1st ed.; Northwestern University: Illinois, IL, USA, 1988.
24. Shlens, J. A Tutorial on Principal Component Analysis. *Cornell Univ.* **2014**, *1404*, 1101.
25. Fischhoff, B. Value elicitation: Is there anything in there. *Am. Psychol.* **1991**, *46*, 835–847. [CrossRef]
26. Keeny, R.L.; von Winterfeldt, D.V.; Eppel, T. Eliciting public values for complex policy decisions. *Manag. Sci.* **1990**, *36*, 1011–1030. [CrossRef]
27. Joo, H.J.; Kim, S.J.; Lee, M.J.; Kim, H.S. A study on determination of investment priority of flood control considering flood vulnerability. *J. Korean Soc. Hazard Mitig.* **2017**, *18*, 417–429. [CrossRef]
28. Saaty, R.W. The Analytic Hierarchy Process–What It Is and How It Is Used. *Math. Model.* **1987**, *9*, 161–176. [CrossRef]
29. Dudek, F.J.; Baker, K.E. The constant method applied to scaling subjective dimensions. *Am. J. Psychol.* **1956**, *69*, 616–624. [CrossRef] [PubMed]
30. Ozkul, S.; Harmancioglu, N.B.; Singh, V.P. Entropy-based assessment of water quality monitoring networks. *J. Hydrol. Eng.* **2000**, *5*, 90–100. [CrossRef]
31. Jensen, F.V. *An Introduction to Bayesian Networks*; Springer: New York, NY, USA, 1966.
32. Kim, S.J.; Rarhi, P.; Jun, H.D.; Lee, J.H. Evaluation of drought severity with a Bayesian network analysis of multiple drought indices. *J. Water Resour. Manag.* **2017**, *144*, 1–10. [CrossRef]
33. Pearl, J. *Probabilistic Reasoning in Intelligent Systems: Networks of Plausible Inference*; Elsevier: San Francisco, CA, USA, 2014.
34. Kaiser, H.F. An index of factorial simplicity. *Psychometrika* **1974**, *39*, 31–36. [CrossRef]
35. Bartlett, M.S. Properties of sufficiency and statistical tests. *Proc. R. Soc. Lond. Ser. A* **1937**, *160*, 268–282.
36. Castelletti, A.; Soncini-Sessa, R. Bayesian networks and participatory modelling in water resource management. *Environ. Model. Softw.* **2007**, *22*, 1075–1088. [CrossRef]
37. Henriksen, H.J.; Barlebo, H.C. Reflections on the use of Bayesian belief networks for adaptive management. *J. Environ. Manag.* **2008**, *88*, 1025–1036. [CrossRef] [PubMed]
38. Ministry of the Interior and Safety. *Natural Disaster Investigation and Recovery Planning Guidelines*; MOIS: Sejong, Korea, 2017.

Article

The Importance of a Dedicated Monitoring Solution and Communication Strategy for an Effective Management of Complex Active Landslides in Urbanized Areas

Daniele Giordan, Aleksandra Wrzesniak * and Paolo Allasia

Consiglio Nazionale delle Ricerche-Istituto di Ricerca per la Protezione Idrogeologica, Strada delle Cacce 73, 10135 Torino, Italy; daniele.giordan@irpi.cnr.it (D.G.); paolo.allasia@irpi.cnr.it (P.A.)

* Correspondence: aleksandra.wrzesniak@irpi.cnr.it; Tel.: +39-011-3977-829

Received: 25 January 2019; Accepted: 11 February 2019; Published: 13 February 2019

Abstract: Over the last decades, technological development has strongly increased the number of instruments suitable for landslide monitoring. For large landslides, monitoring systems are organized in complex and multi-instrumental networks aimed at controlling several representative physical variables. The management of these networks is often a complicated task that must consider technological aspects, data-sets processing, and results publication. We developed a new hybrid system focused on capturing and elaborating data-sets from monitored sites and on disseminating monitoring results to support decision makers. With respect to other available monitoring solutions, we emphasized the importance of technological aspects and a correct communication strategy, which represents the last fundamental step for a correct use of collected data. Monitoring results are often published in a difficult and not user-friendly way because they are intended for technicians with adequate background. Such an approach may be inefficient, especially during emergencies, when also non-expert people are involved. Additionally, this system consists of early warning application, which integrates a threshold-based approach and a failure forecasting modeling. The presented approach represents a possible improvement for a more sustainable management of active landslides that could have a strong impact on population and infrastructures in particular in highly urbanized areas.

Keywords: landslide monitoring; early warning; emergency management; dissemination; urbanized areas

---

1. Introduction

Today, the use of monitoring systems to control the evolution of active landslide and their hazard assessment is very diffused [1,2]. In the last decades, the number of available technical solutions and complexity of landslide monitoring networks have been progressively increasing to fulfill the growing necessity of managing large slope instabilities in urbanized areas.

Urban pressure, amplified by climatic changes, has progressively increased the cases in which the evolution of a landslides can have a direct impact on inhabited areas [3]. Several approaches can be adopted to reduce this impact. They may vary according to different factors like the dimension and the typology of landslide and the distribution of infrastructures. The most critical combination is the presence of large landslides in highly populated areas. In these cases, the delocalization can be hard to implement, and the landslide stabilization can be very difficult and not immediate. Often, the only solution is the use of monitoring systems to improve the knowledge of the landslide behavior and to support decision makers during emergencies related to a partial or total collapse of the instable area.

These systems, if correctly managed, may also create favorable conditions for the population at risk in terms of life comfort.

The adopted monitoring networks are based on the high frequency measurement of the most representative physical variables: (1) superficial displacement, (2) deep displacement, (3) hydro-geological (water table, pore pressure), and (4) meteorological. In the last years, several case studies were analyzed and published, and they represent good examples of this approach. For example in Italy, complex monitoring networks have been developed to control the evolution of landslides such as Ancona [4], Montaguto [5,6], Mt.de La Saxe [7], Rotolon [8], and Ruinon [9]. In other countries, similar approaches have been applied to monitor complex landslides such as Randa (Switzerland—[10]), La Vallette (France—[11]), Super Sauze (France—[12,13]), Sado Island (Japan—[14]), and Slumgullion (USA—[15]).

If we analyze these monitoring systems, we can notice that in all cases the monitoring network workflow consists of (1) on-site data acquisition, (2) data processing, and (3) monitoring result publication and dissemination.

It is evident that technological aspects are predominant in existing monitoring applications, especially in on-site acquisition of representative variables. Starting from Angeli et al. [16], the number of publications dedicated to the technological development of monitoring systems was progressively covering all available technologies, e.g., GPS (Global Positioning System), RTS (Robotized Total Station), GB-InSAR (Ground-Based Synthetic Aperture Radar), and in-place and robotized inclinometers. In the literature one can find the examples of the use of GPS [17,18], RTS [19,20], GB-InSAR [21], and in-place and robotized inclinometers (respectively, [22,23]), among many others.

After on-site acquisition, monitoring systems manage captured data-set and provide representation of the landslide evolution. Allasia et al. [24] and Frigerio et al. [8] present a possible approach for the management of such data-sets. In both cases, informatics tools that are able to acquire, process, validate and store acquired data-sets compose the presented systems. The sequence of these tools makes the on-site acquisition procedures available in a centralized and remote-controlled system. This also allows simultaneous management of several monitoring systems and control of different landslides.

As mentioned before, complex landslides are usually monitored by multi-instrumental networks. These networks provide many types of time series, which are later integrated in a combined database. The integration of different instruments in a unique centralized system, which bypasses applications developed by the producers of monitoring instruments, is a key point for an efficient management of complex monitoring networks [24].

Additionally, the data processing is usually linked to the management of early warning thresholds [25,26]. The adaptation of monitoring systems to early warning purposes is an optimal solution that may facilitate a correct management of landslide acceleration phases [27–30].

Often, early warning systems (EWSs) are associated with superficial displacement, which is probably one of the most frequently used parameter to describe movement tendency and overall knowledge of landslides [31,32]. However, EWSs design and in particular a priori definition of related thresholds is one of the most important and critical aspect of landslides monitoring systems [33]. This is mainly due to the complex and unpredictable behavior of natural hazards, which often results in large number of variables involved [34].

Thresholds are often established by performing back analysis of the available monitoring data. They can be also determined by considering similarities to other events or by following numerical approaches as proposed by Crosta and Agliardi [34]. However, this action may be hampered by insufficient knowledge of the monitored landslide and by the numerous factors (e.g., geotechnical parameters of the shear surface or real effect of pore pressure) [35].

Carlà et al. [36] proposes to use inverse velocity method for setting alarm thresholds and forecasting landslides and structure collapses. In fact, the time of failure forecasting became a major subject of many researchers. The successful work of Saito [37] and Fukuzono [38] inspired many other authors to develop different approaches in order to estimate time of failure (ToF). A summary of the different methods proposed in literature can be found in Federico et al. [39]. These studies introduce simplified empirical and/or graphical methods based on the assumption that material, under constant stress condition, follows a creep mechanism before rupture.

In the past, the monitoring data-sets were exclusively dedicated to technicians with a specific background. This was because the common representations of measured quantities were not user friendly. According to our experience, the use of complex representation limits the number of people that can handle this information [40].

As described in Giordan et al. [41], the large landslides emergencies are often managed by multidisciplinary teams. These teams are usually composed not only of technicians, but also of local authorities who often do not have an adequate background on landslide monitoring. The presence of people with various backgrounds obliges the developer of monitoring systems to consider how to share available information. Often, the importance of a monitoring results communication strategy is underestimated. During emergencies, when time is limited, an easy access to available information could help taking adequate decisions about the safety of people and goods. In this context, the dissemination of monitoring results has become a crucial element of the emergency management.

The experience acquired during the development of the software for management of landslide monitoring data called ADVICE (ADVanced dIsplaCement monitoring system for Early warning) [24] lets us realize that a correct use of monitoring results cannot be limited to a simple near–real time application that publishes the monitoring results on a dedicated webpage. To obtain an effective system that can be used for a more sustainable management of large slope instabilities in urbanized areas, it is mandatory to develop a more complex approach that merges the management of data with the use of thresholds for EW purposes and dissemination procedures that assure a correct communication to stakeholders. In order to fulfill these requirements, we developed a multidisciplinary solution for an effective management of landslide monitoring systems called LANDMON (LANDslides MOnitorng Network). Over the years, we have developed a set of automatic modules that governs in near–real time the management process from data acquisition to results dissemination. Additionally, this system consists of early warning application, which integrates threshold-based approach and failure forecasting modeling. LANDMON emphasizes a proper monitoring data representation and dissemination, due to the heterogeneity of actors involved in the hazard management process. Therefore, LANDMON is an example of innovative solution that integrates technical and modeling features with communication strategy aimed at supporting decision-makers and population at risk, in particular during emergencies. This approach represents a possible improvement for a more sustainable management of monitored active landslides that could have a strong impact on population and infrastructures.

2. Methods

Figure 1 summarizes the workflow of LANDMON, which is divided in three main phases. The first phase is composed of data acquisition procedures, the second one is composed of early warning applications and the third one regards monitoring results representation and dissemination.

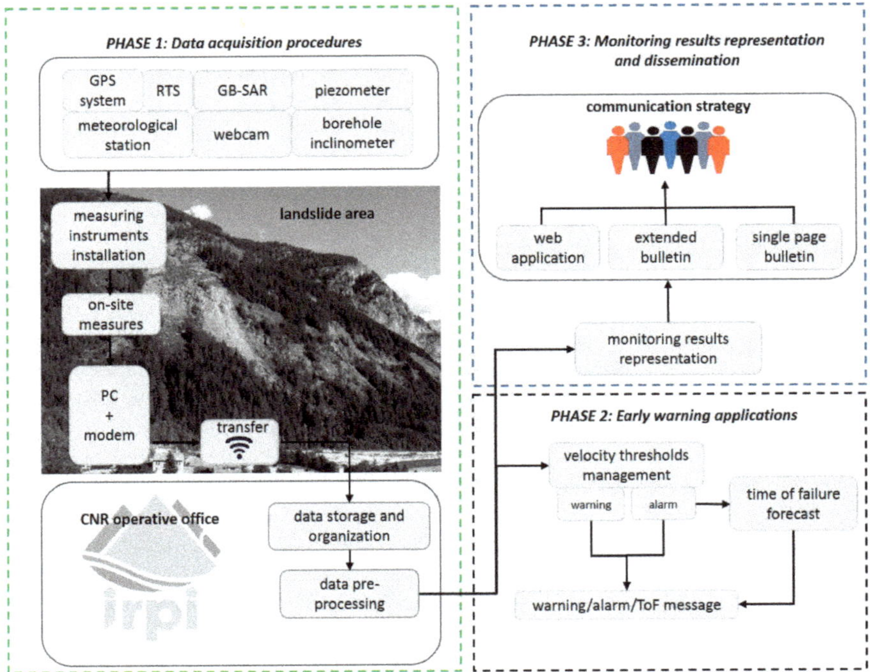

Figure 1. LANDMON (LANDslides MOnitorng Network) structure. The workflow is divided in three main phases: (1) data acquisition procedures, (2) early warning applications, and (3) monitoring results representation and dissemination. Photograph: Mont de La Saxe landslide (Aosta Valley, northern Italy).

Every phase has a sequence of procedures that corresponds to LANDMON modules. Each module has a specific purpose and application. The organization in modules, not in a single unique procedure, is important because the system can be freely adapted and activated according to the landslide features and the aims of the monitoring system. One of the most important element of the developed system is that we combined a sequence of tools able to efficiently acquire, analyze, and represent the monitoring data-set from complex networks in a single, complete, and self-contained software with an appropriate dissemination procedure of monitoring results.

The developed system exploits the experience of ADVICE [24] adding new modules and a dedicated communication strategy. ADVICE was a near–real time software for landslide monitoring and early warning threshold management.

In this new version, we improve the classical concept of early warning by integrating a failure forecast module. This module employs inverse velocity approach to forecast time to failure in near–real time [42], and it is automatically activated when a predefined displacement threshold is exceeded.

Beside technological and computational aspects, LANDMON satisfies a predefined communication strategy, which considers a specific approach for monitoring results dissemination [41,43]. The experience in the use of monitoring systems to manage emergencies emphasizes the need for a more effective communication approach to inform stakeholders and involved population correctly. We consider monitoring results dissemination as an essential aspect in the development of monitoring systems. Such an approach improves the effectiveness of these systems during emergencies. At present, most of the commercial applications provide exclusively data acquisition, elaboration and classical results publication (i.e., tables or time series plots). Additionally, only some of them merge data-set from different monitoring systems and produce bulletins or other solutions aimed to inform population about landslide evolution and related risk level.

Our approach, unlike existing possibilities, is a multidisciplinary solution for efficient hazard management based on series of informatics tools and on particular strategy for the results communication. The proposed system is also human-centered, because it unifies science, monitoring technology, levels of government and pubic [44–46].

We were progressively developing LANDMON and we applied it to several emergency scenarios [47]. In particular, the most important applications are the Montaguto landslide [5], the Costa Concordia wreck [48] and the Mont de La Saxe landslide [49]. In the following, we present the main features of the three phases.

2.1. Phase 1: Data Acquisition Procedures

As introduced in Figure 1, the first phase concerns data acquisition procedures i.e., measuring instruments installation, on-site data acquisition, data transfer, storage and organization, data pre-processing. This structure should be operative for every monitored site and be ready to acquire data from different instruments.

It is fundamental to point out that an innovative system should be able to work with data-sets obtained from different types of instruments installed in the monitored area: RTS, GPS, GB-InSAR, piezometers, meteorological stations, webcam, borehole inclinometers and other geotechnical instrumentation as clinometers and extensimeter. The network composition depends on monitoring scenarios, needs and economic issues. Usually, an accurate planning of the monitoring network is needed, especially in the case of phenomena with heterogeneous directions and velocities of the moving mass. As a consequence, appropriate network design leads to well-organized and efficient monitoring system precisely representing the kinematic of the monitoring area, and to success in hazard management process.

Once the network is ready, the measurements are performed cyclically. For every cycle, the measurements are acquired at certain date and time. The management of instruments acquisition is usually managed via ad-hoc proprietary software arranged in a base station positioned directly at the monitored site. These on-site applications control: (1) frequency of cycles acquisition, (2) acquisition parameters of the instrument, (3) on-site database, (4) export process to obtain a correct data output. The measurement output (its structure and content) depends on used instrument. This output usually is text file containing principal data that describes motion, e.g., displacement in three-dimensional space (x, y and z) in the case of GPS, or vertical and horizontal angles and distance in the case of RTS. The local export of new measurement cycle terminates the on-site acquisition.

Every time a new data-set is available, the system transfers it from local monitoring system to the central server. An internet connection is required in order to transfer and guarantee a continuous flux of these data. This connection may be provided through GPRS (General Packet Radio Service), UMTS (Universal Mobile Telecommunication System), LTE (Long Term Evolution), dial-up or dedicated backbones. The transfer can be done using a dedicated software that uploads the newest file from the local station to the remote server or the local control station may be queried directly from the remote server, and transmitted automatically by secure transfer protocol.

Transferred data are uploaded and stored on a dedicated database server for further elaboration. This step is composed of several actions. First, the raw data is converted into correct metric-scale and coordinates system depending on the technical specification of the instrumentation used. For every instrument a dedicated import tool was developed to transform the original data output from monitoring system in a correct data input for LANDMON. This transformation is fundamental for multi-instrument interoperability. The use of the same time format and coordinate system facilitates the comparison of data-sets registered by different instruments. Then, the data-set is filtered and validated by removing inconsistent data, noisy measurements, spikes, and by handling missing values.

The elaborated data-set (in the form of input file) is then stored in an unique database that contains all the data acquired at the same monitored area by different instruments. The most recent measurements are integrated with the previous ones in the database. When the sequence of measurements is ready, the data can be analyzed by the modules dedicated to displacement threshold management, failure forecast and visualization.

2.2. *Phase 2: Early Warning Applications*

The early warning application developed in LANDMON is composed of two modules. The first one uses velocity thresholds and defines explicitly the warning level based on the historical monitoring data. While the second one uses inverse velocity approach to forecast the probable temporal distance before the collapse. The first module analyses the velocity in a predefined lapse of time with respect to predefined thresholds [50]. The second module evaluates the propensity to the failure using the inverse velocity approach proposed by Fukuzono [38]. These modules, in near–real time, acquire the last measurement session, process the available data-set and produce adequate warnings.

2.2.1. Velocity Threshold Management Module

Design of early warning systems usually starts from defining three levels of attention [26]: (i) ordinary level, when there is no emergency due to the displacement level below predefined threshold, (ii) warning level, when the mass movements oscillate above seasonal perturbations, (iii) alarm level, for critical activation of the unstable area. In our approach, these stages are associated to predefined velocity thresholds. When the velocity limit of warning/alarm level is exceeded, the integrated algorithm of EWS automatically sends alert/alarm messages, e.g., via SMS and/or email to the responsible authorities. After warning, the authorities have often to decide about the activation of specific civil protection procedures. According to the location and possible evolution of the monitored phenomenon, these procedures may include interruption of roads and railway traffic, evacuation of facilities located in the risk zone, and every kind of actions aimed at providing overall safety. The check of thresholds exceeding is executed in near-real time every time new measurements are available. The algorithm checks the warning status based on the last monitoring result, and if the threshold level is exceeded, it activates the predefined procedures to alert decision makers about the current landslide status. This module has been tested under various critical conditions. For example, we applied it during the critical phase of the Mont de La Saxe rockslide occurred in 2014 to support the landslide management team [49]. For further details about Mont de La Saxe landslide see Crosta et al. [7,51] and Alberti et al. [33]. This module was regularly updating decision makers on the evolution of the slide during the most active phases via SMS and with displacement graphs via email. Thresholds were verified for each new measurement session (every hour). SMS and emails were sent every time landslide activity changed. Bulletins were issued weekly, but in the most critical period, they were issued daily.

2.2.2. Near-Real Time Failure Forecasting Model

As described before, EWSs are usually composed of three levels of attention. These levels are determined according to directly measured values of the moving mass. Consequently, this imposes underestimated values for the alarm level, in order to maintain a proper margin of safety. Unfortunately, this approach does not provide any kind of prediction about possible evolution of the landslide, making a classical EWS incomplete and sometimes inefficient. In such a case, the time between the alarm level and the eventual failure (partial or total) is unknown. This fact may be inconvenient for authorities and decision-makers who are in charge of management of emergency scenarios, and who are responsible for the safety of humans and/or public wealth. In this context, the use of modeling techniques seems to be a reasonable solution to estimate eventual ToF of landslide phenomena.

Starting from Fukuzono method, we proposed a new approach to forecast ToF by considering near-real time monitoring [42]. Fukuzono approach, also known as the inverse-velocity method, uses the evolution of inverse value of surface velocity (v) as an indicator of the ToF by correlating the failure with v^{-1} tending to zero [38].

Large landslides during their final stage of failure often follow abrupt evolution that correspond to the tertiary creep behavior. Under such conditions, especially when human lives come into play, it is very important to choose a method with quick implementation, calculation and use. Fukuzono method has an easy and efficient approach, and therefore, due to its simplicity, it is suitable for EWSs and it was integrated in LANDMON to model ToF.

Our model takes the advantage of near-real time data acquired from the monitoring network system. We combine inverse-velocity method, interpolation techniques and statistical methods. In particular, we apply linear regression analysis to describe the function of inverse velocity Iv by fitting its discrete points using the least square approach. According to the mentioned Fukuzono method, the root of the resulting equation of the linear regression provides an estimated ToF. This procedure is then repeated for N-times by means of bootstrap re-sampling procedure [52]. Therefore, we can define the confidence intervals, considering 5% and 95% tiles among N-times estimations of ToF. In addition, the accuracy of linear relation between the model best fit and the data is verified by calculating the Pearson correlation coefficient CC. For more details, the reader may refer to Manconi and Giordan [42,49].

Presented failure time estimation approach uses confidence intervals (5% and 95% tiles) to describe time range for the highest probability of failure, and normalized CC values to describe reliability R of the computed forecast model. We have used this reliability value (in percentage) to define additional threshold for LANDMON EWS as an auxiliary support for early warning purposes. For example, when R is sufficiently high, automatic message for expert operators can be generated.

Figure 2 introduces the adopted concepts for EW, which couples threshold-based EW and failure forecasting. A hypothetical velocity evolution of the landslide point (see the curve on Figure 2) presents a commonly observed landslide behavior toward failure, and it is plotted together with the elements of developed EW (alarm and warning limits, model reliability thresholds).

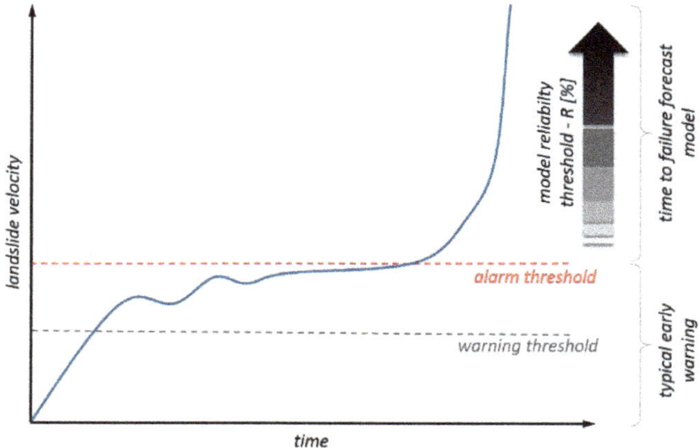

Figure 2. Schematic representation of the adopted approach for early warning applications: coupling threshold-based EW and failure forecast model. The curve presents a hypothetical velocity evolution of the landslide point, which represents usually observed behavior toward failure. The figure adopted from [49].

To exploit ToF results and to make possible the use of thresholds, we adopted the value of reliability R in the EW system. R is a good indicator of the forecast estimation and for this reason it can be used to check if the forecast gives a high probability of failure. For example, during Mont de La Saxe emergency scenario in April 2014 [49], we considered a number of R ranges, as follows: (i) for $50\% < R < 60\%$ the model reliability is low, failure is unlikely but the situation must be surveyed;

(ii) for 60% ≤ R < 75% model reliability is medium, a failure within the estimated ToF range starts to be likely; (iii) for 75% ≤ R < 90% model reliability is high, a failure within the estimated ToF range is likely; (iv) for R ≥ 90% model reliability is very high, a failure within the estimated ToF range is highly probable.

The method application is presented in Figure 3. It is an example of real computation that was performed on 16 April 2014 during the Mont de La Saxe emergency. On Figure 3A we present a one month velocity evolution of five prisms, and on the following plots (B, C and D) the forecasting results for the prism B4. The model reliability R can be obtained for several calculation time windows (CTWs, data acquired over last 12 h, 24 h, 48 h, 1 week, etc.). Plots B, C and D show the results of our methodology for three CTWs, respectively 7-days, 2-days and 1-day.

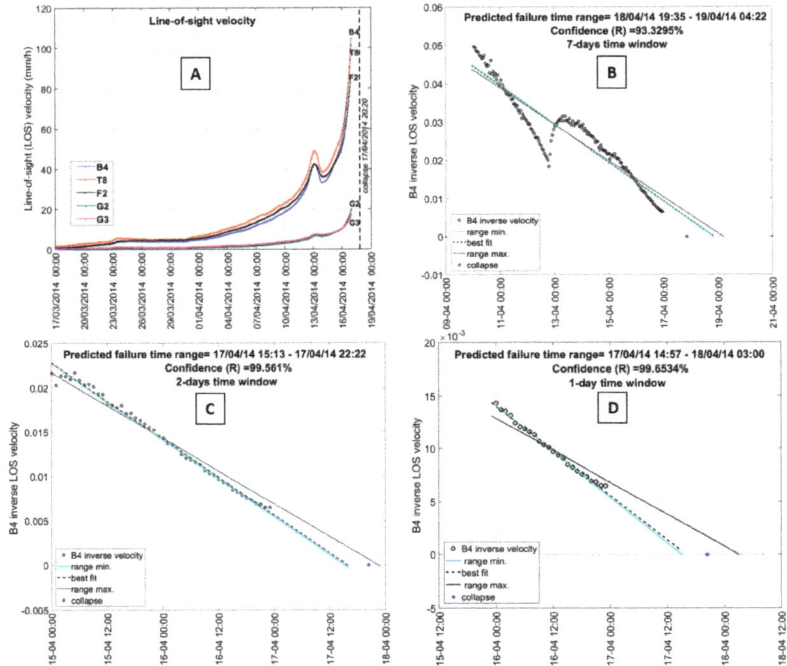

Figure 3. Mont de La Saxe landslide inverse velocity computations. One-month velocity evolution of five prisms (image A), failure time range forecasting and corresponding model confidence R obtained for seven-days CTW (image B), for 2-days CTW (image C) and 1-day CTW (image D).

As output, the algorithm provides predicted failure time range, best fit and confidence of the model. Additionally, we marked the moment of the real collapse of the landslide, in order to present the efficiency of this approach. In conclusion, the results of the presented procedure can be summarized as follows: if the landslide velocity keeps increasing as in the last calculation time window, the probability of a failure within the estimated time of failure is equal to the model reliability.

If the landslide evolution follows the law of creep, the use of ToF can be a great support for decision makers because it gives not only a description of the actual status as velocity thresholds provide but also a forecast. The indication of possible evolution can be an important information for managing the civil protection procedure. The resulting time of failure is shared exclusively with experts, who are qualified to interpret this result, to evaluate the level of reliability and to exploit it during the most critical phases of emergency.

2.3. *Phase 3: Monitoring Results Representation and Dissemination*

Usually, monitoring data are presented in the form of tables or time series plots. These typologies of representation may be clear for experienced people who are familiar with monitoring data. However, natural hazard management process involves directly also non-expert people and such data representations may be difficult to interpret, especially for population who is the terminal of this process. For this reason, in addition to time series plots of measured quantities (displacement, precipitation, temperature), we have developed an illustrative approach (©3DA) to represent the monitoring results, i.e., surface displacement representations in 2-and 3-dimensional space and three-dimensional arrows showing real direction of motion [40].

©3DA represents surface displacement as magnitude maps by interpolating the punctual monitoring data-set over the given Region Of Interest (ROI). Moreover, ©3DA visualizes the measured displacement by means of 3-dimensional arrows at initial measurement points. Such a representation shows magnitude, intensity and direction of displacement measured at this particular point. The referenced results of ©3DA algorithm can be projected on various representations of the monitored area, e.g., digital terrain model (DTM), ortho-photograph, perspective photograph, image obtained from Google Earth/Google Maps or from any other service. The advantage of having many possible backgrounds is the possibility to identify the best representation.

©3DA, integrated within LANDMON, can automatically produce the magnitude maps few minutes after each measurement session, which is an important characteristic especially during emergency conditions. Thanks to the configurability of ©3DA subroutine, this approach can be easily adapted to the hazard conditions under study. ©3DA can be considered as an incipient of the developed and later adopted in LANDMON communication strategy. The aim of this strategy, described in the following section, is a correct and effective dissemination of monitoring results.

The main element that we consider for the definition of our dissemination strategy is the different background of stakeholders, which requires a different approach for the communication of available monitoring results. The dissemination of monitoring results, which is often underestimated by engineering geologists, is extremely important in particular during emergencies, when the time for making a decision (like evacuation or other important actions oriented to damage limitation) is limited.

All monitoring network management solutions consider, after the measurement session, in near–real time a data post-processing and the on-line publication on dedicated website. Accredited users may consult the monitoring results few minutes after the measurement session was completed.

In our system, the latest update of monitoring results is highlighted on the website in a synthetic form, providing a rapid and effective overview of the status of the monitored phenomenon. Content of presented results depends on the characteristics of installed monitoring network. However, the most common data are webcam view, meteorological data (e.g., precipitation and temperature plots), ©3DA elaborations, and time series plot of quantities measured by installed instrument (e.g., RTS, GB-InSAR, GPS). The users may also explore results history by selecting a predefined period (day, week, month, or total) or to use an interactive application to plot desired period.

Our dissemination strategy considers the on-line publication of monitoring results and the use of bulletins. According to United Nations studies, each EWS must be human-centered [44,45], considering either emergency condition or educating population at risk, so that response to warning is efficient. For this purpose, we developed a single-page informative bulletin dedicated to non-experts. The aim of this bulletin is supporting decision-makers and informing the population. Inside LANDMON, the bulletin is generated automatically through the bulletin generator tool, which is compatible with every point measurement data set e.g., target displacement provided by RTS, GPS, or GB-InSAR. An example of such a bulletin is shown in Figure 4. For more details about synthetic bulletin and automatic generation procedure, the reader may refer to Wrzesniak and Giordan [43].

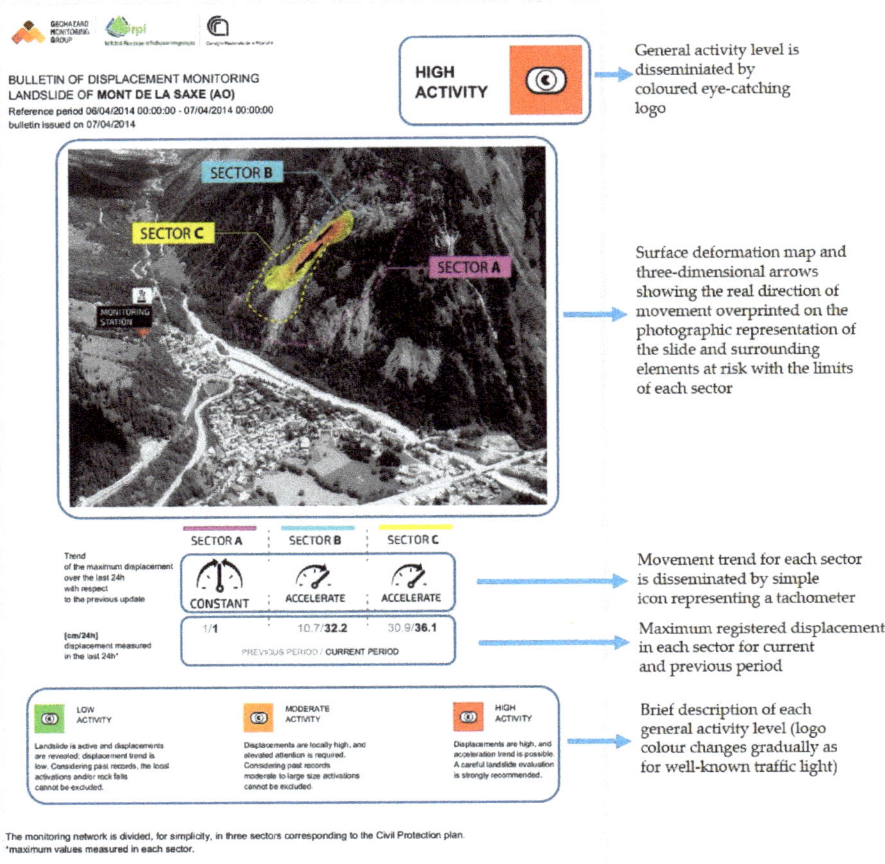

Figure 4. Example of daily single page bulletin. This bulletin was automatically generated based on the real data-set, relative to RTS (Robotized Total Station) measurements of the period 6–7 April 2014. The figure adopted from [43].

The possibility to configure input data and parameters of the algorithm makes this tool flexible and quickly applicable for various scenarios. This tool can operate in near–real time generating the bulletin either continuously with desired frequency (every month, week, day or every 12 h) or within EWS (when the threshold is exceeded the bulletin is generated and attached to the warning message).

In general, the main difference between bulletins and time series of other graphics published on-line is that bulletins are not a simple representation of last monitored data, but an analyzed and commented representation of the landslide activity.

2.4. Communication Strategy of Landslide Monitoring Results

As mentioned before, one of the most important element that we considered for the definition of a correct communication strategy are needs and background of stakeholder. In order to adopt a "usable science" approach [53], owners of monitoring data are responsible not only for a correct acquisition and analysis, but also for a correct dissemination of obtained results. The correct dissemination makes the difference between a simple monitoring data management and an effective support to decision-makers and population in particular during emergencies. In Giordan et al. [41] and in Wrzesniak and Giordan [43], we proposed the identification of different groups of people, according to their role in the emergency management and their background. In particular, we determined four group-types: GROUP-1, GROUP-2, GROUP-3, and POPULATION. The identification of characteristics of stakeholders in terms of background and needs is fundamental for a correct definition of the best communication procedure. In the following a brief description of different defined groups.

GROUP-1 is dedicated to government members (mayor, council members, politicians, etc.). They are usually responsible of the final decision regarding the overall safety. However, their technical background on natural hazards can be limited, so they need a clear overview of monitoring results and current risk level. They also speak to the population, they alert the public about the immediate risk, and they run educational programs to increase risk awareness among the population. The need of this group is a simplified description of the recent landslide evolution and the actual risk level.

Engineers, geologists, and technicians (for example Civil Protection Agency members) are assigned to the GROUP-2. They have a necessary background and experience in order to perform technical analysis and to support GROUP-1. The need of GROUP-2 is a detailed description of landslide evolution, actual warning level, and last results of failure forecast model.

GROUP-3 is composed of experts, mainly engineers and geologists, who are highly specialized in hazard monitoring and EWSs. They are responsible of development and efficiency of monitoring networks and of data analysis, validation, and elaboration. For this reason, they need direct access to raw data to check that the system is working correctly.

As POPULATION, we consider all people indirectly involved in the emergency management. The correct involvement of population is mandatory for they correct perception of the risk level due to the landslide evolution. Only a well-informed population will response correctly during the most critical phases of the emergency.

After defining the main groups' characteristics, we identified one of the most important goal of our communication strategy, which is how we can communicate monitoring results to stakeholders.

Figure 5 shows a schematic representation of the relationship between dissemination, complexity, and comprehension of monitoring results representations with respect to the identified groups. For example, it is shown that there is a direct relationship between the ability of each group to understand monitoring data and the number of members of this group. In this image, GROUP-3 is an end-member because it is often composed by a few people with a dedicated background and the ability to understand even the most complex representations. On the opposite side of the scheme, the other end-member is the POPULATION, which is the largest group of people. Even if the background may vary in this group, we have to consider them to have a limited ability to comprehend landslide monitoring results. For this reason, this group needs to be informed with simplified representations. If such simplification of results communication is not provided, the complexity of adopted representations could hamper the comprehension of the real level of risk and reduce the propensity of the population to cooperate.

To fulfill these necessities, we adopted in our strategy the ©3DA approach [40] that has been developed to represent of the same data-set in many ways with different levels of complexity. In Figure 6, three possible representations of the same RTS data-set are compared. In particular, in images A and B, the data are presented as surface deformation maps with three-dimensional arrows, and as time-series plot of line-of-sight displacement in image C. In image A, a perspective photograph of the landslide is used as a background. This representation is particular effective for population because it represents a simplified representation of the slope and of the actual instable area. Image A also provides a geographic location of the active parts of landslide, which may facilitate overall understanding of the situation. For this reason, the representation in image A is destined to GROUP-1 and POPULATION.

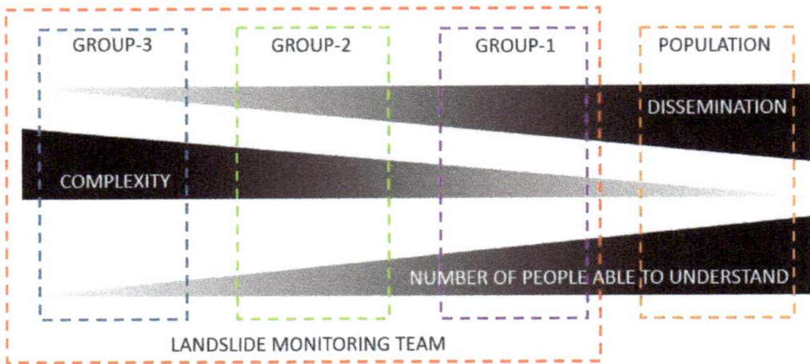

Figure 5. Schematic representation of the relationship between the amount of people to be informed (dissemination process), the complexity of monitoring results representation, the number of people able to understand it and Groups. GROUP-1, -2 and -3 are directly involved in the landslide management and they compose landslide monitoring team. Population is indirectly involved in this process.

Representation presented in image B is more suited for GROUP-1 and GROUP-2. It provides the distribution of landslide activity and the elements at risk (e.g., civil houses, public buildings, transportation structures) thanks to a broader ortho-photograph used as a background. Image C is mainly dedicated to GROUP-2 and GROUP-3 because it requires specific knowledge and experience in landslide monitoring to understand the real meaning of this representation. However, if correctly interpreted, this image is the most representative for the comprehension of the evolution of each single measured target and for the relation between the recent movement rate and the threshold.

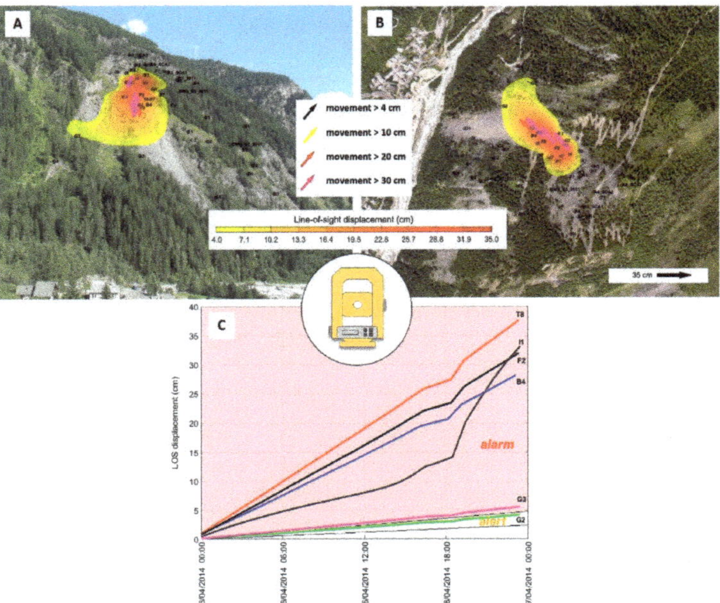

Figure 6. Examples of different representations of the same data-set. The used data are relative to RTS measurements for the period 6–7 April 2017 for Mont de La Saxe landslide. Surface deformation maps (based on line-of-sight displacement) and three-dimensional arrows (based on displacement measured in three directions) are overprinted over frontal view of the landslide (image A) and over ortho-photograph of the landslide with neighborhood area (image B). Image C presents time-series of line-of-sight displacement of the measured targets with warning and alarm zones. These plots are generated by ©3DA.

Another important element that we adopted in our communication strategy is the simultaneous use of near–real time results representation on the website and periodical bulletins, as presented in Figure 7. Our system is able to manage automatically a large flux of monitoring data, from their acquisition to the on-line publication. From the communication point of view, this possibility plays an important role in hazard management, because it can provide a fast and effective update of the latest monitoring results. LANDMON is able to provide an update of monitoring results few minutes after the measurement and processing session is completed. The main limitation of infographics and plots published on the website is the lack of comments and interpretations, which are a key element of adopted bulletins.

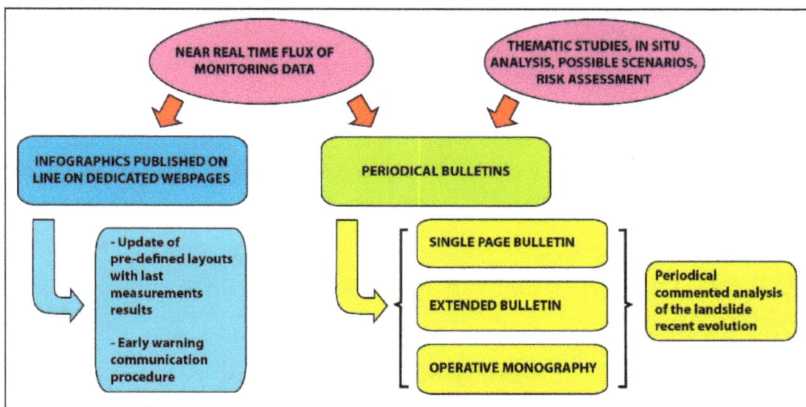

Figure 7. Flowchart of proposed communication strategy based on infographics published on the website and on bulletins.

Decision-makers usually need also a detailed analysis and interpretation of available dataset and that is why we adopted in our communication strategy also different periodical bulletins. We defined three different types of bulletins that contains a different quantity of information and have different communication targets: (i) single page bulletin, (ii) extended bulletin, and (iii) operative monography.

The single page bulletin was described in the previous chapter and it is mainly dedicated to people with limited experience in landslide monitoring (GROUP-1 and POPULATION). This simplified version is the most effective way to communicate the recent level of monitored landslide evolution.

The graphic layout of single page bulletin was designed with special focus on communication aspects so the monitoring results are presented in clear, illustrative and quickly interpretative manner. To achieve this goal, we used infographic techniques to integrate variety of information relative to landslide evolution. Infographics visualize complex data or information in illustrative form, which can be easily understood by the audience [54]. This bulletin presents current displacement status in the form of magnitude map of analyzed physical quantity with three-dimensional arrows associated at the targets location showing the real direction of the motion; all these components are overprinted on the photograph representing monitored area (see Figure 4 for an example). Additionally, we provide information about movement tendency and general activity of monitored phenomenon, and we present them by means of infographics. The bulletin algorithm analyzes the numerical data-set and translates the results into appropriate graphical representation, which is an optimal solution for straight forward communication, especially to non-expert people, e.g., three-dimensional displacement into an arrow, displacement trend into a simple tachometer icon, activity level into an eye-catching logo colored with traffic light graduation (see Figure 4 for these elements).

The extended bulletin is a more detailed and complex version of the single page bulletin. This highly detailed bulletin contains all the available data collected in a defined time interval.

The complexity level is elevated because it is aimed at people with a specific background. This bulletin is prepared in the style of a technical report that describes and comments on the latest update of the monitoring results.

The main purposes of such bulletin are to (i) supply a complete description of the results obtained by the monitoring network in the considered period, (ii) identify and validate eventual trend changes with respect to the previous period and to the same period in previous years, and (iii) to audit the results in terms of eventual updates of the current monitoring network.

The third proposed bulletin is the Operative Monography. This document was designed to host with a standard format all the available information about the monitored landslide. The ICAO (International Civil Aviation Organization) operation manual inspires the Operative Monography structure. A complete description of this document is described in Giordan et al. [53].

3. LANDMON Improvement and Application to Several Case Studies

The presented approach is the product of many years of experience in the management of emergencies of slope instabilities, as well as other natural and anthropic hazards. Table 1 summarizes some of the most representative case studies were our system was applied and the improvement that we added step by step. The following case studies are listed in the table: Paganica fault [55], Montaguto earthflow [5,24,40], Costa Concordia wreck [48,56], Mont de La Saxe rockslide [7,41–43,49], Calatabiano landslide [21], and Ponzano landslide [20,57]. The main characteristics of each emergency (year, type, elements at risk) are presented. Furthermore, we listed the LANDMON modules applied to each case. The evolution of LANDMON can be noticed, as more and more modules were implemented over the years and applied on the new cases.

The emergency of Mont de La Saxe (2014) represents the first complete test bench for our system, as the complexity of monitoring network and the availability of various types of monitoring data allowed the development of the complete set of modules.

Mont de La Saxe is a large and complex slope instability located in the region of the Aosta Valley in northern Italy. Its unstable volume is of ca. 8×10^6 m^3 [7,51], and it hazards a part of Courmayeur municipality and access road to the Mont Blanc Tunnel, an important connection for trans-Alpine transport. Due to these characteristics, the continuous and automatic monitoring of surface displacement has started in 2009.

Some examples of application for La Saxe case can be found in the present manuscript. Figure 3 shows the results of failure range forecasting module. Figure 4 presents, as an example of communication strategy, a daily single-page bulletin (developed in 2014, validated, and automatized in 2017). Figure 6 shows the application of automatic data processing and communication strategy.

Table 1. Temporal development of the system.

	Emergency Name						
	Paganica Fault	**Montaguto**	**Costa Concordia Wreck**	**Mont de La Saxe**		**Calatabiano**	**Ponzano**
Year	2009	2010	2012	2013	2014	2015	2017
Type	Post-earthquake fault movement	Earthflow	Vessel wreck	Rockslide		Landslide	Landslide
Elements at risk	Main water pipeline of L'Aquila city	National road and railway	Coastal environment and seawater	Inhabited zones, access road to Mont Blanc Tunnel		Main water pipeline of Messina city	Old town of Ponzano village
			Applied modules				
automatic data acquisition and transfer	x	x	x	x	x	x	x
automatic data processing		x	x	x	x	x	x
on-line data publication		x	x	x	x	x	x

Table 1. *Cont.*

	Emergency Name						
	Paganica Fault	**Montaguto**	**Costa Concordia Wreck**	**Mont de La Saxe**		**Calatabiano**	**Ponzano**
threshold based EW			x	x	x	x	x
time failure forecasting				x	x	x	x
failure range forecasting					x	x	x
Communication strategy [1]					x	x	x

[1] Since 2017, communication strategy includes automatic single-page bulletin.

4. Discussion

In the last decades, the development of complex monitoring systems was focused on the technological improvements, which enabled quick automatic data acquisition, and dissemination of warnings and monitoring results. RTS, GPS, and GB-InSAR are only a few examples of possible instruments that can be adopted for the acquisition of physical variables and used for monitoring network architecture. Currently, monitoring systems are also used to support decision-makers and to inform population at risk.

The large availability of automatic and reliable monitoring sensors triggered the diffusion of EWSs. Modern monitoring systems provide a large amount of accurate measures at high sampling rates, so they are widely used to set early warning thresholds. However, the use of threshold-based approaches is characterized by the well-known difficulty to establish the threshold levels. These values are often conservative, and their underestimated values can produce reiterated alerts/alarms without a real failure. If the analyzed variable (e.g., superficial displacement or its derivative) exceeds the predefined threshold without real changes in the stability of the slope, the management procedure can reach a critical phase, which can be worsened by continuous alarms over a long period. The solution to avoid such a critical phase may be the revision of threshold values, or the use of supplementary models that can be useful in these particular conditions. As introduced before, LANDMON incorporates failure-forecasting model dedicated to early warning purposes (see Section 2.2.2).

To forecast ToF, we combined inverse velocity and statistic methods, thus the results should be always interpreted with regard to probability. This is why, in our opinion, describing a time range when the failure may occur is more adequate than trying to define an exact time of failure. Our goal was to develop an efficient module for LANDMON, which helps in near–real time the management of the early warning systems in the most critical phases and provides as many information as possible about landslide status during the emergencies. However, the aim of such EWSs is to support decision-makers, not to replace them. Technicians must first consult the results of the monitoring network and later associate them with other factors as weather conditions, previous landslide behavior, and experience in order to adopt an adequate decision.

The availability of large amount of near–real time monitoring data can be an advantage on one side, but on the other, it can create issues related to the management and to the processing of these data. LANDMON has been developed to deal with this large influx of data. Moreover, LANDMON recognizes another important aspect that often underestimated by classical technological solutions: the dissemination of results and the management of the available data.

The management of hazard monitoring networks is an interdisciplinary activity due to variety of involved people (engineers, geologists, technicians, politicians, council members, population, etc.). All these people participate (directly and indirectly) in this process, thus the collaboration among them is necessary. Our experience, in particular in the management of landslide emergencies as scientific advisor of National Civil Protection Agency, evidenced that the lack of correct communication strategy for monitoring results dissemination can be critical. For this purpose, beside hardware and software developments, we focused on communication aspects. We developed an appropriate communication approach, which we adopted in LANDMON. The assumption of our communication strategy is that monitoring systems can be considered efficient only if all involved people are able to understand the results produced by these systems. From this point of view, LANDMON is not just a system composed of technological features but also an expression of the communication strategy that supplies different monitoring results representations according to the stakeholders' background.

As presented in Table 1, LANDMON is the result of many years of experience in the management and monitoring of various hazardous scenarios. We developed LANDMON mainly for applications in the field of geohazards like landslides [24,41], but we also considered other experiences as the hazard management in anthropic environment like open pit mines [58] or cultural heritage sites [59].

In many applications, LANDMON considers not only the natural process but also the stability of elements at risk like buildings and infrastructures, which are usually integrated within the instable areas. For example, the water pipeline of Paganica fault [47,55] and Calatabiano case studies, as well as the buildings and retaining wall of the old town of Ponzano [20], were subject to monitoring together with main landslide evolution. Another example of unusual application of our system is the monitoring of Costa Concordia wreck [48].

The first case study shown in the Table 1 is Paganica fault (2009), when only the "automatic data acquisition and transfer" module was implemented. In the following years, due to growing social requirements and technological possibilities, we developed more modules in occasion of Montaguto, Costa Concordia and Mont de La Saxe (2013) case studies. The first complete LANDMON system was finally implemented for Mont de La Saxe (2014) case study, where we successfully applied our communication strategy during the emergency. Our complete system was later used to manage Calatabiano and Ponzano landslides.

5. Conclusions

In this paper, we presented LANDMON (LANDslides MOnitorng Network), a multidisciplinary solution for effective management of monitoring networks used for the complex landslides evolution control. The developed modular approach can be adopted also for others emergencies that should be managed using complex monitoring systems. We developed and tested this approach and its modules in several landslide-related emergencies like Montaguto or Mont de La Saxe.

The proposed method is a sequence of computerized tools and a more structured procedure that considers technological, computational, and communication aspects.

The first aspect of LANDMON considers a modular organization of applications, which can be configured according to the characteristics of the monitored site. These features guarantee flexibility in application for various scenarios, not only related to landslide monitoring, but also to the monitoring of other hazard types, like open pit mines, buildings, and dams. All developed modules may be executed automatically and queried in the main process to perform the cycle in automatic mode from acquisition to dissemination.

Following the technological aspect, the system development focused on a more complex management of early warning by integrating a failure forecast modeling. This module provides a prediction about possible evolution of the landslide when the "classical" threshold-based alarm level is exceeded. Additionally, the supplementary alarm threshold is determined based on reliability of the implemented model.

The last aspect of LANDMON is the communication strategy. The proposed approach considers the needs and backgrounds of users, as it follows the principles of usable science by translating complex data sets into adequate representations for the receivers. This communication strategy carefully considers the population at risk, which is the real end-user of monitoring systems, by providing a dedicated solution. In particular, the last monitoring results are disseminated by means of a single-page informative bulletin where the landslide status is presented using infographic techniques to obtain clear, illustrative, and quickly interpretative overview. The use of user-friendly representations of complex data is a key point for a correct communication of real risk level that should be easily transmitted to the population, especially during emergencies.

Therefore, LANDMON is a complex, automatic, complete, and self-contained system designed for efficient acquisition, analysis, and appropriate representation and dissemination of the monitoring data. This system merges engineering with communication aspects, improving the effectiveness of the management of monitoring network and providing a sustainable risk management approach for active landslides in urbanized areas.

Author Contributions: This paper represents a result of collaborative teamwork. Conceptualization, methodology, software development and validation, D.G., A.W. and P.A.; manuscript writing and revising D.G. and A.W.

Funding: This research received no external funding.

Conflicts of Interest: The authors declare no conflict of interest.

References

1. Wieczorek, G.F.; Snyder, J.B. Monitoring slope movements. In *Geological Monitoring*; Young, R., Norby, L., Eds.; Geological Society of America: Boulder, CO, USA, 2009; pp. 245–271.
2. Frodella, W.; Ciampalini, A.; Bardi, F.; Salvatici, F.; Di Traglia, F.; Basile, G.; Casagli, N. A method for assessing and managing landslide residual hazard in urban areas. *Landslides* **2018**, *15*, 183–197. [CrossRef]
3. Vranken, L.; Vantilt, G.; Van Den Eeckhaut, M.; Vandekerckhove, L.; Poesen, J. Landslide risk assessment in a densely populated hilly area. *Landslides* **2015**, *12*, 787–798. [CrossRef]
4. Cotecchia, V. The second Hans Cloos lecture. Experience drawn from the great Ancona landslide of 1982. *Bull. Eng. Geol. Environ.* **2006**, *65*, 1. [CrossRef]
5. Giordan, D.; Allasia, P.; Manconi, A.; Baldo, M.; Santangelo, M.; Cardinali, M.; Corazza, A.; Albanese, V.; Lollino, G.; Guzzetti, F. Morphological and kinematic evolution of a large earthflow: The Montaguto landslide, southern Italy. *Geomorphology* **2013**, *187*, 61–79. [CrossRef]
6. Ferrigno, F.; Gigli, G.; Fanti, R.; Intrieri, E.; Casagli, N. GB-InSAR monitoring and observational method for landslide emergency management: The Montaguto earthflow (AV, Italy). *Nat. Hazards Earth Syst. Sci.* **2017**, *17*, 845. [CrossRef]
7. Crosta, G.B.; Lollino, G.; Paolo, F.; Giordan, D.; Andrea, T.; Carlo, R.; Davide, B. Rockslide monitoring through multi-temporal LiDAR DEM and TLS data analysis. In *Engineering Geology for Society and Territory-Volume 2*; Springer: Cham, Switzerland, 2015; pp. 613–617.
8. Frigerio, S.; Schenato, L.; Bossi, G.; Cavalli, M.; Mantovani, M.; Marcato, G.; Pasuto, A. A web-based platform for automatic and continuous landslide monitoring: The Rotolon (Eastern Italian Alps) case study. *Comput. Geosci.* **2014**, *63*, 96–105. [CrossRef]
9. Del Ventisette, C.; Casagli, N.; Fortuny-Guasch, J.; Tarchi, D. Ruinon landslide (Valfurva, Italy) activity in relation to rainfall by means of GBInSAR monitoring. *Landslides* **2012**, *9*, 497–509. [CrossRef]
10. Jaboyedoff, M.; Ornstein, P.; Rouiller, J.-D. Design of a geodetic database and associated tools for monitoring rock-slope movements: The example of the top of Randa rockfall scar. *Nat. Hazards Earth Syst. Sci.* **2004**, *4*, 187–196. [CrossRef]
11. Squarzoni, C.; Delacourt, C.; Allemand, P. Differential single-frequency GPS monitoring of the La Valette landslide (French Alps). *Eng. Geol.* **2005**, *79*, 215–229. [CrossRef]
12. Malet, J.-P.; Maquaire, O.; Calais, E. The use of Global Positioning System techniques for the continuous monitoring of landslides: Application to the Super-Sauze earthflow (Alpes-de-Haute-Provence, France). *Geomorphology* **2002**, *43*, 33–54. [CrossRef]
13. Travelletti, J.; Malet, J.-P. Characterization of the 3D geometry of flow-like landslides: A methodology based on the integration of heterogeneous multi-source data. *Eng. Geol.* **2012**, *128*, 30–48. [CrossRef]
14. Ayalew, L.; Yamagishi, H.; Marui, H.; Kanno, T. Landslides in Sado Island of Japan: Part I. Case studies, monitoring techniques and environmental considerations. *Eng. Geol.* **2005**, *81*, 419–431. [CrossRef]
15. Parise, M.; Coe, J.A.; Savage, W.Z.; Varnes, D.J. The Slumgullion landslide (southwestern Colorado, USA): Investigation and monitoring. In Proceedings of the International Conference FLOWS, Sorrento, Italy, 11–13 May 2003; pp. 11–13.

16. Angeli, M.-G.; Pasuto, A.; Silvano, S. A critical review of landslide monitoring experiences. *Eng. Geol.* **2000**, *55*, 133–147. [CrossRef]
17. Gili, J.A.; Corominas, J.; Rius, J. Using Global Positioning System techniques in landslide monitoring. *Eng. Geol.* **2000**, *55*, 167–192. [CrossRef]
18. Tagliavini, F.; Mantovani, M.; Marcato, G.; Pasuto, A.; Silvano, S. Validation of landslide hazard assessment by means of GPS monitoring technique? a case study in the Dolomites (Eastern Alps, Italy). *Nat. Hazards Earth Syst. Sci.* **2007**, *7*, 185–193. [CrossRef]
19. Arbanas, Ž.; Sassa, K.; Marui, H.; Mihalić, S.; Eberhardt, E. Comprehensive monitoring system on the Grohovo Landslide, Croatia. In *Landslides and Engineered Slopes: Protecting Society through Improved Understanding, Proceedings of the 11th International & 2nd North American Symposium on Landslides, Banff, Canada, 2–8 June 2012*; Taylor & Francis: London, UK; pp. 1441–1447.
20. Allasia, P.; Baldo, M.; Giordan, D.; Godone, D.; Wrzesniak, A.; Lollino, G. Near Real Time Monitoring Systems and Periodic Surveys Using a Multi Sensors UAV: The Case of Ponzano Landslide. In Proceedings of the IAEG/AEG Annual Meeting, San Francisco, CA, USA, 2019.
21. Lombardi, L.; Nocentini, M.; Frodella, W.; Nolesini, T.; Bardi, F.; Intrieri, E.; Carlà, T.; Solari, L.; Dotta, G.; Ferrigno, F.; et al. The Calatabiano landslide (southern Italy): Preliminary GB-InSAR monitoring data and remote 3D mapping. *Landslides* **2017**, *14*, 685–696. [CrossRef]
22. Simeoni, L.; Mongiovì, L. Inclinometer monitoring of the Castelrotto landslide in Italy. *J. Geotech. Geoenviron. Eng.* **2007**, *133*, 653–666. [CrossRef]
23. Allasia, P.; Lollino, G.; Godone, D.; Giordan, D. Deep displacements measured with a robotized inclinometer system. In Proceedings of the 10th International Symposium on Field Measurements in Geomechanics, Rio de Janeiro, Brasil, 16–20 July 2018.
24. Allasia, P.; Manconi, A.; Giordan, D.; Baldo, M.; Lollino, G. ADVICE: A new approach for near-real-time monitoring of surface displacements in landslide hazard scenarios. *Sensors* **2013**, *13*, 8285–8302. [CrossRef] [PubMed]
25. Eberhardt, E.; Watson, A.; Loew, S. Improving the interpretation of slope monitoring and early warning data through better understanding of complex deep-seated landslide failure mechanisms. In *Landslides and Engineered Slopes*; Taylor & Francis: London, UK, 2008; pp. 39–51.
26. Intrieri, E.; Gigli, G.; Mugnai, F.; Fanti, R.; Casagli, N. Design and implementation of a landslide early warning system. *Eng. Geol.* **2012**, *147*, 124–136. [CrossRef]
27. Bell, R.; Glade, T.; Thiebes, B.; Jäger, S.; Krummel, H.; Janik, M.; Holland, R. Modelling and Web Processing of Early Warning. Available online: https://homepage.univie.ac.at/thomas.glade/Publications/BellEtAl2009.pdf (accessed on 12 February 2019).
28. Michoud, C.; Bazin, S.; Blikra, L.H.; Derron, M.-H.; Jaboyedoff, M. Experiences from site-specific landslide early warning systems. *Nat. Hazards Earth Syst. Sci.* **2013**, *13*, 2659–2673. [CrossRef]
29. Glade, T.; Nadim, F. Early warning systems for natural hazards and risks. *Nat. Hazards* **2014**, *70*, 1669–1671. [CrossRef]
30. Thiebes, B.; Glade, T. Landslide early warning systems—Fundamental concepts and innovative applications. In *Landslides and Engineered Slopes: Experience, Theory and Practice, Proceedings of the 12th International Symposium on Landslides, Napoli, Italy, 12–19 June 2016*; Aversa, S., Cascini, L., Picarelli, L., Scavia, C., Eds.; Taylor & Francis: London, UK; pp. 12–19.
31. Di Biagio, E.; Kjekstad, O. Early Warning, Instrumentation and Monitoring Landslides. In Proceedings of the 2nd Regional Training Course, RECLAIM II, Phuket, Thailand, 29 January–2 February 2007. Available online: http://www.adpc.net/V2007/Programs/UDRM/PROGRAMS_PROJECTS/RECLAIMIII/Downloads/RECLAIMProceedings(1).pdf (accessed on 12 February 2019).
32. Balis, B.; Kasztelnik, M.; Bubak, M.; Bartynski, T.; Gubała, T.; Nowakowski, P.; Broekhuijsen, J. The urbanflood common information space for early warning systems. *Procedia Comput. Sci.* **2011**, *4*, 96–105. [CrossRef]
33. Alberti, S.; Crosta, G.B.; Rivolta, C. Statistical analysis of displacement rate for definition of EW thresholds applied to two case studies. In Proceedings of the Workshop on World Landslide Forum, Ljubljana, Slovenia, 2017; pp. 285–292.
34. Crosta, G.; Agliardi, F. How to obtain alert velocity thresholds for large rockslides. *Phys. Chem. Earth Parts ABC* **2002**, *27*, 1557–1565. [CrossRef]

35. Bhandari, R. Some lessons in the investigation and field monitoring of landslides. In Proceedings of the 5th International Symposium on Landslides, Lausanne, Switzerland, 10–15 July 1988; pp. 1435–1457.
36. Carlà, T.; Intrieri, E.; Di Traglia, F.; Nolesini, T.; Gigli, G.; Casagli, N. Guidelines on the use of inverse velocity method as a tool for setting alarm thresholds and forecasting landslides and structure collapses. *Landslides* **2017**, *14*, 517–534. [CrossRef]
37. Saito, M. Forecasting the Time of Occurrence of a Slope Failure. 1965. Available online: https://www.issmge.org/uploads/publications/1/39/1965_02_0116.pdf (accessed on 12 February 2019).
38. Fukuzono, T. A new method for predicting the failure time of a slope. In Proceedings of the 4th International Conference and Field Workshop on Landslide, Tokyo, Japan, 23–31 August 1985; pp. 145–150.
39. Federico, A.; Popescu, M.; Elia, G.; Fidelibus, C.; Internò, G.; Murianni, A. Prediction of time to slope failure: A general framework. *Environ. Earth Sci.* **2012**, *66*, 245–256. [CrossRef]
40. Manconi, A.; Allasia, P.; Giordan, D.; Baldo, M.; Lollino, G.; Corazza, A.; Albanese, V. Landslide 3D Surface Deformation Model Obtained Via RTS Measurements. In Proceedings of the Second World Landslide Forum, Rome, Italy, 3–7 October 2011.
41. Giordan, D.; Manconi, A.; Allasia, P.; Bertolo, D. Brief Communication: On the rapid and efficient monitoring results dissemination in landslide emergency scenarios: The Mont de La Saxe case study. *Nat. Hazards Earth Syst. Sci.* **2015**, *15*, 2009–2017. [CrossRef]
42. Manconi, A.; Giordan, D. Landslide failure forecast in near-real-time. *Geomat. Nat. Hazards Risk* **2016**, *7*, 639–648. [CrossRef]
43. Wrzesniak, A.; Giordan, D. Development of an algorithm for automatic elaboration, representation and dissemination of landslide monitoring data. *Geomat. Nat. Hazards Risk* **2017**, *8*, 1898–1913. [CrossRef]
44. UNISDR. Global Survey of Early Warning Systems. 2006. Available online: https://www.unisdr.org/2006/ppew/info-resources/ewc3/Global-Survey-of-Early-Warning-Systems.pdf (accessed on 12 February 2019).
45. UNISDR. The Sendai Framework for Disaster Risk Reduction 2015–2030. 2015. Available online: https://www.preventionweb.net/files/43291_sendaiframeworkfordrren.pdf (accessed on 12 February 2019).
46. Garcia, C.; Fearnley, C.J. Evaluating critical links in early warning systems for natural hazards. *Environ. Hazards* **2012**, *11*, 123–137. [CrossRef]
47. Lollino, G.; Manconi, A.; Giordan, D.; Allasia, P.; Baldo, M. Infrastructure in geohazard contexts: The importance of automatic and near-real-time monitoring. In *Environmental Security of the European Cross-Border Energy Supply Infrastructure*; Springer: Dordrecht, The Netherlands, 2015; pp. 73–89.
48. Manconi, A.; Allasia, P.; Giordan, D.; Baldo, M.; Lollino, G. Monitoring the stability of infrastructures in an emergency scenario: The "Costa Concordia" vessel wreck. In *Global View of Engineering Geology and the Environment*; Taylor & Francis Group: London, UK, 2013.
49. Manconi, A.; Giordan, D. Landslide early warning based on failure forecast models: The example of the Mt. de La Saxe rockslide, northern Italy. *Nat. Hazards Earth Syst. Sci.* **2015**, *15*, 1639–1644. [CrossRef]
50. Crosta, G.; Agliardi, F. Failure forecast for large rock slides by surface displacement measurements. *Can. Geotech. J.* **2003**, *40*, 176–191. [CrossRef]
51. Crosta, G.; Di Prisco, C.; Frattini, P.; Frigerio, G.; Castellanza, R.; Agliardi, F. Chasing a complete understanding of the triggering mechanisms of a large rapidly evolving rockslide. *Landslides* **2014**, *11*, 747–764. [CrossRef]
52. Efron, B. Bootstrap methods: Another look at the jackknife. 1979. Available online: http://jeti.uni-freiburg.de/studenten_seminar/stud_sem_SS_09/EfronBootstrap.pdf (accessed on 12 February 2019).
53. Giordan, D.; Cignetti, M.; Wrzesniak, A.; Allasia, P.; Bertolo, D. Operative Monographies: Development of a New Tool for the Effective Management of Landslide Risks. *Geosciences* **2018**, *8*, 485. [CrossRef]
54. Smiciklas, M. *The Power of Infographics: Using Pictures to Communicate and Connect with Your Audiences*; Que Publishing: Indianapolis, IN, USA, 2012.
55. Manconi, A.; Giordan, D.; Allasia, P.; Baldo, M.; Lollino, G. Surface displacements following the Mw 6.3 L'Aquila earthquake: One year of continuous monitoring via Robotized Total Station. *Ital. J. Geosci.* **2012**, *131*, 403–409.
56. Raspini, F.; Moretti, S.; Fumagalli, A.; Rucci, A.; Novali, F.; Ferretti, A.; Prati, C.; Casagli, N. The COSMO-SkyMed constellation monitors the Costa Concordia wreck. *Remote Sens.* **2014**, *6*, 3988–4002. [CrossRef]

57. Solari, L.; Raspini, F.; Del Soldato, M.; Bianchini, S.; Ciampalini, A.; Ferrigno, F.; Tucci, S.; Casagli, N. Satellite radar data for back-analyzing a landslide event: The Ponzano (Central Italy) case study. *Landslides* **2018**, *15*, 773–782. [CrossRef]
58. Allasia, P.; Giordan, D.; Lollino, G.; Cravero, M.; Iabichino, G.; Bianchi, N.; Monticelli, F. *Monitoring and Computations on a Landslide in an Open Pit Mine*; American Rock Mechanics Association: Alexandria, VA, USA, 2009.
59. Margottini, C.; Corominas, J.; Crosta, G.B.; Frattini, P.; Gigli, G.; Iwasaky, I.; Lollino, G.; Marinos, P.; Scavia, C.; Sonnessa, A. Landslide hazard assessment, monitoring and conservation of Vardzia monastery complex. In *Engineering Geology for Society and Territory-Volume 8*; Springer: Cham, Switzerland, 2015; pp. 293–297.

Article

An Earthquake Fatalities Assessment Method Based on Feature Importance with Deep Learning and Random Forest Models

Hanxi Jia *, Junqi Lin and Jinlong Liu

Key Laboratory of Earthquake Engineering and Engineering Vibration, Institute of Engineering Mechanics, China Earthquake Administration, Harbin 150080, China; linjunqi@iem.net.cn (J.L.); liujinlong@iem.ac.cn (J.L.)
* Correspondence: jiahanxi@iem.ac.cn; Tel.: +86-0451-866-73509

Received: 27 March 2019; Accepted: 8 May 2019; Published: 14 May 2019

Abstract: This study aims to analyze and compare the importance of feature affecting earthquake fatalities in China mainland and establish a deep learning model to assess the potential fatalities based on the selected factors. The random forest (RF) model, classification and regression tree (CART) model, and AdaBoost model were used to assess the importance of nine features and the analysis showed that the RF model was better than the other models. Furthermore, we compared the contributions of 43 different structure types to casualties based on the RF model. Finally, we proposed a model for estimating earthquake fatalities based on the seismic data from 1992 to 2017 in China mainland. These results indicate that the deep learning model produced in this study has good performance for predicting seismic fatalities. The method could be helpful to reduce casualties during emergencies and future building construction.

Keywords: earthquake fatalities; deep learning; random forest; feature importance; structure type

---

1. Introduction

Earthquakes impose a large number of threats to the Chinese (Table 1). If there is a proper rapid estimation of the number of casualties in an earthquake, the impact and losses of the disaster could be decreased [1]. The human and material resources of emergency management can be allocated by predicting the death toll [2]. We use the surface-wave magnitude (Ms) in the study. According to the current emergency response regulations of relevant Chinese departments, the following categories of emergency personnel and materials are obtained: (1) When the magnitude is less than 6 and the predicted number of deaths is 0–10. The government will need 10–50 emergency personnel and 200–300 tents; (2) When the magnitude is greater than or equal to 6 and less than 6.5, and the predicted number of deaths is 0–10. The number of emergency personnel is 50–100, and the number of tents is 1000–3000. (3) When the magnitude is greater than or equal to 6.5 and less than 7, and the predicted death toll is 0–10, 200–500 emergency personnel and 3000–5000 tents will be needed. If the predicted number more than 10, 500–1000 emergency personnel and 5000–10000 tents will be needed. (4) When the magnitude is more than 7 and the death toll is less than 10, 500–1000 emergency personnel and 5000–10000 tents will be required. When the death toll is between 10–100, 1000–5000 emergency personnel and 10000–20000 tents will be required. When the death toll is 100–1000, 5000–10000 emergency personnel and more than 20,000 tents will be needed and (5) when the number of deaths is greater than 1000, it is necessary to draw the necessary emergency personnel and material distribution according to the specific economic and political conditions in the local area.

Table 1. Earthquake disaster losses in China mainland from 1992 to 2017, including the number of death and injured people and the economic costs. The unit of the economic loss is Chinese Renminbi Yuan (CNY).

Year	Deaths (People)	Injured (People)	Economic Costs (CNY)
1992	5	480	1.60×10^8
1993	9	381	2.84×10^8
1994	4	1378	3.29×10^8
1995	85	15024	1.16×10^9
1996	365	17956	4.60×10^9
1997	21	150	1.25×10^9
1998	59	13631	1.84×10^9
1999	3	137	4.74×10^8
2000	10	2977	1.47×10^9
2001	9	741	1.48×10^9
2002	2	360	1.48×10^8
2003	319	7147	4.66×10^9
2004	8	688	5.78×10^8
2005	15	867	2.63×10^9
2006	25	204	8.00×10^8
2007	3	419	2.02×10^9
2008	69283	377010	8.59×10^{11}
2009	3	404	2.74×10^9
2010	2705	11088	2.36×10^{10}
2011	32	506	6.01×10^9
2012	86	1331	8.29×10^9
2013	294	15671	9.95×10^{10}
2014	624	3688	3.56×10^{10}
2015	33	1217	1.80×10^{10}
2016	2	103	6.68×10^9
2017	37	638	1.48×10^{10}

However, there are many factors which may affect fatalities, and not every factor has a decisive impact on earthquake casualties. Therefore, it is also necessary to select a suitable method to evaluate the importance of each factor.

Linear models are the most constantly used methods for assessing feature correlation [3]. In reference [4], the research has given the relationship between human losses and factors such as population density and the intensity and magnitude of the earthquakes based on the linear models. Nevertheless, due to the uncertainties and fuzziness in the data of the factors [5], integrated ensemble models were proposed and applied to the feature importance assessment models [6–9] for the purpose of improving accuracy and generalization ability of the traditional linear models [10]. In the present studies, the excellent performance of ensemble algorithms on prediction ability and generalization capacity has been proven better than the linear models [3]. However, so far, no research has been conducted to evaluate the importance of influencing factors and different structure types on earthquake casualties using machine learning methods. Previous studies of earthquake casualties based on experience directly gave influencing factors [2,11] and structural types [12] or based on the statistical methods gave [13].

Different methods were developed to estimate the casualties in earthquakes. Most studies used empirical analysis methods [12,14,15] and some software systems to assess casualties. For instance, geographic information system (GIS) [11,16], the U.S. Geological Survey's Prompt Assessment of Global Earthquakes for Response (PAGER) system [17] and the Disaster Management Tool (DMT) software [18]. In reference [18], the authors present the casualty estimation model, which is part of the DMT software. The model is based on the evaluation of laserscanning data that collected by the airborne sensors and it also can be used to detect collapsed buildings, to assess their damage type, and to compute the number of the trapped victims for each collapsed building. The PAGER system, made recourse to the EERI World Housing Encyclopedia (WHE) project (including the non-engineered building) [19], can estimate the fatality for large earthquakes in the two hours [17]. However, these systems cannot assess losses in a few minutes. Empirical methods usually established linear models that were evaluated by fitting one or more functions [20]. These models have many disadvantages: the workloads tend to be large and the amount of data small; the abnormal points were usually deleted instead of calculating fit together within the models; and they have strong subjectivity. These shortcomings can be compensated by neural networks in the field of machine learning [21].

With the rise of machine learning algorithms, some studies of estimating fatalities based on back propagation neural network (BPNN) method have begun to emerge [21]. Because of different earthquakes of intensity, population density, and different structure types, it is extremely perplexing to define a certainty relevance to evaluate fatalities caused by an earthquake. Hence, deep learning method, with its abilities to estimate perplexed relevances, could be an outstanding method to evaluate fatalities. However, BPNN method is not a very perfect network, it has many shortcomings: (1) The convergence speed is too slow and it takes hundreds or more than hundreds of times to learn to converge [22]; (2) it cannot guarantee convergence to a global minimum point [23,24]; (3) there are a number of hidden layers and neurons in that are not theoretically guided, but are determined empirically, thus, the network tends to be large [22]—the redundancy invisibly increases time of network learning [25]; and (4) learning and memory of the network are unstable. Deep learning optimization algorithms can improve the shortages of BPNN method.

Therefore, we assessed the importance of the factors based on three machine learning methods and selected the random forest algorithm as the optimal classifier. At the same time, we evaluated the contribution degree of 43 different structure types based on the random forest algorithm. Finally, the deep learning assessment model was established with the factors of population density, magnitude, focal depth, epicentral intensity, and time. Figure 1 shows the flowchart of the entire assessment process.

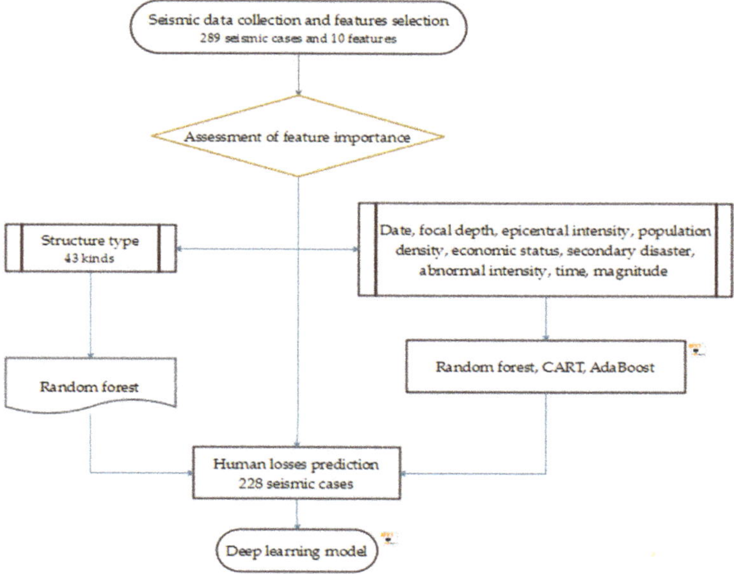

Figure 1. Flowchart for assessing human losses with machine learning methods. CART: classification and regression tree.

2. Features

2.1. Data

An important question, in human losses expected studies, is whether or not a conditioning variable is actually useful and needed for the assessment and prediction. There are many factors affecting the casualties of earthquakes, such as the intensity of earthquakes, the vulnerability of houses and the economic development in earthquake areas. However, some factors do not have data for every earthquake. We chose the following ten features: date, time, magnitude, epicentral intensity, abnormal intensity, focal depth, secondary disasters, population density, economic situation, and damage ratio of different structure types.

1. The regularity of people's working and resting time makes the differences of earthquake occurrence time have a great impact on the fatalities [4,13]. Time was divided into two parts of this study: daytime (7:00–21:00) and sleeping time (21:00–next7:00).
2. Magnitude is a parameter to measure the energy release by the seismic wave, generally, the greater the magnitude, the greater the disaster caused. The magnitude is expressed by the surface wave magnitude Ms commonly used in China.
3. Considering the seismic source as a point, the vertical distance from the point to the ground is called the focal depth. Often, the smaller the focal depth is, the closer to the epicenter and the greater the damage caused.
4. In general, the higher the intensity, the greater the casualties [4]. This study used the epicenter intensity listed in the Earthquake Disasters and Losses Assessment Report in Chinese Mainland.
5. Population density has a great impact on the number of earthquake deaths [4]. There are obvious differences between densely populated and more sparsely populated areas. For example, the population density in Tibet province is low and there are even existing places where no people, hence, the casualties in the earthquake are bound to be small. Conversely, high population densities contribute to an increase in the number of deaths [2]. The location of the epicenter can be obtained from the China Earthquake Networks Center after an earthquake, and the population density can be obtained from the data published by the Statistical Bureau.
6. Generally, seismic intensity decreases with increasing distance from the epicenter. But high intensity points appear in low intensity areas, or, conversely, for reasons such as geological structure, topography, and the superposition of deep seismic reflection waves, which is called seismic intensity abnormally. The abnormal intensity in this paper refers to the occurrence of high intensity points in low intensity areas, which often aggravates disaster losses. Abnormal intensity was expressed in two cases: yes and no.
7. The economic situation has a great impact on disaster losses [4]. Usually, the better the economic situation, the lighter the disaster will be in the same earthquake intensity, and the higher the population density will be in the concentrated areas of social wealth. According to the situation mentioned in the Earthquake Disasters and Losses Assessment Report in Chinese Mainland, the economic situation was divided into seven categories, from top to bottom, the economic situation is getting better: (1) national poverty region; (2) special poverty area, deep poverty area, remote and poor mountainous area, remote and poor area, border poverty areas, provincial poverty region, remote area; (3) minority poverty area, general poverty area; (4) financial deficit area; (5) economically backward area, minority area; (6) general area; (7) western medium-developed areas.
8. Secondary disasters will cause secondary damage to the disaster area. The impacts cannot be underestimated: they mostly manifest as mountain collapse, landslide and debris flow, and very few are the result of fires. They can only be divided into two cases in this study due to the small number of generations: yes and no.

9. Most of the fatalities are caused by building damage [2] and this factor is vital to the number of deaths [13]. Therefore, this paper chooses the damage ratio of houses as the feature. Damage ratios consist of collapsed structures, heavy damage, moderate damage, and slight damage.
10. Different earthquake occurrence dates can sometimes lead to aggravation of earthquake damage, for instance, rain and snowy weather will affect rescue efforts. Dates were processed quarterly, and a year is divided into four quarters in this study.

2.2. Importance Assessments of 9 Features

The selected supervised classifiers are the random forests (RF), adaptive boosting (AdaBoost) and classification and regression tree (CART). The CART algorithm, a decision tree model and a non-parametric data mining method, has many advantages including ease of handling numerical and categorical data and multiple outputs situations [25]. The CART algorithm is a component learner with gini features, such as division standard, while ensemble learning combines multiple weak classifiers using different methods. The most common methods of ensemble learning are bootstrap aggregating (bagging) method and boosting method, in which bagging is a parallel algorithm and boosting is a sequential algorithm. The RF algorithm, the expansion of bagging method, exploits random binary trees to discriminate and classify data [7]. The AdaBoost approach, a boosting algorithm, constructs a strong classifier with weak classifiers and updates the weight of samples based on learning error, as shown in Figure 2. Each sample in the training data is given equal weight α at first. A weak classifier is trained on the training data and the error rate ε of the classifier is calculated. Then, the classifier is trained again on the unified data. The ε will be increased while αwill be reduced to the first classification on the second training classifier. AdaBoost calculates ε of each weak classifier and assigns α to each classifier. In Figure 2, the first row is the data set, where the different width of histograms represents different weight on each sample. The data set will be weighted by α in the third row after passing through the classifier [26]. The final output is obtained by summing the weighted results. The the error rate ε (Equation (1)) and the weight α (Equation (2)) are calculated as follows:

$$\varepsilon = \frac{N1}{N2} \tag{1}$$

$$\alpha = \frac{1}{2}\left(\frac{1-\varepsilon}{\varepsilon}\right) \tag{2}$$

where N1, N2 are the number of incorrectly classified and classified samples, respectively.

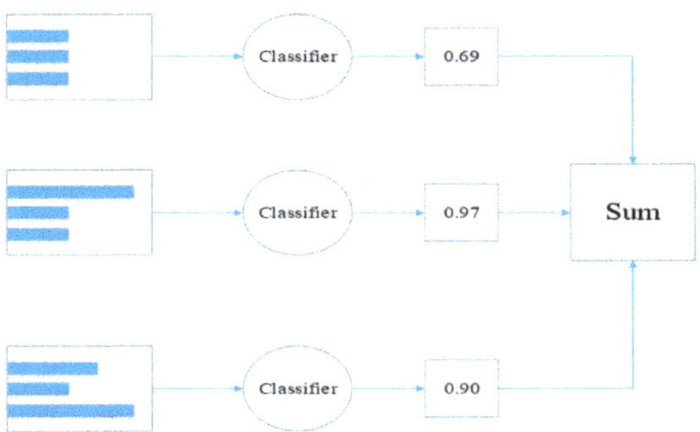

Figure 2. Simple example showing that AdaBoost can construct strong classifier from set of weak classifiers.

We worked with the CART, RF, and AdaBoost models implemented in the jupyter of the Anaconda Navigator. Data was generally required to be standardized in machine learning. We used the

StandardScaler preprocessing method of the sklearn function library to process the magnitude, focal depth, epicentral intensity, and population density. Equation (3) presents the process:

$$x = (x - \mu)/\sigma \quad (3)$$

where x is the features, μ is the mean of the data and δ is the variance of the data. The parameters selection of each model is shown in Table 2. Table 3 presents the result of verifying the model with cross validation score function in sklearn function library. And the importance of nine features in casualty assessment is shown in Figure 3.

Table 2. Parameters of random forest (RF), CART, and AdaBoost (Unset parameters were used as default parameters in sklearn, AdaBoost has two classes of classification algorithms SAMME and SAMME.R, among which SAMME.R is better based on class probability).

Random Forest	CART	AdaBoost
number of estimators = 82	max depth = 10	base estimator = decision trees
number of jobs = −1	max features = none	classifier algorithm = SAMME.R
max features = none	number of estimators = 500	learning rate = 0.5
criterion = gini	criterion = gini	number of estimators = 379

Table 3. Results of testing the models on unseen validation set of 9 features.

Algorithm	Random Forest	CART	AdaBoost
Mean accuracy	0.820	0.745	0.766

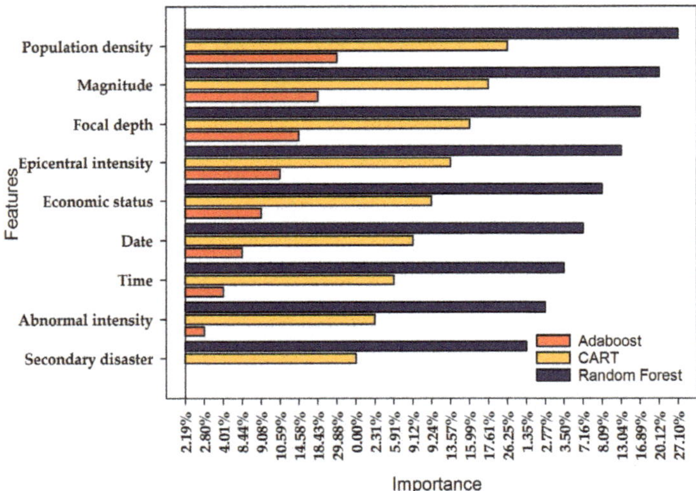

Figure 3. Comparing the relative importance of input variables with RF, CART, and AdaBoost algorithms. To make the contrast clearer and more readable, use the bar chart in the sigmaplot software for drawing.

2.3. Importance Assessments of Structure Types

The importance of different structure types was assessed alone because of the complexity of the structure types and the great contribution to the death [27]. For the accuracy and comprehensiveness of the assessment, this study listed 43 universal and special structure types from the Earthquake Disasters and Losses Assessment Report in Chinese Mainland from 1992 to 2007: reinforced concrete frame structure, masonry structure, brick-wood structure, civil structure, national brick-wood structure, brick-concrete structure, bucket-piercing frame structure, brick-concrete structure (building of two or more floors), shed, brick-concrete structure(building of only one floor), brick-column civil structure, wing-room, national civil structure, simple house with dry fortified earth wall, brick-column adobe structure, dry brick building, brick structure, timber structure, timber stack structure, timber framework, brick-masonry structure, frame structure, adobe structure, wood-column adobe structure, reinforced frame structure, mixed house, brick adobe structure, general houses (owned by citizens), stone-wood structure, stone structure, simple house, old Tibetan house, stone-grass structure, stone-concrete structure, alunite house, earth rock house, industrial plant, steel frame structure, soil tamper structure, cave dwelling, wooden frame house, flag stone house, and general house.

Structural damages were divided into five grades: collapse, heavy damage, moderate damage, slight damage, and basically undamaged. We chose the collapse, heavy damage of buildings, and population density as input parameters. Firstly, death was almost being caused by the collapse and heavy damage of the architecture. Another, population density, was closely related to the seismic casualty and the number of buildings. We only selected random forest algorithm to assess the importance of structure types for the mean accuracy of the RF model was higher than the other algorithm as seen from Table 3. Table 4 presents the procedure of RF in feature assessment. The importance of structure types can be seen in Figure 4.

Table 4. Steps of the random forest algorithm.

Step by Step Procedure of Random Forests Algorithm
Inputs: a, b, c, d a: Damage ratio of collapse of different structures b: Damage ratio of heavy damage of different structures c: Population density d: Death **Parameters**: number of estimators = 500, criterion = gini, numbers of jobs = -1, max features = None, criterion = MSE, max depth = None, max leaf nodes = None, min impurity decrease = 0.0, min impurity split = None, min samples leaf = 1, min samples split = 2, min weight fraction leaf = 0.0, presort = False, random state = None, splitter = best **Process:** step1: Bootstrap sampling is used to extract sub-training sets from the training set. step2: Generate the feature subsets by randomly selecting features before node splits. step3: Establish decision trees step4: Obtain the results for the sample to be tested step5: Vote on the results and got the results. **Output:** Importance of structure types

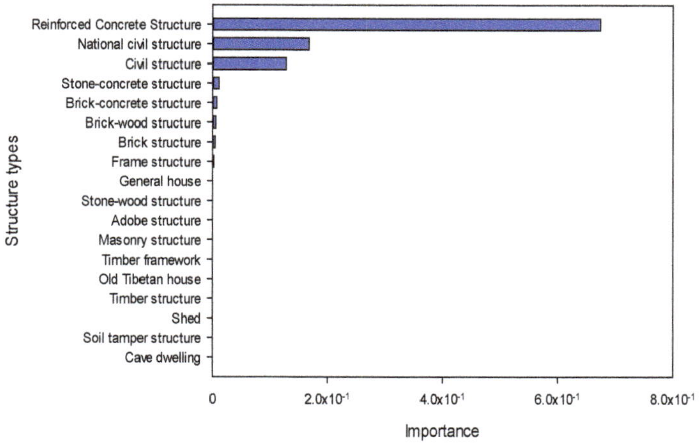

Figure 4. Importance of structure type (The structure types that contribute little to the casualty not shown. The importance of general houses and structure types below it close to zero).

3. Model

We chose the population density, magnitude, focal depth, epicentral intensity, and time as the input parameters. The criteria and reasons for selecting five input parameters among ten features were as follows:

(1) Features were selected from high to low in the order of features importance. According to their subjectivity, the number of occurrences and obtained time after the earthquake.
(2) The results of the importance assessment show the relative values of the nine factors that contribute to the assessment of the casualty, rather than the absolute value of each of the factors alone. Time ranked seventh instead of higher in the importance assessment results because we only divided it into the two parts. In the deep learning model, we did not to classify the time. Therefore, we chose the time because it could be almost obtained at the same time as an earthquake and it had not subjectivity in the deep learning model.
(3) The division of the economic status had a strong subjectivity and the data selected in the test set was the economic situation of the year of the occurrence of the earthquake. With the annual inflation and the depreciation of the coin, the situation of that year may not be applicable to the future.
(4) Although the date was more important than the time, the subjectivity of the date was also great, which was only divided into four seasons.
(5) The secondary disasters and abnormal intensity were only divided into yes and no. The combined number of the two phenomena was small and not every earthquake was accompanied by them.
(6) The structure types were too complex. Structure destroyed during each earthquake were different and every destruction was divided into five parts.

3.1. Data

We collected 289 destructive earthquakes occurred in the Chinese mainland from 1992 to 2017 in the Earthquake Disasters and Losses Assessment Report in Chinese Mainland (Table A1). The excel table data were pre-processed by openpyxl module in deep learning method, and the data set of 228 earthquake cases were returned without a missing value. Among these data, we selected 180 data as the training set and 38 data as the test set. The remaining ten were used as validation sets. The time was recorded in minutes, and the hours are converted into minutes, which are calculated at 1440 min per day. When the time is x: y, the data will be processed as (60x+y)/144. For example, if the time is 04:32, the change is 0.19. Other parameters need not be specially processed.

3.2. Deep Learning Model

3.2.1. Hyper Parameter

A deep learning model is a multilayer stack of simple modules and many of which compute non-linear input to output mappings. The hidden layers of 3 to 18 in the deep learning model can implement extremely complex functions of the original variables that are sensitive to minute details [23]. Given the small data set in this study, we chose a four-layers back propagation network. The number of neurons in the hidden layers was optimized for getting accurate output [21]. To select the number of two hidden layer's neurons, the training began with ten and three neurons separately and was repeated for more neurons. The number of neurons in the hidden layers were determined to be 40 and 5 separately for reducing the complexity and running time of the model.

There are different methods to avoid over fitting during the training process and to obtain models that are expert in generalizing the problem to be explained. For instance, stopping iteration in advance, regularization, dropout, and data set expansion. Stopping iteration in advance and data set expansion are not applicable to this model. The dropout method is a random deletion of half of the hidden layer nodes, hence, it is also not applicable. Finally, the regularization method was chosen.

Regularization in deep learning includes coefficients L1 and L2. We chose the most commonly used L2 regularization. L2 regularization is to add an additional regularization term to the loss function, presented in Equation (4).

$$c = c_0 + \frac{\lambda}{2n}\sum_{w} w^2 \tag{4}$$

where c and c_0 are the new and old loss functions, separately, λ is the L2 regularization rate, n is the number of the sample and w is the weight. Through a large number of tests, the optimal parameters were obtained as shown in Table 5. Figure 5 demonstrates the model with two hidden layers, consisting of forty and five neurons respectively, five input layers, and one output layer.

Table 5. Hyper parameters.

Batch Size	Learning Rate Base	Learning Rate Decay	Regularization Rate	Training Steps	Moving Average Decay
16	0.8	0.99	0.00005	500000	0.99

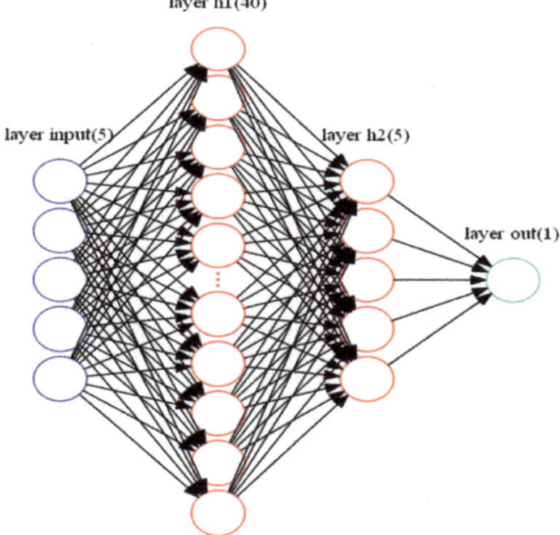

Figure 5. Scheme of deep learning network.

3.2.2. Optimization Algorithm

The optimization algorithms used in the model including the adaptive moment estimation (Adam) [28], mini-batch gradient descent [29], and moving average model [30]. We used the adam algorithm to make the learning rates updating independent for parameters by calculating the first and second moment estimation of gradients. The mini-batch is to reduce the randomness for the data in a pool determine the direction of the gradient together and reduce the possibility of deviation during the descent process [31]. The model computer was a workstation with a good performance and the data volume of the model was less than 2000. The size of the training pool was usually chosen as the n-th power of 2. Hence, we chose 16 data as a training pool through experiments. The moving average model is to prevent the sudden change when updating parameters [32].

3.2.3. Process of the Model

Definition structure and forward propagation used tensorflow framework [30]. The parameters of deep learning network needed only define weights w and b, and the normal distribution was chosen as the weight generating function. The activation function was relu function [23] and was used in the first hidden layer. The second hidden layer and the output layer are basically equivalent to linear regression. We selected Adam algorithm as back propagation algorithm. The learning rate is optimized. The mean square error loss function was selected for the data set was little (Equation (5)).

$$c = \frac{1}{n}\sum (y - t)^2 \quad (5)$$

where c presents loss function, n is the number of the samples, y is the predictive value and t is the true value.

4. Results

We chose ten seismic cases with different parameters as the test data, of which the eighth, ninth, and tenth cases were Ludian earthquake in 2014, Yushu earthquake in 2010 and Wenchuan earthquake in 2008. Selection of specific test sets is shown in Table 6. Comparing with the intensity-based mortality estimation method in *Assessment of Earthquake Disaster Situation in Emergency Period* [33] (China-National Standard), presented in Equation (6) and (7). The Standard is an important basis for the government's decision-making on earthquake relief. After an earthquake, the primary tasks of the emergency work are to conduct disaster investigation and assessment in a short period of time and to provide the government with timely and necessary information on the disaster situation. Therefore, relevant departments have formulated the Standard in a comprehensive summary of the research methods and field practices of earthquake disaster assessment in China. Focus on rapid and dynamic disaster recovery and assessment.

$$N_D = \sum_{j=6}^{I_{max}} A_j \rho R_j \quad (6)$$

$$\ln R_j = -44.365 + 7.516 I_j - 0.329 I_j^2 \quad (7)$$

where N_D is the number of deaths; I_{max} is the intensity of the meizoseismal area; A_j is the distribution area of intensity value *j*; ρ is the population density; R_j is the death rate corresponding to the intensity value *j*; I_j is the intensity value j.

Table 6. Test sets.

Number	Time	Magnitude (Ms.)	Epicentral Intensity (Degree)	Focal Depth (km)	Population Density (People/km^2)
1	0.33	5.1	6	12	58.28519
2	0.84	5.0	7	33	105.7979
3	0.25	5.5	7	8	16.63824
4	0.38	5.1	6	9	169.8517
5	0.96	6.9	8	11	3.384553
6	0.69	6.1	8	10	131.524
7	0.47	5.7	8	14	226.8326
8	0.69	6.5	9	12	216.6675
9	0.33	7.1	9	14	8.686792
10	0.60	8.0	11	14	238.3409

Table 7 and Figure 6 show the comparison results of the deep learning model, China-National Standard and true value. Table 8 presents the accuracy of the two methods in cases from1 to 7 and cases from 8 to 10, separately. The results demonstrate as follows:

(1) Adam algorithm and moving average model based on deep learning can accelerate the convergence speed and improve the accuracy of model prediction.
(2) The accuracy was higher, and the factors considered were more than the Standard.
(3) The prediction accuracy of the last three earthquake cases is generally not high.

(4) The China-National Standard model selected fewer parameters. The model was based on the experience of many experts and had been tested for many years. In the cases 1–7, the predicted values and the true values were all on the order of magnitude, but the performance on the cases 8–10 was far worse than the deep learning model, which was far from the true value and has no reference for the actual earthquakes.

Table 7. Comparison results.

Number	Deep Learning	China-National Standard	True Value
1	0	0	0
2	0	0	1
3	7	0	8
4	26	0	22
5	24	0	24
6	36	10	41
7	80	18	81
8	245	173	617
9	2625	7	2698
10	37406	191	69227

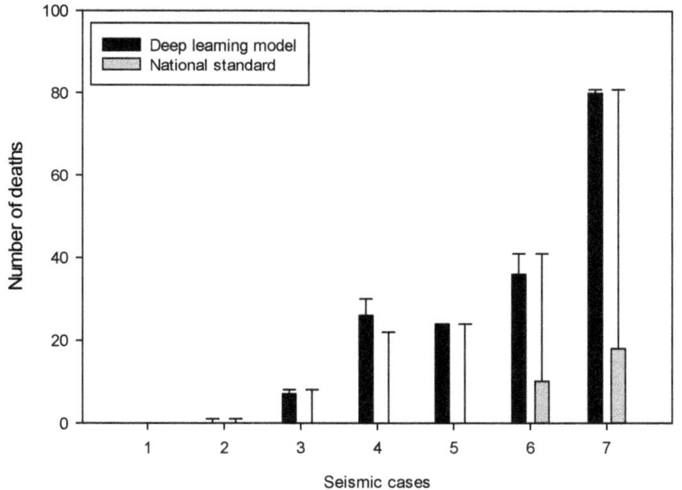

Figure 6. The number of deaths of the deep learning model and the national standard on test data. And the errors of two models with actual death toll are also presented. (The results of earthquake cases 8–10 are not shown for the large difference will affect the results of other earthquake cases).

Table 8. The accuracy of two models.

Method	Cases 1–7	Cases 8–10	Cases 1–10
China-National Standard	52.27%	25.32%	44.19%
Deep learning model	93.61%	45.85%	79.28%

5. Discussion

Three issues were considered in this study: the importance of nine features to the human losses in China mainland, the importance of 43 structure types of the fatalities, and human losses assessment. The first issue was addressed by adopting three traditional machine learning method domains to investigate different features' influence on the seismic fatalities. The results can provide a basis for further seismic assessment. The second issue was addressed by RF algorithm to assess the importance of 43 structure types. The third issue was addressed by establishing a deep learning model domain to predict the seismic death. A detailed investigation of these two issues is presented as follows.

5.1. Importance of Different Features

We propose an automatic classifier based on RF, CART, and AdaBoost, and some attributes to describe the feature importance of seismic. Random forest is the most stable among three methods. Figure 7 exhibits the results of three tests on that three methods. RF provides probability estimates on the classification that are useful to accept or reject a new classification [8]. Thus, we chose the RF algorithm as the main method. It fully proved the importance of the population density, magnitude, focal depth, epicentral intensity. The sum of the four features is 74.68%, which is far more important than other factors.

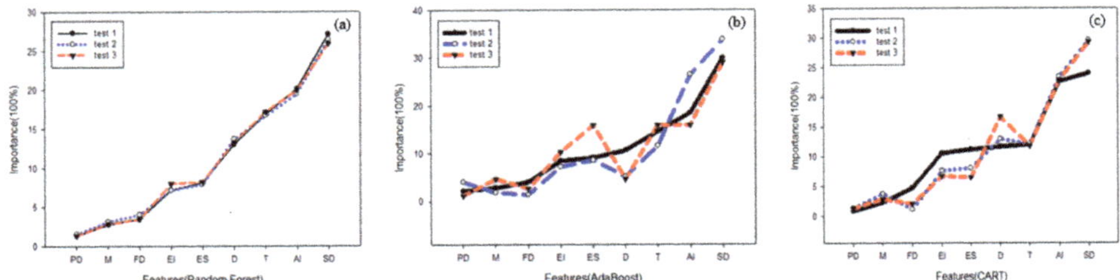

Figure 7. Results of three tests by Random Forest (**a**), AdaBoost (**b**), and CART (**c**) algorithms. The horizontal axis is the first letter of the feature, followed by: PD (population density), M (magnitude), FD (focal depth), EI (epicentral intensity), ES (economic status), D (date), T (time), AI (abnormal intensity) and SD (secondary disaster).

Time is vital to seismic, for example, the Tangshan earthquake occurred at 3:24 on July 28, 1976, when people slept in the house, and time aggravated the disaster. Losses will be relatively mitigated in daytime for most people are not sleeping and could escape quickly. The reason that time ranks seventh in this study is that time was only divided into two possibilities: daytime and sleeping time. More classes are discriminated by an attribute, the more important the attribute is [8]. In the same way, the ratio of abnormal intensity and the secondary disaster are only 2.63% and 1.18% separately for the fewer classes. However, time is divided into different levels, not modeled at an actual time, so there is interference with the results. The secondary disasters and the abnormal intensity are classified because they occur less frequently. In summary, the features selected by the RF algorithm experiments are consistent with those used by other scholars to study earthquake casualties (population density [2], magnitude [2], focal depth [34], epicentral intensity [16], and time [13]).

5.2. Importance of Different Structure Types

We chose all 43 structure types in China mainland to assess, which are far more than the number of the types in the previous studies [35]. Although the structure types of the WHE project are suitable for the most region of the world, it also lacks some structures that Chinese characteristic structure, such as the national civil structure, the old Tibetan house and the national brick-wood structure. Hence, the data from the Earthquake Disasters and Losses Assessment Report in Chinese Mainland of every earthquake in China mainland is more suitable for the study comparing the data of the WHE project. The HAZUS system (HAZad United States of Multi-Hazard) only has twelve structure types and estimates the casualties for collapsed buildings and not for the heavy damaged buildings [18]. In reference [18], the casualties are estimated at the level of single buildings and not for an entire zone in an adapted HAZUS system. The population density varies greatly in different parts of China, so the HAZUS system and the adapted system are not suitable for the casualties estimation of China.

The importance of different structural types was obtained based on RF method first. In the part of data collecting, there are no integration of structural forms and some may be slightly repeated in order not to omit any structure type. For instance, brick-column civil structure, brick-concrete structure (building of two or more floors) and brick-concrete structure (building of only one floor). However, repeated structures, even when combined, the importance was close to zero.

Figure 4 displays that the contribution to the casualties of reinforced concrete structure is the largest, followed by civil structure, stone-concrete structure, brick-concrete structure, brick-wood structure, brick house and the frame house. The other structures are not shown in Figure 4 and could be neglected. We concluded that the structure with large effect may not be good at seismic behavior. The result manifests that once the building is destroyed, the damage will be greater.

5.3. Human Losses Assessment Model

We proposed a human losses rapid prognostic assessment model based on the above feature engineering. The results indicate the predictions of large earthquakes with magnitude 6.5 or more are lower than the others. For example, the actual death toll of the Wenchuan earthquake is 69227, but the estimation of the deep learning model is 37406, the PAGER model is 50,000 and the other empirical model [20] is 30,000. It can be seen that the accuracy of the assessment of casualties for a large destructive earthquake is not high according to the traditional accuracy calculation (Equation (8)). The reason may be that there are more factors affecting large earthquakes than small earthquakes [2], and the uncertainty is greater than that of small earthquakes [34]. For example, the mountainous area of Ludian County is as high as 87.9% and the high incidence of secondary disasters (e.g., debris flow and landslide) cause large number of casualties. The accuracy of general prediction models is reduced due to the neglect of geological conditions [12].

$$\text{accuracy} = \left(1 - \frac{|\textit{true value} - \textit{estimate value}|}{\textit{true value}}\right) \times 100\% \tag{8}$$

Different geological conditions in different parts of China. In China, the areas with the most earthquakes are Sichuan Province, Yunnan Province, and Xinjiang Province. According to the earthquake cases and the results of the deep learning model, we can obtain conclusions as follows: (1) The basin region with soft rock-soil can aggravate earthquake disaster in Yunnan Province and Sichuan Province, such as Ludian earthquake in 2014 and Wenchuan earthquake in 2008; (2) In the event of an earthquake, the number of casualties in the area of the fault zone will be more serious in Xinjiang Province and Yunnan Province, such as Jinghe earthquake in 2017, Hutubi earthquake in 2016, and Ninger earthquake in 2007, and (3) Qinghai Province and Sichuan Province are close to each other, and the geological conditions are similar. Areas with crumbly strata have higher seismic vulnerability, such as Yushu earthquake in 2010. Above all, the error of the deep learning most comes from the geological problems.

There are some important reasons that affect the differences of the accuracy between the case 1–7 and case 8–10 besides the geological conditions. Earthquake casualties come from the direct and the secondary disasters, and the assessment mode of the latter is more difficult. However, there is currently no professional method to distinguish the human losses from the direct and the secondary disasters after the earthquake. Therefore, although the secondary disaster was evaluated at the time of importance assessment, it was not selected as an input variable. For destructive earthquakes, such as case 8–10, the types of secondary disasters are more abundant, so the impact is greater. Moreover, the traffic network is very important for post-earthquake rescue. Small earthquakes with magnitudes less than 6.5, such as cases 1–7, usually have limited damage to the traffic network and rarely completely block traffic. However, in the case of a large earthquake with a magnitude more than 6.5, such as the cases 8–10, the traffic network is usually interrupted, which seriously hinders rescue and greatly increases the number of deaths. The study did not consider the damage of the traffic network is also the main reason for the accuracy of the case 8–10.

The purpose of this study is to rapidly assess the human losses. Hence, the input features should be obtained in a short time after an earthquake. In reference [21], the train set was only from the Bam earthquake in 2003, thus, the results were not representative. Compared with empirical methods [36], the data set is larger [4], covering almost all the data from 1990 to 2017 without default. The accuracy of the results is higher; the data set is larger, and the factors considered are more than the China-National Standard [33]. It proved the deep learning technical can be used to estimate the causalities without any assumptions as compared with statistical methods, and can be directly processed by inner functions.

6. Conclusions and Future Works

6.1. Conclusions

We proposed a method to assess the importance of the nine factors affecting casualties in earthquakes and rank the importance of each feature based on the random forest algorithm. At the same time, 43 structural types were evaluated by this method, and the contribution of different structural forms to the death toll were obtained, which provides a basis for the future construction of structural forms. Based on the above evaluation of importance, we have reached the following conclusions:

(1) The works of the features importance fully prove the importance of the population density, magnitude, focal depth, and epicentral intensity on contribution to the death. The importance of the time less than the date and economic status because it is divided into only two parts.
(2) The Random Forest algorithm performs better than the AdaBoost and the CART algorithms, both in terms of stability and accuracy.
(3) Reinforced concrete structure, national civil structure, and civil structure have the highest contribution to the death toll. The contribution of stone-concrete structure, brick-wood structure, brick structure, and frame structure are small. The other structures are smaller, and even cannot be displayed in the figure.

A deep learning model for estimating human losses based on the results of the Random Forest algorithm was established. We selected five important features and compared the results with the China-National Standard because the Standard is suitable for the rapid assessment works. The results demonstrate that the accuracy is higher than the other methods and the running time are suitable for the emergency rescue work. Therefore, this method can be used to evaluate the fatalities for future earthquakes in China mainland. This model can serve the China Earthquake Administration and the Chinese government.

6.2. Extension of the Works

Further extensive studies are needed and some recommendations for future research are given as follows. First, this research is based on the evaluation of factors importance. Future studies can extend the study by adding the number of factors. Second, this paper estimates the importance of different structures to death based on the Random Forest algorithm. Future studies can add the sparse learning to process the data for the classifier getting better results. Third, the deep learning model assesses the human losses with some optimization algorithms. Future studies can add the hidden layers and continue to optimize the algorithms. It will be of great interests to focus on death prediction in the future works.

Author Contributions: Conceptualization, H.J. and J.L. (Junqi Lin); methodology, H.J.; software, H.J.; validation, all three authors; formal analysis, H.J.; investigation, H.J.; resources, J.L. (Junqi Lin); data curation, J.L. (Junqi Lin); writing—original draft preparation, H.J.; writing—review and editing, H.J.; visualization, H.J.; supervision, J.L. (Junqi Lin) and J.L. (Jinlong Liu); project administration, J.L. (Junqi Lin) and J.L. (Jinlong Liu); funding acquisition, J.L. (Junqi Lin) and J.L. (Jinlong Liu)

Funding: This research was funded by NATIONAL KEY R&D PROGRAM OF CHINA, grant number 2018YFC1504503.

Acknowledgments: We would like to thank Chen Zhao for guidance and help with applying machine learning algorithms.

Conflicts of Interest: The authors declare no conflict of interest.

Appendix A

Table A1. The 289 earthquake cases of the deep learning model with the default value from 1992 to 2017 in China mainland.

	Date	Time	Magnitude (Ms)	Epicentral Intensity	Focal Depth (km)	Population Density (People Per Square)	Deaths
1	1992/4/23	11:32 PM	6.9	7		64.98	4
2	1992/12/18	7:21 PM	5.4	6		311.11	1
3	1993/1/27	4:32 AM	6.3	8	14	60.68	0
4	1993/2/1	3:33 AM	5.3	6	10	35.63	0
5	1993/3/20	10:52 PM	6.6	8		11.76	2
6	1993/5/24	7:57 AM	5	6		21.28	0
7	1993/5/30	3:26 PM	4.9	6	10	122.38	0
8	1993/7/17	5:46 PM	5.6	6	16	11.06	1
9	1993/8/14	10:30 PM	5.6	7	8	105.60	0
10	1993/9/5	4:22AM	5.1	6	40	32.33	0
11	1993/12/1	4:37 AM	6	7	28	94.26	2
12	1994/1/11	8:51 AM	6.7	7	10	16.45	0
13	1994/9/16	2:20 PM	7.3	7	20	700.00	4
14	1994/9/19	11:28 PM	5.2	6		197.35	0
15	1995/2/18	8:14 AM	5.1	6		101.51	0
16	1995/4/25	12:13 AM	5.6	7	20	62.42	0
17	1995/7/12	5:46 AM	7.3	8	10	55.50	11
18	1995/7/22	6:44 AM	5.8	8	10	80.31	12
19	1995/9/20	11:14 AM	5.2	6	12		0
20	1995/9/26	12:39 PM	5.3	6	27	11.62	0
21	1995/10/6	6:26 AM	4.9	6	10	945.95	0
22	1995/10/24	6:46 AM	6.5	9	12.5	81.94	59
23	1996/2/3	7:14 PM	7	9	10	57.44	309
24	1996/2/28	7:21 PM	5.4	7	15	412.94	1
25	1996/3/19	11:00 PM	6.9	8	11	3.38	24
26	1996/5/3	11:32 AM	6.4	8	24	183.19	26
27	1996/6/1	8:49 PM	5.4	6	10	91.97	0
28	1996/7/2	3:05 PM	5.2	6	10	125.55	2
29	1996/9/25	3:24 AM	5.7	7	15	42.56	1
30	1996/12/21	4:39 PM	5.5	7		5.16	2
31	1997/1/21	9:47 AM	6.4	8		180.59	12
32	1997/1/25	10:38 AM	5.1	6	10	3.03	0
33	1997/1/30	5:59 PM	5.5	7	10	36.82	0
34	1997/3/1	2:04 PM	6	7		131.98	1
35	1997/4/11	1:34 PM	6.6	8	17	58.78	8
36	1997/5/31	2:51 PM	5.2	6		76.28	0
37	1997/8/13	4:13 PM	5.3	7	7.9	545.09	0
38	1997/9/26	11:19 AM	4.2	6	4.8	314.60	0
39	1997/10/23	8:28 PM	5.3	6	10	587.57	0
40	1997/11/3	10:29 AM	5.6	7	14	0.93	
41	1998/1/10	11:50 AM	6.2	8	10	82.11	49
42	1998/3/19	9:51 PM	6	7	11	6.75	0
43	1998/4/14	10:47 AM	4.7	6	10.7	0.00	0
44	1998/5/29	5:11 AM	6.2	7	11	19.09	0
45	1998/6/25	2:39 PM	5.2	5	11	926.69	1
46	1998/7/11	7:04 PM	5	6	15.5	17.03	0
47	1998/7/20	9:05 AM	6.1	7	14	3.59	
48	1998/7/28	12:51 PM	5.5	6	11	25.28	0
49	1998/8/27	5:03 PM	6.6	8	11	42.09	3
50	1998/9/18	11:53 AM	4.8	6	15	361.84	0

Table A1. *Cont.*

	Date	Time	Magnitude (Ms)	Epicentral Intensity	Focal Depth (km)	Population Density (People Per Square)	Deaths
51	1998/10/2	8:49 PM	5.3	7	13	83.46	0
52	1998/11/19	7:38 PM	6.2	8	10	40.22	6
53	1998/12/1	3:37 PM	5.1	7	10	322.07	0
54	1999/3/11	9:18 PM	5.6	7	11	116.08	0
55	1999/3/15	6:42 PM	5.6	6	11	16.93	0
56	1999/4/15	2:29 PM	4.7	6	8	88.60	1
57	1999/5/17	11:29 AM	5.2	6	20	471.60	0
58	1999/6/17	11:02 PM	5.3	5	11	25.28	0
59	1999/8/17	6:47 PM	5	7	12	5413.56	0
60	1999/9/14	8:54 PM	5	6	9	91.67	0
61	1999/9/27	7:49 PM	5.1	6	20	14.14	0
62	1999/11/1	9:24 PM	5.6	7	9	271.87	0
63	1999/11/25	12:40 AM	5.2	6	10	314.93	1
64	1999/11/26	4:51 AM	5	6	13	15.07	0
65	1999/11/29	12:10 PM	5.6	7	15	27.03	
66	1999/11/30	4:24 PM	5	6	26	651.86	1
67	2000/1/15	7:37 AM	6.5	6	30	123.04	7
68	2000/1/27	4:55 AM	5.5	7	10	115.12	0
69	2000/4/15	5:32 PM	5.3	7	13	1.19	0
70	2000/4/29	11:54 AM	4.7	6	16	10.33	1
71	2000/6/6	6:59 PM	5.9	8	15	247.06	0
72	2000/8/21	9:25 PM	5.1	6	8	87.61	2
73	2000/9/12	8:27 AM	6.6	8	13	4.55	0
74	2000/10/6	8:05 PM	5.8	6	18	147.33	0
75	2001/2/23	8:09 AM	6			4.70	3
76	2001/3/12	4:57 PM	5	6	10	103.06	0
77	2001/3/24	7:23 AM	5	5		3.63	0
78	2001/4/12	11:13 AM	5.9	8	6	196.15	2
79	2001/5/24	5:10 AM	5.8	7	5	39.07	1
80	2001/5/24	5:10 AM	5.8	7	5	99.64	0
81	2001/6/8	2:03 AM	5.3	6	5	208.09	1
82	2001/6/23	10:48 AM	4.9	6	15.4	4736.80	0
83	2001/7/10	7:51 AM	5.3	6	13	125.50	0
84	2001/7/11	5:41 AM	5.3	6	10	8.32	0
85	2001/7/15	2:36 AM	5.1	6	8	337.43	0
86	2001/9/4	12:05 PM	5	6	8	102.91	0
87	2001/10/27	1:35 PM	6	7	15	138.41	1
88	2001/11/14	5:26 PM	8.1	10			0
89	2002/8/8	7:42 PM	5.3	7		7.63	0
90	2002/9/5	12:18 PM	4	6		185.70	0
91	2002/10/20	11:46 PM	5	6	15	19.15	0
92	2002/12/14	9:27 PM	5.9	7	15	7.35	2
93	2002/12/25	8:57 PM	5.7	6	20	5.28	0
94	2003/1/4	7:07 PM	5.4	6	17	124.75	0
95	2003/2/14	1:34 AM	5.4	6	25		0
96	2003/2/24	10:03 AM	6.8	9	25.2	30.67	268
97	2003/4/17	8:48 AM	6.6	8	14	3.84	0
98	2003/4/24	6:37 AM	4.5	6	12	56.53	0
99	2003/5/4	11:44 PM	5.8	7	26.7	49.29	1
100	2003/6/17	10:46 PM	4.8	5	5	172.20	0
101	2003/7/10	9:54 AM	4.8	6	10	241.69	0
102	2003/7/21	11:16 PM	6.2	8	6	101.91	16
103	2003/8/16	6:58 PM	5.9	8	15	36.46	4
104	2003/8/18	5:03 PM	5.7	7	9	3.08	2
105	2003/8/21	10:16 AM	5	6	10	2287.92	0
106	2003/9/2	7:16 AM	5.9	6	10	1.77	0
107	2003/9/27	7:33 PM	7.9	11	15	5.48	0
108	2003/10/16	6:28 PM	6.1	8	5	84.30	3
109	2003/10/25	8:41 PM	6.1	8	18	58.87	10
110	2003/11/13	10:35 AM	5.2	8	12	165.95	1

Table A1. *Cont.*

	Date	Time	Magnitude (Ms)	Epicentral Intensity	Focal Depth (km)	Population Density (People Per Square)	Deaths
111	2003/11/15	2:49 AM	5.1	7	10	415.91	4
112	2003/11/25	1:40 PM	4.9	6	20	27.87	0
113	2003/11/26	9:38 PM	5	7	8	411.22	0
114	2003/12/1	9:38 AM	6.1	8	18	20.68	10
115	2004/3/7	9:29 PM	5.6	6	15	2.43	0
116	2004/3/24	9:53 AM	5.9	7	30	2.91	1
117	2004/5/11	7:27 AM	5.9		20	2.34	0
118	2004/6/17	5:25 AM	4.7	6	5	322.13	1
119	2004/7/12	7:08 AM	6.7		33	0.45	0
120	2004/8/10	6:26 PM	5.6	8	10	353.50	4
121	2004/8/24	6:05 PM	5.8		25	1.11	0
122	2004/9/7	8:15 PM	5	7	33	105.80	1
123	2004/9/17	2:31 AM	4.9	6	12	295.74	0
124	2004/10/19	6:11 AM	5	6	6	899.16	0
125	2004/12/26	3:30 PM	5	6	7	72.44	1
126	2005/1/5	6:05 AM	4.7	6	5	6.32	0
127	2005/1/26	12:30 AM	5	6	6	69.30	0
128	2005/2/15	7:38 AM	6.2	7	32	38.81	0
129	2005/4/8	4:04 AM	6.5	6	10	21.14	0
130	2005/6/2	4:06 AM	5.9	6		6.37	0
131	2005/7/25	11:43 PM	5.1	6	15	88.85	1
132	2005/8/5	10:14 PM	5.3	6	21	178.52	0
133	2005/8/13	12:58 PM	5.3	6	15	143.40	0
134	2005/8/26	5:08 AM	5.2	6		2.27	0
135	2005/10/27	7:18 PM	4.4	6	16	152.50	1
136	2005/11/26	8:49 AM	5.7	7	10	638.10	13
137	2006/1/12	9:05 AM	5	6	16	83.14	0
138	2006/3/27	3:20 AM	4.3	6	14	0.00	0
139	2006/3/31	8:23 PM	5	6	15	54.31	0
140	2006/6/21	12:52 AM	5	6	15	93.70	1
141	2006/7/4	11:56 AM	5.1	5	20	437.74	0
142	2006/7/19	5:53 PM	5.6	7	15	4.90	0
143	2006/7/22	9:10 AM	5.1	6	9	169.85	22
144	2006/8/25	1:51 PM	5.1	7	7	194.47	2
145	2006/11/23	10:04 AM	5.1	5	17		0
146	2007/3/13	10:22 AM	4.9	6	6	2175.44	0
147	2007/6/3	5:34 AM	6.4	8	5	103.63	3
148	2007/6/23	4:17 PM	5.8	6	16	94.32	0
149	2007/7/20	6:06 PM	5.7	7	25	27.43	0
150	2008/2/1	5:06 AM	4.6	6	6	496.13	0
151	2008/3/21	6:33 AM	7.3	7	33	14.39	0
152	2008/3/21	8:36 PM	5	6	11	157.98	0
153	2008/3/24	11:24 PM	4.1	6	10	1000.00	0
154	2008/3/30	4:32 PM	5	6	33	26.15	0
155	2008/4/20	9:14 PM	5.1	6	22	62.46	0
156	2008/4/21	5:42 AM	4.2	6	33	26.15	0
157	2008/5/12	2:28 PM	8	9	14	238.34	69227
158	2008/6/10	2:05 PM	5.2	6	14	4.20	0
159	2008/8/21	8:24 PM	5.9	8	7	78.78	5
160	2008/8/30	4:30 PM	6.1	8	10	131.52	41
161	2008/8/30	8:46 PM	5.3	6	25	2.85	0
162	2008/10/5	11:52 PM	6.8	8	27	3.62	0
163	2008/10/6	4:30 PM	6.6	8	8	10.17	10
164	2008/11/10	9:22 AM	6.3	7	10	9.25	0
165	2008/11/22	4:01 PM	4.1	6	8	200.00	0
166	2008/12/26	4:20 AM	4.9	6	5	399.36	0
167	2009/1/25	9:47 AM	5	6	7	14.41	0
168	2009/2/20	6:02 PM	5.2	6	6	15.22	0
169	2009/4/19	12:08 PM	5.5	6	7	8.45	0
170	2009/4/22	5:26 PM	5	6	7	6.24	0

Table A1. *Cont.*

	Date	Time	Magnitude (Ms)	Epicentral Intensity	Focal Depth (km)	Population Density (People Per Square)	Deaths
171	2009/7/9	7:19 PM	6	8	10	115.44	1
172	2009/8/8	9:26 PM	4	6	11	1885.35	2
173	2009/8/28	9:52 AM	6.4	7	8	2.36	0
174	2009/11/2	5:07 AM	5	6	10	117.18	0
175	2010/1/17	5:37 PM	3.4		7		6
176	2010/1/31	5:36 AM	5	7	10	1661.60	1
177	2010/2/22	9:32 PM	4.2	6	10	51.52	0
178	2010/2/25	12:56 PM	5.1	6	16	105.26	0
179	2010/4/4	9:46 PM	4.5	5	8	38.20	0
180	2010/4/14	7:49 AM	7.1	9	14	8.69	2698
181	2010/6/5	8:58 PM	4.6	5	5	5.76	1
182	2010/6/10	2:38 PM	5.1	6	8	0.93	0
183	2010/8/29	8:53 AM	4.8				0
184	2010/10/24	4:58 PM	4.7	6	8	714.98	0
185	2011/1/1	9:56 AM	5.1		10		
186	2011/1/8	7:34 AM	5.6		560		
187	2011/1/12	9:19 AM	5		10		
188	2011/1/19	12:07 PM	4.8	6	9	1749.16	0
189	2011/2/1	4:16 PM	5.3		7		
190	2011/2/15	3:18 PM	5.1		10		
191	2011/3/10	12:58 PM	5.8	8	10	84.55	25
192	2011/3/20	4:00 PM	5.2		30		
193	2011/3/24	9:55 PM	7.2	6	20	58.67	0
194	2011/4/10	5:02 PM	5.3	7	7	10.96	0
195	2011/4/16	9:11 AM	6		130		
196	2011/4/30	4:35 PM	5		60		
197	2011/5/10	11:41 PM	6.1		560		
198	2011/5/22	9:34 AM	5.2		10		
199	2011/6/8	9:53 AM	5.3	6	5	0.02	0
200	2011/6/20	6:16 PM	5.2	6	10	139.08	0
201	2011/6/26	3:48 PM	5.2		10	235.26	0
202	2011/7/25	3:05 AM	5.2	6	10	6.02	0
203	2011/8/2	3:40 AM	5.1		10		
204	2011/8/9	7:50 PM	5.2		11	134.75	0
205	2011/8/11	6:06 PM	5.8	7	8	19.71	0
206	2011/9/15	11:27 PM	5.5	6	6	1.75	0
207	2011/9/18	8:40 PM	6.8	7	20	3.51	7
208	2011/10/16	9:44 PM	5	6	4	11.29	0
209	2011/10/30	11:23 AM	5.7		223		
210	2011/11/1	5:58 AM	5.4		20		
211	2011/11/1	8:21 AM	6	7	28	45.00	0
212	2011/12/1	8:48 PM	5.2	6	10	104.13	0
213	2012/1/8	2:20 PM	5	6	27	24.52	0
214	2012/3/9	6:50 AM	6		30	1.18	0
215	2012/5/3	6:19 PM	5.4	7	8	7.78	0
216	2012/6/15	5:51 AM	5.4	6	20	9.38	0
217	2012/6/24	3:59 PM	5.7	7	11	52.39	4
218	2012/6/30	5:07 AM	6.6	8	10	13.60	0
219	2012/7/20	8:11 PM	4.9	6	6	640.63	1
220	2012/8/12	6:47 PM	6.2	7	30	2.62	0
221	2012/9/7	11:19 AM	5.7	8	14	226.83	81
222	2012/11/26	1:33 PM	5.5	6	8	93.75	0
223	2012/12/7	10:08 PM	5.1	6	9	7.29	0
224	2013/1/18	8:42 PM	5.4	7	15	11.90	0
225	2013/1/23	12:18 PM	5.1	6	7	625.00	
226	2013/1/24	2:01 AM	4.2	5			
227	2013/1/29	12:38 AM	6.1	7	20	16.34	0
228	2013/3/3	1:41 PM	5.5	7	9	68.04	0
229	2013/3/11	11:01 AM	5.2	6	8	42.24	0
230	2013/3/29	1:01 PM	5.6	6	13	5.23	0

Table A1. *Cont.*

	Date	Time	Magnitude (Ms)	Epicentral Intensity	Focal Depth (km)	Population Density (People Per Square)	Deaths
231	2013/4/17	9:45 AM	5	7	9	68.87	0
232	2013/4/20	8:02 AM	7	9	13	116.88	196
233	2013/4/22	5:11 PM	5.3	7	6	106.99	2
234	2013/7/22	7:45 AM	6.6	8	20	139.97	95
235	2013/8/12	5:23 AM	6.1	8	10	13.04	
236	2013/8/28	4:44 AM	5.1	8	9	15.75	3
237	2013/8/31	8:04 AM	5.9	8	10	3.57	3
238	2013/11/23	6:04 AM	5.5	7	9		
239	2013/12/1	4:34 PM	5.3	6	9	29.01	0
240	2013/12/16	1:04 PM	5.1	7	5	147.91	0
241	2014/2/12	5:19 PM	7.3	9	12	3.39	0
242	2014/4/5	6:40 AM	5.3	6	13	274.05	0
243	2014/5/30	9:20 AM	6.1	8	12	87.12	
244	2014/8/3	4:30 PM	6.5	9	12	216.67	617
245	2014/8/17	6:07 AM	5	6	7	325.71	
246	2014/10/1	9:23 AM	5	6	15	30.23	0
247	2014/10/7	9:49 PM	6.6	8	5	48.25	1
248	2014/10/25	1:20 PM	4.2	6	5	1611.89	0
249	2014/11/22	4:55 PM	6.3	8	18	16.72	5
250	2014/12/6	4:20 PM	5.9	8	10	48.25	1
251	2015/1/10	2:50 PM	5	5	10	2.69	0
252	2015/1/14	1:21 PM	5	6	14	164.58	0
253	2015/2/22	2:42 PM	5	6	14	171.79	0
254	2015/3/1	6:24 PM	5.5	7	11	70.81	0
255	2015/3/14	6:14 AM	4.3	6	10	7625.50	2
256	2015/3/30	9:47 AM	5.5	7	7	119.03	0
257	2015/4/15	7:08 AM	4.5	6	9	144.31	1
258	2015/4/15	3:39 PM	5.8	7	10	4.56	0
259	2015/4/25	2:11 PM	8.1	9	20	4.33	27
260	2015/5/22	12:05 AM	4.6	5	7		0
261	2015/7/3	9:07 AM	6.5	8	10	44.83	3
262	2015/10/30	7:26 PM	5.1	6	10	62.39	0
263	2016/1/14	5:18 AM	5.3	6	5	8.22	0
264	2016/1/21	1:13 AM	6.4	8	10	0.64	0
265	2016/2/11	9:10 PM	5	6	8	3.81	0
266	2016/3/12	11:14 AM	4.4	6	5		0
267	2016/5/11	9:15 AM	5.5	7	6.9		0
268	2016/5/18	12:48 AM	5	6	15	7.49	0
269	2016/5/22	5:08 PM	4.6	6	6	162.50	0
270	2016/7/31	5:18 PM	5.4	7	10	21.98	0
271	2016/8/11	11:49 AM	4.4	5	10	13.15	0
272	2016/9/23	1:23 AM	5.1	6	16	35.77	0
273	2016/10/17	3:14 PM	6.2	7	9	21.83	1
274	2016/11/25	10:24 PM	6.7	8	10	0.65	1
275	2016/12/8	1:15 PM	6.2	8	6	3.10	0
276	2016/12/14	4:14 PM	5		5		0
277	2016/12/20	6:04 PM	5.8	7	9	0.73	0
278	2016/12/27	8:17 AM	4.8	6	10		0
279	2017/1/28	2:46 AM	4.9	6	11		0
280	2017/2/8	7:11 PM	4.9	7	10		0
281	2017/3/27	7:55 AM	5.1	6	12	58.29	0
282	2017/5/4	1:40 PM	4.9	6	10		0
283	2017/5/11	5:58 AM	5.5	7	8	16.64	8
284	2017/6/16	7:48 PM	4.3	7	5	20.77	0
285	2017/8/18	9:19 PM	7	9	20	28.79	30
286	2017/8/9	7:27 AM	6.6	8	11	169.66	0
287	2017/9/30	2:14 PM	5.4	6	13	3.62	0
288	2017/11/18	6:34 AM	6.9	8	10		0
289	2017/11/23	5:43 PM	5	6	10		0

References

1. Erdik, M.; Şeşetyan, K.; Demircioğlu, M.B.; Zülfikar, C.; Hancilar, U.; Tüzün, C.; Harmandar, E. Rapid Earthquake Loss Assessment After Damaging Earthquakes. *Soil Dyn. Earthq. Eng.* **2011**, *31*, 247–266. [CrossRef]
2. Samardjieva, E.; Badal, J. Estimation of the Expected Number of Casualties Caused by Strong Earthquakes. *Bull. Seismol. Soc. Am.* **2002**, *92*, 2310–2322.
3. Chen, W.; Sun, Z.; Han, J. Landslide Susceptibility Modeling Using Integrated Ensemble Weights of Evidence with Logistic Regression and Random Forest Models. *Appl. Sci.* **2019**, *9*, 171. [CrossRef]
4. Chen, Q.F.; Mi, H.; Huang, J. A simplified approach to earthquake risk in mainland China. *Pure Appl. Geophys.* **2005**, *162*, 1255–1269. [CrossRef]
5. Chen, W.; Shirzadi, A.; Shahabi, H.; Ahmad, B.B.; Zhang, S.; Hong, H.; Zhang, N. A novel hybrid artificial intelligence approach based on the rotation forest ensemble and naïve Bayes tree classifiers for a landslide susceptibility assessment in Langao County, China. *Geomat. Nat. Hazards Risk* **2017**, *8*, 1955–1977. [CrossRef]
6. Hu, H.Y.; Lee, Y.C.; Yen, T.M.; Tsai, C.H. Using BPNN and DEMATEL to modify importance-performance analysis model—A study of the computer industry. *Expert Syst. Appl.* **2009**, *36*, 9969–9979. [CrossRef]
7. Park, S.; Kim, J. Landslide Susceptibility Mapping Based on Random Forest and Boosted Regression Tree Models, and a Comparison of Their Performance. *Appl. Sci.* **2019**, *9*, 942. [CrossRef]
8. Provost, F.; Hibert, C.; Malet, J.P. Automatic classification of endogenous landslide seismicity using the Random Forest supervised classifier. *Geophys. Res. Lett.* **2017**, *44*, 113–120. [CrossRef]
9. Sung, A.H.; Mukkamala, S. Identifying Important Features for Intrusion Detection Using Support Vector Machines and Neural Networks. In Proceedings of the 2003 Symposium on Applications and the Internet, Orlando, FL, USA, 28 February 2003; pp. 3–10.
10. Altmann, A.; Toloşi, L.; Sander, O.; Lengauer, T. Permutation importance: A corrected feature importance measure. *Bioinformatics* **2010**, *26*, 1340–1347. [CrossRef]
11. Karimzadeh, S.; Miyajima, M.; Hassanzadeh, R.; Amiraslanzadeh, R.; Kamel, B. A GIS-based seismic hazard, building vulnerability and human loss assessment for the earthquake scenario in Tabriz. *Soil Dyn. Earthq. Eng.* **2014**, *66*, 263–280. [CrossRef]
12. Wilson, B.; Paradise, T. Assessing the impact of Syrian refugees on earthquake fatality estimations in southeast Turkey. *Nat. Hazards Earth Syst. Sci.* **2018**, *18*, 257–269. [CrossRef]
13. Maqsood, S.T.; Schwarz, J. Estimation of Human Casualties from Earthquakes in Pakistan—An Engineering Approach. *Seismol. Res. Lett.* **2011**, *82*, 32–41. [CrossRef]
14. Jaiswal, K.; EERI, M.; Wald, D. An empirical model for Global Earthquake fatality estimation. *Earthq. Spectra* **2010**, *26*, 1017–1037. [CrossRef]
15. So, E.; Spence, R. Estimating shaking-induced casualties and building damage for global earthquake events: A proposed modelling approach. *Bull. Earthq. Eng.* **2013**, *11*, 347–363. [CrossRef]
16. Hashemi, M.; Alesheikh, A.A. A GIS-based earthquake damage assessment and settlement methodology. *Soil Dyn. Earthq. Eng.* **2011**, *31*, 1607–1617. [CrossRef]
17. Earle, P.S.; Wald, D.J.; Allen, T.I.; Jaiswal, K.S.; Porter, K.A.; Hearne, M.G. Rapid Exposure and Loss Estimates for The May 12, 2008 Mw 7.9 Wenchuan Earthquake Provided by The U.S. Geological Survey's Pager System. In Proceedings of the 14th World Conference on Earthquake Engineering, Beijing, China, 12–17 October 2008.
18. Schweier, C. Geometry Based Estimation of Trapped Victims After Earthquakes. *Int. Symp. Strong Vranc. Earthq. Risk Mitig.* **2007**, *9*, 4–6.
19. Goretti, A.; Bramerini, F.; Di Pasquale, G.; Dolce, M.; Lagomarsino, S.; Parodi, S.; Iervolino, I.; Verderame, G.M.; Bernardini, A.; Penna, A.; et al. The Italian Contribution to the USGS PAGER Project. In Proceedings of the 14th World Conference on Earthquake Engineering, Beijing, China, 12–17 October 2008.
20. Jaiswal, K.; Wald, D.J.; Hearne, M. *Estimating Casualties for Large Earthquakes Worldwide Using an Emperical Approach*; Open-File Rep. 2009-1136; U.S. Geological Survey: Reston, VA, USA, 2009; p. 78.
21. Aghamohammadi, H.; Mesgari, M.S.; Mansourian, A.; Molaei, D. Seismic human loss estimation for an earthquake disaster using neural network. *Int. J. Environ. Sci. Technol.* **2013**, *10*, 931–939. [CrossRef]
22. Wang, F.; Niu, L. An Improved BP Neural Network in Internet of Things Data Classification Application Research. In Proceedings of the 2016 IEEE Information Technology Networking Electronic and Automation Control. Conference. (ITNEC 2016), Chongqing, China, 20–22 May 2016; pp. 805–808.

23. Lecun, Y.; Bengio, Y.; Hinton, G. Deep learning. *Nature* **2015**, *521*, 436–444. [CrossRef]
24. Xu, L.; Cui, Y.; Song, Y.; Ma, X. The Application of an Improved BPNN Model to Coal Pyrolysis. *Energy Sources Part. A Recovery Util. Environ. Eff.* **2015**, *37*, 1805–1812. [CrossRef]
25. Chand, J.; Singh Chauhan, A.; Kumar Shrivastava, A. Review on Classification of Web Log Data using CART Algorithm. *Int. J. Comput. Appl.* **2013**, *80*, 41–43. [CrossRef]
26. Tian, H.X.; Mao, Z.Z. An Ensemble ELM Based on Modified AdaBoost.RT Algorithm for Predicting the Temperature of Molten Steel in Ladle Furnace. *IEEE Trans. Autom. Sci. Eng.* **2009**, *7*, 73–80.
27. Nadim, F.; Andresen, A.; Bolourchi, M.J.; Mokhtari, M.; Tvedt, E.; Moghtaderi-Zadeh, M.; Lindholm, C.; Remseth, S. The Bam Earthquake of 26 December 2003. *Bull. Earthq. Eng.* **2005**, *2*, 119–153. [CrossRef]
28. Kingma, D.P.; Ba, J. A Method for Stochastic Optimization. In Proceedings of the 3rd International Conference for Learning Representations, San Diego, CA, USA, 7–9 May 2015; pp. 1–15.
29. Gimpel, K.; Das, D.; Smith, N.A. Distributed Asynchronous Online Learning for Natural Language. In Proceedings of the 14th Conference on Computational Natural Language learning (CoNLL 2010), Uppsala, Sweden, 15–16 July 2010; pp. 213–222.
30. Abadi, M.; Barham, P.; Chen, J.; Chen, Z.; Davis, A.; Dean, J.; Devin, M.; Ghemawat, S.; Irving, G.; Isard, M.; et al. TensorFlow: A system for large-scale machine learning. In Proceedings of the 12th USENIX Symposium on Operating Systems Design and Implementation (OSDI '16), Savannah, GA, USA, 2–4 November 2016.
31. Danner, G.; Jelasity, M. Fully Distributed Privacy Preserving Mini-Batch Gradient Descent Learning. In Proceedings of the 15th IFIP International Conference on Distributed Applications and Interoperable Systems, Grenoble, France, 2–4 June 2015; pp. 30–44.
32. Said, E.S.; Dickey, D.A. Testing for unit roots in autoregressive-moving average models of unknown order. *Biometrika* **1984**, *71*, 599–607. [CrossRef]
33. China Earthquake Administration. *Assessment of Earthquake Disaster Situation in Emergency Period*; GB/T 30352-2013; China Standard Press: Beijing, China, 2014.
34. Wyss, M. Human losses expected in Himalayan earthquakes. *Nat. Hazards* **2005**, *34*, 305–314. [CrossRef]
35. Nichols, J.M.; Beavers, J.E. Development and Calibration of an Earthquake Fatality Function. *Earthq. Spectra* **2003**, *19*, 605–633. [CrossRef]
36. Wyss, M.; Zuñiga, F.R. Estimated casualties in a possible great earthquake along the Pacific Coast of Mexico. *Bull. Seismol. Soc. Am.* **2016**, *106*, 1867–1874. [CrossRef]

 *sustainability*

Article

A Proposed Theoretical Approach for the Estimation of Seismic Structural Vulnerability of Wastewater Treatment Plants

Ploutarchos N. Kerpelis [1,2,*], Spyridon K. Golfinopoulos [2] and Dimitrios E. Alexakis [1]

[1] Department of Civil Engineering, School of Engineering, Ancient Olive Grove Campus, University of West Attica, 250 Thivon Str., GR-12244 Egaleo, Greece; d.alexakis@uniwa.gr

[2] Department of Financial and Management Engineering, School of Engineering, University of Aegean, 41 Kountourioti Str., GR-82132 Chios, Greece; s.golfinopoulos@fme.aegean.gr

* Correspondence: kerpelis@uniwa.gr; Tel.: +30-2105381364

Abstract: The assessment of seismic vulnerability is critical for lifelines such as wastewater treatment plants (WTPs) because failures may result in environmental degradation, deterioration of water quality and human diseases development. The main scope of this research is the testing and application of a rapid, simple methodology for assessing the seismic structural vulnerability (SSV) of WTPs (according to the qualitative method Rapid Visual Screening), using structural variables as indices of these infrastructures. An original new method involving the assessment of the SSV of thirteen steps (four for a sample set of WTPs and nine for an individual one) is introduced following systematic literature retrieval. The analysis highlights twenty one factors that may determine the SSV of WTPs: three factors involving general characteristics, five factors involving seismicity and geotechnical data, six factors involving technical data (including structural data) and seven additional factors about WTPs' materials (concrete and the steel reinforcement of concrete frames). The structural data is analyzed to six additional factors. The implementation of the proposed methodology constitutes a simple, rapid methodological approach for assessing the SSV of WTPs using unique factors that were pinpointed and identified for the first time in this study.

Keywords: structural vulnerability; seismicity; Wastewater Treatment Plants; sustainability

Citation: Kerpelis, P.N.; Golfinopoulos, S.K.; Alexakis, D.E. A Proposed Theoretical Approach for the Estimation of Seismic Structural Vulnerability of Wastewater Treatment Plants. *Sustainability* **2021**, *13*, 4835. https://doi.org/10.3390/su13094835

Academic Editors: George D. Bathrellos, Hariklia D. Skilodimou, Konstantinos G. Chousianitis, Charalampos Vasilatos and Tommaso Caloiero

Received: 13 March 2021
Accepted: 20 April 2021
Published: 25 April 2021

1. Introduction

The increased incidence of natural and technological disasters in recent years poses threats to infrastructures. Climate change, construction complexity, and urbanization can have more extended impacts on WTPs than earthquakes [1], especially if a multicriteria analysis related to urban geology and urban geomorphology is not included in proactive planning against natural hazards [2]. The vulnerability of sewage systems is increased if other factors (such as contamination of surface waters by organochlorine pesticides) exist in addition to potential fluid leakages [3]. Earthquakes affecting wastewater treatment plants (WTPs) may trigger other hazards such as liquefaction, landslides, tsunamis, fires, and odours, as well as karst collapse [4]. The effects of these hazards remain in the environment for many years, posing a risk to health and public safety and negatively influencing sustainability development. Wildfires caused by earthquakes or urban fires can release toxic elements into the soil and water resources [5].

The "Sendai Framework for Disaster Risk Reduction 2015–2030" established the principles of the prevention and mitigation of disasters through the Sendai World Conference [6]. Researchers in earthquake-prone countries such as Greece [7] have recorded earthquake events and construction assets during vulnerability assessments after significant [8,9] or catastrophic earthquakes [10]. HAZUS methods have been implemented to assess infrastructure in EU projects [11]. Methods of estimating the impacts of earthquakes have been raised [12]. Practically, the most important buildings and infrastructures must be checked as a priority due to the created costs and huge effects of seismic disasters. Another priority is the health and safety of people and the protection of the environment.

According to the United Nations, Sustainability Development (SD) is one of society's main priorities. The goals of SD include "Good Health and Well-being", "Clean Water and Sanitation", "Sustainable Cities and Communities", and "Industry, Innovation, and Infrastructure" [13]. Earthquake damage to WTPs may influence infected individuals' health, because it may push infected people to maintain unsanitary practices, such as open defecation. Resilient infrastructure is an urgent need for sustainable cities to ensure human and environmental safety. Previous research highlights that "Novel sustainability concepts, approaches, methods and tools need to be developed" [14]. The European Union is committed to the implementation of SD, as documented in Agenda 2030 of the United Nations [15]. Previously conducted surveys have focused on environmental sustainability, aiming to minimize the economic cost of treatment. Tools for supporting management have been developed [16].

WTPs are a critical type of infrastructure that operates as a lifeline. The increase in the total development of the conduit length of pipelines in Japan has occurred due to an increase in the population and has resulted in considerable financial costs and extended periods of restoration (after seismic impacts) [17]. Sewage facilities such as sewers, pumping stations, and WTPs may be damaged by an earthquake [18]. The more vulnerable facilities are the worst designed [19]. The assessment of their vulnerability requires a disciplinary scientific approach [20]. The evaluation of the seismic structural vulnerability of WTPs in earthquake-prone countries or regions using concrete is fascinating. The impacts of an earthquake on WTPs may cause malfunctioning or non-operation, resulting in increased risks for both human health [21] and the environment [22]. In case of a disaster, the psychology of the citizens, families and children may also be affected at anywhere, e.g., at home and at work. Additional effects will be created as many other lifelines would be out of control.

The effects of seismic impacts have a direct relationship with sustainability principles. Following the Kobe earthquake in Japan (1995), many WTPs were quake-stricken. The restoration of these WTPs was part of the reconstruction actions [23]. Twenty out of the twenty-three pumping stations in the area were damaged and rendered useless, primarily due to power outages. The impacts of 15 great earthquakes from 1989 to 2011 have been recorded [24,25]. It is mentioned that WTPs were part of the affected areas. Many facilities, such as pumping stations, were destroyed during the New Zealand earthquake. Following the 1999 Taiwan and the 1995 Northridge earthquakes, a power outage involving pumping stations influenced the nonstructural vulnerability of WTPs. During the 1999 Turkey earthquake, heavy damage to mechanical equipment occurred. Similar damage to large-diameter interceptor pipes and small-diameter connection pipes was also observed during the Chile earthquake in 2010. The 2007 Gisborne earthquake caused minor damage to mechanical parts, and all the sewage systems collapsed due to the tsunami in Thailand in 2004. Changes in oceanographic characteristics near WTPs' outlets (such as ocean pollution in Izmit bay after the 1999 Turkey earthquake and the accumulation of sewage solids in the harbor during the 1931 New Zealand earthquake) can cause serious problems [26]. During the 1989 earthquake in LomePrieta, California, liquefaction and damage to older buildings (in areas without antiseismic design) influenced steel tanks, a post-tensioned and pre-stressed concrete tank, and mechanical equipment. These seismic impacts led to the loss of commercial power [27]. According to the Oregon Resilience Plan for Water and Wastewater Systems, essential facilities (such as intakes, treatment plants, pump stations, and outfalls) are vulnerable to damage from liquefaction [28]. The sewage pipeline network is susceptible to failure from permanent ground deformations, causing an increasing degree of damage over time that also affects society [17,29]. Pipelines are prone to failure at connections with these essential facilities. In Des Moines, Iowa, due to the 1993 earthquake, floods led to a loss of water [30], which is useful for WTP facilities.

Europe adopted rules (called Eurocodes) to construct structures, aiming to harmonize the Member States' other design traditions [31]. Nationally Determined Parameters (NDPs) are included in them. The Eurocodes consist of 10 European Standards (EN1990–1999) covering areas such as the basis of structural design (EN1990), the actions of structures (EN1991), the design of concrete structures (EN1992), the design of steel structures (EN1993), the design of composite steel and concrete frames (EN1994), geotechnical design (EN1997), and the creation of structures for earthquake resistance (EN1998). Each Code (except EN 1990) is divided into 58 Parts covering specific aspects in detail. In Greece, the Eurocodes were completed in 2007 and legislated in 2014, with implementation occurring in parallel with the implementation of the National building regulations (GGG 1457/B/2014). In the United States, the implementation of building codes such as the International Building Code (IBC) and the International Existing Building Code (IEBC) have helped to strengthen WTPs.

The United States has not adopted rules on the structural vulnerability of WTPs such as the "Water and Wastewater Systems Sector-Specific Plan" [32] or the "Roadmap to a Secure and Resilient Water and Wastewater Sector". Assessments of the seismic risk of WTPs are not required by the Environment Protection Agency, but some WTPs prepare these without structural or nonstructural vulnerabilities [33]. Instead, the Federal Emergency Management Administration (FEMA) has tried to enhance resilience and SD by using fragility curves and new technologies (Geographical Information Systems, GIS) for seismic structural assessment, although this may result in many failures [34]. The project HAZUS, which operates under the above mentioned holistic view of structures, evaluates vulnerability in terms of losses and financial terms [35]. The European Union produced an overview of the existing methodologies used worldwide for practical risk assessment of critical infrastructures [36].

The European Programme for Critical Infrastructure Protection (EPCIP) and the US Critical Infrastructure Protection (CIP) operate with different views. These programs are based on SD in order to address the key factors of financial and human losses. The American Society of Civil Engineers (ASCE) specifies Guidelines for the seismic evaluation of water transmission facilities such as WTPs [37]. Structural failures are investigated according to the functions of WTP structures, the materials used (e.g., reinforced concrete, steel, PVC, HDPE, and GRP), and the pipeline network [28].

In Greece, the Syner-G Program named "Systemic Seismic Vulnerability and Risk Analysis for Buildings, Lifeline Networks, and Infrastructures' Safety Gain" was developed to estimate the seismic vulnerability of these structures [11,38]. Structural parameters were investigated by another Pan-Hellenic Project by the Earthquake Planning and Protection Organization (EPPO), which is currently active and is related to the seismic structural vulnerability of public buildings and public welfare institutions. It is a primary-stage pre-earthquake assessment and is based on rapid visual inspection using structural variables [39,40]. Table S1 in the Supplementary Material presents the abbreviations and definitions in this study.

2. Potential Threats to Water and Soil Quality

The importance of sustainability has been increasing [30]. Countries must implement governance for disaster risk management in parallel with environmental control [29]. The response and motivation of the authorities are essential, according to the rules of civil protection. Delays in these mechanisms may lead to the direct or indirect degradation of the environment. The planning of structures must follow the main principle "growth must meet today's needs without jeopardizing the well-being of future generations".

Structures must reduce their energy consumption (or produce energy from wastewater), and their materials must acquire increased strength to withstand increased seismic forces. Best practices advise stakeholders to mitigate against the impacts of disasters and preserve earth sources [30,41]. WTP planning must involve an increase in resilience, the reuse of effluent water, and the management of seismic impacts [42].

A high-intensity earthquake that occurs in an area containing WTPs may lead to disaster. Consequences may be immediate or long-term in terms of time and being local, national, or global. Damage to or malfunctions of infrastructure may trigger the contamination of soil, water, and air [43]. Fluid or air leakage may cause the ignition and explosion of flammable materials. Accidental discharge of wastewater into water resources may increase the contamination of a stressed area of society. Damaged sewer pipes can release sewage or waste, including toxic components.

Furthermore, a fire in petroleum refinery infrastructure that had broken out during the Turkey earthquake in 2000 caused the death of three inhabitants and the release of exhaust gases and particulate matter to the environment. During the 1995 Kobe (Japan) earthquake, asbestos fibers were released. Climate change may aggravate the situation, leading to more frequent fires and the release of dangerous materials [5,44]. Other disasters such as inundations or heavy rainfall may contribute to the dispersal of hazardous materials.

In case that the nutrients and elements concentration in untreated sewage of WTPs exceed the corresponding parametric values proposed by the Directive 91/271/EEC may have a serious impact on human health and ecosystems [45]. Chloride leakages from a WTP deposit occurred after the 1987 California earthquake [1]. The priority of countries regarding SD is to mitigate the impacts of disasters on the environment and human beings [46].

Disinfection byproducts, volatile organic compounds and other materials released from WTPs which contain heavy metals (e.g., Co, Cr, As, Ni, and Pb) may affect seriously ecology, ecosystems, and inhabitants' health [47,48]. Civil protection authorities should investigate areas immediately with increased concentrations of trace elements.

At present, the amount of pollution threatening soil and/or water is much more significant due to climate change, urbanization, and ageing structures [49]. The impacts of these events co-occurring are much more critical than the pollution of each of them independently.

3. Estimation of WTPs' SSV—Future Needs

The assessment of the seismic vulnerability of structures has been researched using categorization methodologies to quantify the level of damage to structural elements [50]. Few surveys about the resilience of WTPs have been conducted due to the great effort and costs involved [42]. Many parameters, such as the vulnerability of materials used in these structures, their contribution to structural vulnerability, and the relationships among different elements of vulnerability, have not been assessed. New studies must identify the seismic vulnerability and resilience of WTPs (using a specific methodology), considering structural factors.

The planning and management of WTPs must meet two of the four fields that SD is focused on, i.e., water management—wastewater treatment and environmental engineering and management [51]. The creation of action plans on a large scale using innovative, practical solutions is required. These infrastructures focus on operational environmental performance but must increase the level of ecological thinking across the whole sector (including vulnerability factors coming from other scientific fields) [46].

Researchers have approached sustainability through the use of conceptual and mathematical assets [52] using indicators of the SD of WTPs [53] and investigating technical assets (such as the monitoring of pollutant removal) [16]. It has been observed that there is a lack of structural asset approaches for these infrastructures. Materials such as concrete are subject to building rules regarding their durability and corrosive exposure (such as carbonation or frost). The result of these is the changing of their stainability degree [54]. This study investigates the criteria and variables that must be identified for the structural vulnerability assessment of WTPs. The structural requirements of factors that affect the structural vulnerability assessment of WTPs are the same as those used for standard structural buildings. Still, additional parameters, such as acid resistance, must be considered [55,56]. An approach focusing on structural vulnerability parameters (in addition to existing methods) could aid in urban and rural planning using new technologies such as GIS [57].

In this study, a new methodology involving rapid visual inspection by technicians with the ultimate goal of sustainable development is proposed. Currently, there is no similar universal fast estimation method for these facilities. The methodology is categorized as an empirical assessment approach, and the next step must be to develop a questionnaire for Rapid Visual Screening [50]. Similar questionnaires have been used in Canada, Japan, Turkey, Greece, New Zealand, and India, as practical methods. An example of this approach used in Greece is a questionnaire about the variation in the individual seismic structural assessment standards of public and public-use buildings described by the EPPO [39]. A future questionnaire based on the proposed methodology would enrich previous research [57] that has attempted to adapt the EPPO's questionnaire and to make it appropriate for the needs of WTPs. The survey highlights the further enrichment and analysis of SSV factors in WTPs made from reinforced concrete, a material with many ambiguities compared with other standard materials used. Another attempt to focus on seismic vulnerability factors related to WTPs was then conducted for the whole infrastructure [58], especially for SSV [59]. The enriched categorization of these factors in a questionnaire that contributes to the investigation of the seismic vulnerability of WTPs is shown in this study.

Other researchers tried to introduce surveys using questionnaires about the variables related to the effects of WTPs as a method of rapid assessment, as time is a crucial factor in this type of assessment [60]. This method facilitated the formation of opinions by elected officials (technical specialists, decision-makers, stakeholders) through a multi-objective decision analysis of the questionnaires. The above approach involved six main questionnaires, including nine criteria for evaluating a WTP. According to local rules, the minimization of construction impacts included only the minimization of erosion and the insurance of operations.

4. Proposed Method

The proposed method for the rapid assessment of SSV includes the following phases [61] and is similar to the planning of the "All-hazard Consequence Management Tool" that was proposed by the Water Environment Federation [62]. The sample set of WTPs involves several representative WTPs. The proposed methodology will be applied initially. The findings of this step will be used for possible revisions of a questionnaire sheet and as an input for Phase B. Phase A investigates the weighting factors of a sample set of WTPs, while the Phase B estimate the SSV score and compare the sample set of Phase A. A qualitative scale is used to visualize the results to protect individual WTPs.

4.1. Investigating the SSV of a Sample Set of WTPs

Phase A uses an obligatory step for the collection of SSV data. This is useful for future assessments of the SSV of an individual WTP, done using Phase B. The data must come from a representative sample set of WTPs using statistical methods. The means for conducting this assessment must be based on reliable and valid investigations. Figure 1 illustrates the steps required for the collection of SSV data of a WTP sample set.

1. An inspectional visit to a sample set of WTPs to check the characteristics of their facilities: The impacts of flows on the structural components can only be investigated by contacting the facilities [63].
2. Collection of data: involves records, studies, interviews, and questionnaires. The assessment of SSV will be more accurate if many significant variables are collected.
    - Collection of the scientific information (seismological, geotechnical, and technical data about its parts): Any data associated with the processes of a WTP [64], the structure itself [9], and the seismic vulnerability, including geotechnical data [25], help assess the vulnerability. Structural variables can be used as indices, similar to empirical methods [50].

- Review of the variables used in the past. Any existing records about structural variables such as concrete tests conducted before or after the repair of seismic impacts and specific variables extracted from the Antiseismic Codes of construction times may be helpful. Geological or technical data (mentioned before) about the structure may be part of the set of data to be utilized.

3. Issue a questionnaire to the survey participants. The questionnaire must include all the SSV variables mentioned, and their validity and reliability must be investigated. The participants in the survey must have relevant education or training and great experience with WTPs.
4. Weighting of factors for each of the above variables: Simple empirical values are often used to estimate structures (even at WTPs), such as equations used related to SD and cost issues [64]. In this study, the nomination of parameters must be followed by the weighting of these factors, because some parameters have more significant impacts (e.g., the foundation of the WTP) than others (e.g., the environmental temperature when the concrete was poured). A similar approach was implemented in another survey conducted by Keeney et al. [60] in WTPs of Seattle (USA).

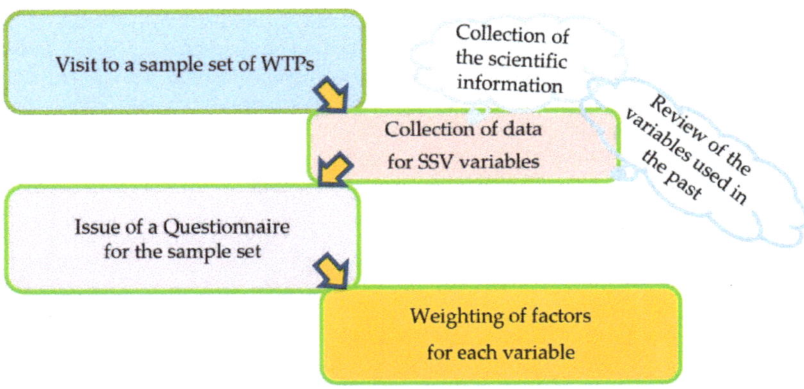

Figure 1. Investigating the SSV weighting factors from a sample set of WTPs (Phase A).

Weighting factors, such as the frequency of occurrence, related to the means of collecting data, e.g., the survey participants' answers to the SSV questionnaire, are calculated. The central tendency and variability of the SSV variables will assist in the definition of weighting factors. The determination of weighting factors completes Phase A.

4.2. Investigating the SSV of an Individual WTP

Phase B follows the following nine steps that are suggested for the estimators for the evaluation of WTPs (Figure 2). First, the estimator decides against conducting a rapid assessment of SSV. The potential risk related to a decision depends on people's perception of risk, the time and means available, and the management of collected data as financial issues as well as the percentage of SD achieved. Economic, social, and environmental considerations may encourage a researcher to investigate the final structural values in a certain way [65]. An absence of qualitative scales or weighting factors for SSV variables from the past will trigger research (as described before) to investigate these weighting factors from a sample set of WTPs.

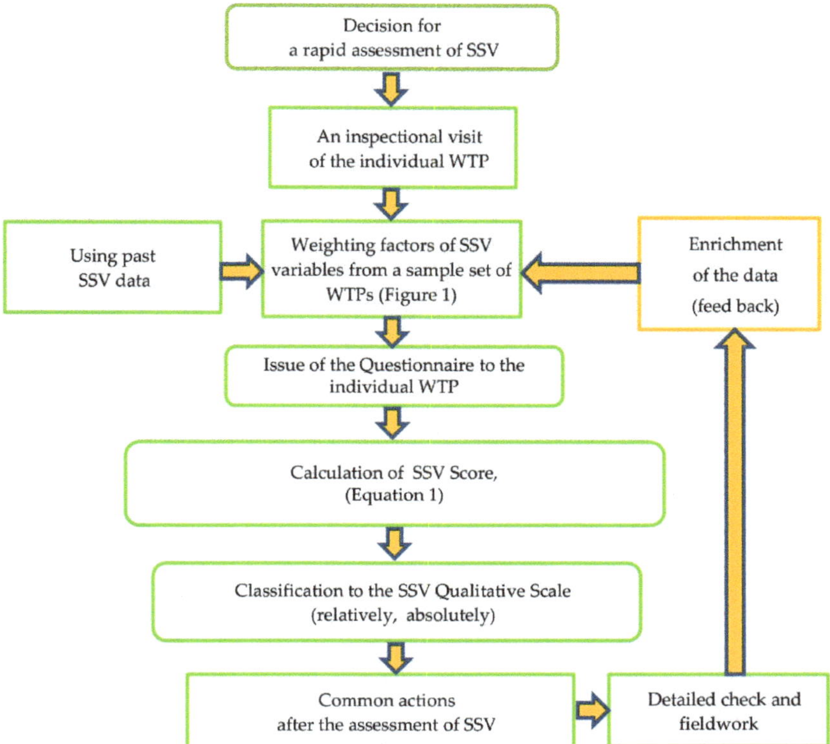

Figure 2. Investigating the SSV of an individual WTP (Phase B).

Additionally, an inspectional visit to the individual WTP is needed. Methods similar to questionnaires are used to collect scientific data about infrastructures. Each factor's weighted value is calculated as described in the previous stage (Phase A). Descriptive statistics assist in the determination of the weighting factors. It is considered that the weighting used for factors in the individual WTP questionnaire is the same as that used when assessing the sample set of WTPs. The score (total value) of a WTP's SSV is equal to the total value of each factor multiplied by its SSV weighting. This is done using Equation (1) as follows [66]:

$$\text{Score} = \Sigma(\text{Weight of factor} * \text{Factor}). \tag{1}$$

A qualitative scale for assessing SSV based on the above quantitative weighted variables of the WTP sample set is developed. The correlation of the scores of individual WTPs with the maximum and minimum values of similar products in the sample set of WTPs can assist in defining the relative qualitative scale for assessing SSV. The relative classification of the SSV of an individual WTP in question compares its score with similar scores in the sample set of WTPs. Scientists have stated that ordinal classification could benefit from the use of available relative information [67].

The final absolute classification is processed by constructing a whole qualitative scale, which includes the average individual WTP scores for the SSV factors compared with the ideal maximum and minimum scores expected for the sample set of WTPs.

This step in Phase B involves the classification of scores on the Qualitative Scale, which estimates SSV according to the SSV of the sample set of WTPs, in both relative (compared with each other) and absolute (corresponding to the desirability results according to the rules) terms. The assessment of the SSV of an individual WTP is performed. The absolute and relative classification of any WTP in question can be achieved using the abovementioned SSV qualitative scales. These results offer inestimable value to SD planning and implementation.

After assessing the SSV of any WTP, another joint action conducted may be detailed checking and fieldwork if the level of vulnerability is significant (or if no structural factors can be evaluated). Depending on the circumstances, everyday actions conducted after assessing a WTP's SSV are the implementation of urgent/emergency protective meters on the structure (to prevent malfunctions or non-function). These situations will cause heavy consequences to the inhabitants of the affected area and the environment. Any detailed in situ check will provide valuable data about the seismic vulnerability of the structures [68].

Everyday actions conducted after the assessment of the SSV of any WTP may involve the enrichment of data using raw data related to the investigation's questionnaire with the results of the above calculation. Valuable data will be used for the restoration of existing problems. Feedback is necessary, as future estimates of SSV will use more updated data. The update of SD planning and actions is imperative.

5. Description of the Variables

After analyzing the collected data, as described previously (Figure 2, Phase B), we concluded that the method's variables must be in accordance with the nominated ones and the limits of the building rules. In Greece, these rules are the Eurocodes and the Greek Code for Seismic Resistant Structures (GCSRS) [69], Greek Code for Reinforced Concrete (GCRC), which are used for usual construction projects (or even the Greek Code of Structural Interventions). Remarkably, most countries' structures were built prior to the SD principles becoming part of the building rules.

Experience with impacts on WTPs should be incorporated into the countries' building rules. Other empirical approaches use a Seismic Priority Index consisting of variables that can estimate the levels of risk and vulnerability [50]. Similarly, a questionnaire may include the following elements that are important for assessing SSV.

Figure 3A–D shows the four main parts that are important for evaluating of SSV, including the general characteristics of WTPs, seismicity and geotechnical settings, technical factors and the material used for the construction of WTPs.

5.1. The General Characteristics of WTPs

5.1.1. Year of Construction and Year of Last Intervention/Addition

This factor is directly related to the structural regulations at the time [28]. Researchers of the great earthquake that occurred in Bhuj in 2001 stated that reinforced concrete (RC) frames built in previous decades were more vulnerable than other structures built in recent years [70]. Sendai in Japan has sewage facilities that are more than 118 years old [18], and the process and auxiliary equipment in the WTP of the Sewer Authority Mid-Coastside have expired [33]. Earlier structure studies produce more safe conditions.

5.1.2. Entire Surface Area of the WTPs

Measurement of the entire surface area is helpful as an estimator to calculate the "serving population". The maximum population that the infrastructure will serve is known as the "serving population". A proportional relationship exists between served populations and the surface area of each WTP [71].

5.1.3. Capacity of the WTPs

The importance of WTPs is more significant than usual buildings and depends on the population served because of the potential seismic impacts on them. A large population needs more facilities for sewage treatment. Domestic sewage needs primary and/or secondary treatment, and wastewater from industries also requires tertiary treatment [72]. A low level of inclusion of SD principles may increase the vulnerability of WTPs.

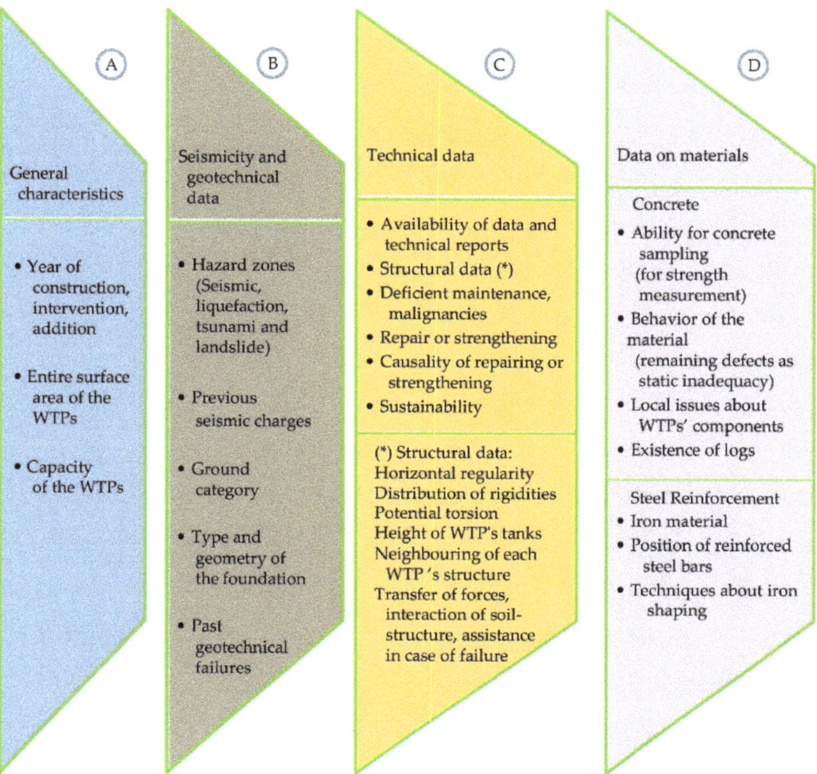

Figure 3. SSV variables of WTPs: (**A**) General characteristics, (**B**) Seismicity and geotechnical data, (**C**) Technical data, (**D**) Data on construction materials.

5.2. *Seismicity and Geotechnical Data*

Earthquakes are an acute stressor of WTP facilities [42]. Seismic and geological data related to WTPs should be classified as critical vulnerability variables that must be recorded.

5.2.1. Seismic Hazard Zone

The seismic data related to a country's areas determines which zones may be useful for vulnerability assessment [28]. Worldwide, in Europe, and in Greece, the formulation of these zones is based on peak ground acceleration (PGA) for a return period of 475 years [73–75]. In Greece, GCSRS categorizes the technical project areas in line with the expected acceleration of the soil [76].

Liquefaction, tsunamis, and landslide effects, as secondary phenomena, may influence structures. WTPs are usually established at lower heights (near rivers and/or near the sea) where alluvial deposits exist and liquefaction danger is serious. Using the wrong WTP settlement site may increase the seismic vulnerability, severely impacting the structure [77]. A possible tsunami may stress the structural framework [28]. Recorded data which include past events of liquefaction, tsunamis and landslides, is also valuable.

5.2.2. Previous Seismic Charges

The existence of non-repaired damage to WTPs may change the structure, increasing its vulnerability. Records or testimonies may provide required data. Meanwhile, the buildings' structural characteristics and ground-motion records following past earthquakes are used to export the fragility functions of typical buildings [78].

5.2.3. Ground Category

Additional information about the ground category may be collected from past data records and observations [79]. Data related to geological formations, water level, geomorphological relief and geological faults is crucial for estimating the related vulnerability factor [80]. Moreover, the ground categories used for common structures are categorized in Greece by GCSRS.

5.2.4. Type and Geometry of the Foundation

Data of seismic vulnerability factors is significant, because it is related to possible subsidence, seismic stress, fluctuation of water level and liquefaction [81].

5.2.5. Past Geotechnical Failures of the Broader Area

It is essential to develop a database record which include information related to past failures (e.g., landslides, liquefaction) in the region [82].

5.3. Technical Data

WTPs are critical infrastructures with special requirements regarding structural strength and sustainability, so special rules must be used to ensure maximum protection and safety for human and the environment. Remarkably, construction codes are used as the standard rules for usual buildings. The importance of introducing SSV parameters is necessary. Sustainability parameters are aggressively embedded in building rules.

In the United States, the IBC is one of the codes established by the International Code Council (ICC) based on its usage. It is the existing regulation for constructing buildings and structures. The IEBC sets out requirements for repairs, alterations, and additions to existing buildings and structures. WTPs are not subject to special building rules, as shown at the IBC, IEBC, and the Eurocodes of the European Union [31]. In Greece, structure laws like the Greek Codes for Reinforced Structures using Concrete [83] or the Code for Earthquake-Resistant Structures [69] (or even the Code of Structural Interventions 2012) do not refer to high-risk special constructions, e.g., WTPs, and their needs. As shown, no particular rules exist for building codes of structures (including WTPs) for the recording of structural values. Instead, these facilities are designed to have longer life cycles.

The technical variables for each part of a WTP are presented at the Section 5.3.1, Section 5.3.2, Section 5.3.3, Section 5.3.4, Section 5.3.5, Section 5.3.6 and should be examined, regardless of the type of treatment involved.

5.3.1. Availability of Data Records and Technical Reports

If the past technical reports are available, it is useful to compare them against the updated reports [84].

5.3.2. Structural Data

Structural data show similarities to the data required for standard buildings, as presented below:

- The horizontal regularity of each independent part of a WTP must be recorded. The shear walls of structures (as tanks) may experience failure after an earthquake [85];
- The distribution of its members' rigidity [86]: An earthquake may cause structural damage to sections with different levels of rigidity. After a strong earthquake, building damage mainly occurs because of the intense beams and weak columns of a building inside the WTP (such as the facilities' control building) [77].
- Potential torsion may affect the structure [87].

- The considerable height of a WTP's tanks may be a significant factor [88]. The large size of tanks leads to a great degree of shaking of the sewage, producing forces beyond its design capacity and increasing its seismic vulnerability [28].
- The close neighboring of each WTP part with another. Mutual collisions or the pounding between parts may be factors that increase seismic vulnerability [89].
- The transferability of forces (in general), soil–structure interactions, and assistance following construction failure [90,91] or associations with geotechnical problems [92]: A possible reason for changing the route of forces to the ground is an earthquake. Knowledge about the existence of other methods of force transfer is valuable for the estimation of this factor.

5.3.3. Deficient Maintenance/Malignancies

It is possible for these situations to decrease the infrastructure's resilience and increase its vulnerability [93], while estimation is mandatory. Due to deficient maintainance, steel oxidation and concrete carbonation might be presented. Subsidence might produce malignancies to the structure.

5.3.4. Repair or Strengthening of Infrastructure

In this case, resilience increases and vulnerability decreases, although there are barriers, including a lack of design, the installation guidelines, and long-term durability studies [94].

5.3.5. Causality of Repairing or Strengthening of WTPs

Another factor is the investigation of the reasons that led restoration scholars to undergo improvements or strengthening. Such reasons may include the presence of soil subsidence or even a fire near the WTPs [95]. The causes must be separated in static and shock-dynamic loads (such as earthquakes) and time damage.

5.3.6. Sustainability

Technical issues should be according the sustainability principles. Researchers examined the relationship among the disaster and the seismic hazard, exposure of population and fragile buildings [20]. Each of these determinants is directly related with WTPs, posed to strong earthquakes.

5.4. Data on Materials

A complete qualitative study must include additional factors such as indices [50] related to the construction materials (RC, Steel, PVC, HDPE, and GRP). Materials often have specific standards, but reinforced concrete has many uncertainties regarding its composition, its method of construction, and its maintenance. The role of materials is essential to provide the strength of the frame [28,96]. Complementary data are primarily technical issues related to reinforced concrete, the primary material used in WTPs, as shown below [97].

- Data related to the concrete frame.
    (a) Ability for concrete sampling. To measure the strength of the construction's structural frame, concrete sampling (where and when it is possible) and measurement of their values must be performed. An existing evaluation of damage caused by earthquakes may be useful [98]. Additional methods may be used to calibrate the concrete strength, including nondestructive methods such as ultrasound measurements, percussion methods, and a nail extractor [99].
    (b) The behavior of the material. Defects can be influenced by static inadequacies from the past. Previous exposure to seismic stress may have caused residual faults (including sloping and sedimentation of structural elements) [100].
    (c) Local issues about WTPs' components. Knowledge on local issues with WTP components may be valuable. Possibly, the presence of holes in piping con-

struction [101] or the permeability of components may affect the structural vulnerability.

(d) Existence of logs. Concrete samples' logs taken by the owner (during construction) are valuable for assessing a structure's vulnerability [102]. Information about the substantial quality can be obtained from logs, for example, the quality of sand used, the place of concrete production (in situ or industrial), the existence of a sieve analysis and proper gradation of aggregates, and the appropriate vibration and curing of concrete and use of low cement values in the creation of structural elements [77]. In recent times, it has been mandatory (e.g., in environmental legislation) for SD parameters to be introduced in studies about vulnerability evaluation.

- Data related to steel reinforcement (of the concrete frame).
    (a) Iron material. Any subject related to the material used in WTPs, such as iron-shaped memory alloys [103], must be checked. Characteristics of these steel bars (e.g., diameters, anchorage, and overlapping) may influence the resistance of structures.
    (b) The positions of reinforced steel bars. Technical studies predict the position of seel bars inside the concrete, according to building codes [83].
    (c) Techniques about iron shaping. Data on the "closure of fasteners" at the columns or walls are valuable, as these factors may decrease the strength of materials in WTPs [104,105]. Structural details, such as the absence of 135° seismic hooks and a lack of transverse ties, may be the causes of an increase in seismic vulnerability [77].

There is a contractual obligation regarding these critical infrastructures because this secures the WTP quality. The only way to investigate potential construction failures is to identify structures with high SSV. Other conditions such as the temperature of the environment, concrete construction, and the corrosion or carbonation of concrete [106] are essential because they may influence the strength of the WTP. Extreme temperature conditions caused by climate change over time may be a significant factor, too.

A sustainable WTP structure, characterized by its ability to act as a green building, mitigates against negative impacts and decreases the respective lifecycle of substances discharged to the natural environment. The wastewater treatment processes require significant energy resources, resulting in elevated emission levels [51,107]. The integration of all the parameters should provide planners for vulnerability with a holistic view, as shown by similar surveys [108,109].

6. Initial Evaluation of the Proposed Method

The origin methodology introduced herein is based on two main sets of actions conducted on totally 13 steps to achieve the objective—determining the SSV of WTPs. For the first set of four steps, previously collected SSV data from a sample set of WTPs is used (Figure 1) to perform the second steps (Figure 2). Proper implementation is based on onsite checking of twenty-seven critical structural variables in the WTP sample set (described by this study) and weighting to view their impacts on vulnerability (Figure 3). The parameters of the valuable structural data serving the SD principles are categorized into general characteristics, seismicity, geotechnical, and technical data (including related materials such as concrete and steel reinforcement).

The three variables that refer to the general characteristics are the following: (a) the year of construction and the year of last intervention/addition, (b) the surface area of the whole WTP, and (c) the WTP capacity. Similarly, there are five factors related to seismicity and geotechnical data: (a) the seismic hazard zone (including the hazard zones for liquefaction, tsunamis, and landslides), (b) damages related to past earthquakes, (c) the ground category, (d) information about the type and geometry of the foundation, and (e) past geotechnical failures.

The six factors related to technical data are the following: (a) the availability of data and technical reports, (b) structural data (such as the horizontal regularity, the distribution of rigidity to its members, potential torsion, the considerable height of the WTP's tanks, the too-close neighboring of WTP parts, and the transfer of forces), (c) deficiencies in maintenance, malignancies (due to the presence of subsidence), (d) repair or strengthening conducted, (e) the reasons provoked repairing or strengthening, and (f) the sustainability of a WTP's structure.

From the point of view of sustainability, an investigation about the materials of which a WTP is made is necessary. When reinforced concrete is used as a material for a WTP, four issues arise: the ability to undergo concrete sampling (proof of concrete samples, use of additional methods to calibrate the concrete strength), the behavior of the material including remaining defects as a past static inadequacy, local issues about WTP components, and the existence of logs of concrete tests conducted on the set of concrete samples during or after construction.

Three issues about the steel reinforcement of a concrete frame include the use of materials such as iron-shaped memory alloys, the positions of reinforced steel bars and techniques about iron shaping (diameters, anchorage, overlapping, and data about the "closure of fasteners" at columns or walls).

Phase A of the methodology suggests that the variables' weightings may be estimated via methods such as descriptive statistics (Figure 1). Historical data, interviews and questionnaires (extracting from experienced estimators), and bibliographies can provide data about SSV. These data are organized through a questionnaire given to participants of the survey. The calculation of the weighting factors of SSV for the sample set of WTPs consists of the collection of the data analyzed previously, e.g., from the answers given by the questionnaire's participants. The sorting of them and discovery of the proportions and frequency relations in the WTPs sample data is the next step in the methodology.

Phase B of the methodology suggests that the estimation of the SSV of an individual WTP requires a set of actions (nine steps; Figure 2). As there are many methodologies regarding the assessment of a structure's vulnerability (which are unreliable according to some researchers) [50], the decision to perform a rapid, qualitative, empirical analysis survey must be adopted (Figure 2). In the first case, an estimation is performed using briefing records, questionnaires, and interviews. A qualitative assessment is conducted by comparing the SSV of an individual WTP with the SSV of the previously calculated WTP sample set. The results are produced by constructing relative and absolute qualitative scales to categorize the SSV of the individual WTP at the survey time. Two qualitative scale categories exist: (a) a relative scale, which uses the values of the sample set, and (b) an absolute scale, which uses ideal maximum and minimum values.

Verification of the results and the relations among elements of vulnerability may be achieved through onsite checks, and protective measures may prevent post-seismic damage related to sustainability issues. Enrichment of the primary data in the sample set is necessary as a type of feedback. Correlations among specific variables and the classification of absolute and relative findings are based on imported data, the time at which the survey is completed, and the people involved, including their perceptions. Very distant future assessments of SSV require renewed study with new rules and restrictions set by the state.

7. Concluding Remarks

WTPs are usually old-structured establishments constructed applying old building codes. Damages in the structure of WTPs are very often in seismic-prone countries, resulting in serious problems. Nowadays, SSV is assessed by applications, such as GIS, and scenarios that produce virtual results about losses and financial assets. As resilience and sustainability are priorities for modern societies, the rapid implementation of a universal program requires the application of a method to examine the values of variables that affect the SSV of these constructions. There is a need for the rapid assessment of SSV following an earthquake, although the empirical approaches which use Rapid Visual

Screening seem to be ineffective tools. This approach needs qualitative analysis. Additionally, earthquakes themselves involve many uncertainties that necessitate the use of qualitative methodological approaches.

A new qualitative, rapid, and comprehensive methodology is available to assess the SSV of WTPs, as well as satisfying the principles of SD. The proposed method involves totally 13 steps to investigate the SSV of WTPs using twenty-one critical variables (where structural data is analyzed to six more variables) and examines the WTP characteristics. There is a need to implement two sets of actions (one for assessing the SSV of the WTP sample set and one for evaluating the SSV of an individual WTP). The investigation takes place through the assessment of general characteristics, seismicity, geotechnical, technical aspects, and factors related to the materials used at the WTPs.

The method used is an empirical methodology due to a lack of time in earthquake-prone areas. It involves using questionnaires containing SSV variables and compares the relative and absolute scores of an individual WTP with a sample set of WTPs. The proposed method by the present study is a qualitative approach with advantages and disadvantages, and it follows the SD principles. The most significant advantage of the method is the time and cost saved in assessing the SSV of an individual WTP; while the Phase A is time-consuming and should be improved. The weighting factors of the variables should be subjected to extensive and thorough research based on various research methods. Many factors may change the results, e.g., the technicians involved, the timing of the survey, and the society's rules and regulatory constraints at the time of the study. The vulnerability assessors using the proposed method must be specialized officers of these facilities with relevant education and experience. There is a need for accurate implementation of the descriptive statistical methods for the sample set of WTPs, such as using the occurrence percentage of each SSV variable. After the SSV assessment of an individual WTP, an update of data is required.

Similar studies about estimating of the structural resilience of WTPs have investigated past failures of pipelines, their lateral spread, the percentage of maintenance hole replacements, a comparison of the age distribution of the existing facilities with the building code requirements, and parameters such as the seismic hazard zone, liquefaction zones, landslide zones, tsunami inundation, and extra forces produced by the shaking of liquids inside the tanks. In contrast, the proposed coherent method introduces and qualitatively handles more parameters related to seismic structural vulnerability. The total number of SSV-evaluating parameters are parallel to the sustainability goals.

The main aim of each type of vulnerability classification is to protect human life and the environment. Beyond that, there are more issues, such as the social and economic values involved in decisions that need to be made following the assessment of a WTP's structural vulnerability. Future investigation may discover other criteria, such as the operators' perceptions of vulnerability, SSV assessment mechanisms, and the importance of used factors. Modern threats such as natural hazards and climate change should be considered seriously using the appropriate evaluation of WTPs' structures, using a total review of SSV.

Supplementary Materials: The following are available online at https://www.mdpi.com/article/10.3390/su13094835/s1, Table S1: Abbreviations and definitions.

Author Contributions: Conceptualization: D.E.A.; methodology: P.N.K.; software, figures development: P.N.K.; formal analysis, inverstigation: D.E.A., S.K.G., P.N.K.; resources: D.E.A., S.K.G., P.N.K.; data curation: D.E.A., S.K.G.; writing original draft preparation: P.N.K.; writing review and editing: D.E.A., S.K.G., P.N.K.; supervision: S.K.G., D.E.A. All authors have read and agreed to the published version of the manuscript.

Funding: This research received no external funding.

Institutional Review Board Statement: Not applicable.

Informed Consent Statement: Not applicable.

Data Availability Statement: Data is contained within the article.

Acknowledgments: The authors would like to acknowledge Nikolaos Kerpelis, graphic designer, for supporting and creating the graphical abstract.

Conflicts of Interest: The authors declare no conflict of interest.

References

1. Alexakis, D. Dispersion of hazardous material (haz-mat) triggered by natural disasters and related impacts on the quality of water and soil resources. Potentially effects on human health. In *Special Volume in Memory of Petros Vythoulkas*; NTUA, Center for Natural Risk Assessment and Preventive Planning: Athens, Greece, 2010; pp. 90–104. (In Greek)
2. Bathrellos, G.D. An overview in urban geology and urban geomorphology. *Bull. Geol. Soc. Greece* **2007**, *40*, 1354–1364. [CrossRef]
3. Golfinopoulos, S.K.; Nikolaou, A.D.; Kostopoulou, M.N.; Xilourgidis, N.K.; Vagi, M.C.; Lekkas, D.T. Organochlorine pesticides in the surface waters of Northern Greece. *Chemosphere* **2003**, *50*, 507–516. [CrossRef]
4. Papadopoulou-Vrynioti, K.; Bathrellos, G.D.; Skilodimou, H.D.; Kaviris, G.; Makropoulos, K. Karst collapse susceptibility mapping considering peak ground acceleration in a rapidly growing urban area. *Eng. Geol.* **2013**, *158*, 77–88. [CrossRef]
5. Alexakis, D.E. Suburban areas in flames: Dispersion of potentially toxic elements from burned vegetation and buildings. Estimation of the associated ecological and human health risk. *Environ. Res.* **2020**, *183*, 109153. [CrossRef]
6. UNDRR Implementing the Sendai Framework. Available online: https://www.undrr.org/.../what-sf (accessed on 30 January 2021).
7. Makropoulos, K.; Kaviris, G.; Kouskouna, V. An updated and extended earthquake catalogue for Greece and adjacent areas since 1900. *Nat. Hazards Earth Syst. Sci.* **2012**, *12*, 1425–1430. [CrossRef]
8. Mousiopoulos, N.; Penelis, G.; Abramidis, I.; Stylianidis, K.; Kalogirou, N.; Arabantinos, D. *30 Years after the Thessaloniki Earthquake Memories and Perspective—Tribute on the Anniversaru of the Thessaloniki Earthquake*; AUTH/ Faculty of Engineering: Thessaloniki, Greece, 2008. (In Greek)
9. EPPO. *Study Group Investigation and Recordings of Reasons That Caused Typical Damages to Buildings, during the Athens Earthquake of 7.9.1999*; EPPO: Athens, Greece, 2000; Available online: https://www.oasp.gr/sites/default/files/%20182.pdf (accessed on 28 April 2020). (In Greek)
10. Karidis, P.; Lekkas, E. *The Hαiti Earthquake Ms7.2R 12 January 2010*; NKUA-NTUA: Athens, Greece, 2010; Available online: http://www.elekkas.gr/images/stories/missions/haiti2010_extras/haiti_booklet_elekkas.pdf (accessed on 28 April 2020). (In Greek)
11. AUTH SYNER-G: Systemic Seismic Vulnerability and Risk Analysis for Buildings, Lifeline Networks and Infrastructures Safety Gain. Available online: http://www.vce.at/SYNER-G/files/downloads.html (accessed on 28 April 2020).
12. Dritsos, S. *Repairs and Reinforcement of Reinforced Concrete Buildings*, 3rd ed. 2005. Available online: http://www.episkeves2.civil.upatras.gr/wp-content/uploads/2015/07/ (accessed on 28 April 2020). (In Greek)
13. United Nations Take Action for the Sustainable Development Goals. Available online: https://www.un.org/sustainabledevelopment/sustainable-development-goals/ (accessed on 17 February 2021).
14. Murgante, B.; Borruso, G.; Lapucci, A. Sustainable development: Concepts and methods for Its application in urban and environment planning. In *Geocomputation, Sustainability and Environmental Planning*; Springer: Berlin/Heidelberg, Germany, 2011; Volume 348, pp. 1–15.
15. European Commission Next Steps for a Sustainable European Future—European Action for Sustainability. Available online: https://eur-lex.europa.eu/legal-content/EN/TXT/HTML/?uri=CELEX:52016DC0739&from=EN (accessed on 17 February 2021).
16. Guerrini, A.; Romano, G.; Ferretti, S.; Fibbi, D.; Daddi, D. A Performance measurement tool leading Wastewater Treatment Plants toward economic efficiency and sustainability. *Sustainability* **2016**, *8*, 1250. [CrossRef]
17. Nishisaka, H. Research towards a long-term restoration plan for sewage pipes. In Proceedings of the 6th EWA/JSWA/WEF Joint Conference "The resilience of the water sector", Munich, Germany, 15–18 May 2018; European Water Association: Hennef, Germany; pp. 44–58.
18. Kato, K. Restoration of Sendai Sewerage Service from the Great East Japan Earthquake and disaster-prevention measures for the future. In Proceedings of the 6th EWA/JSWA/WEF Joint Conference "The Resilience of the Water Sector", Munich, Germany, 15–18 May 2018; European Water Association: Hennef, Germany; pp. 120–144.
19. Wakimoto, H. Building and utilizing the Wastewater Treatment Plant network in Kobe City. In Proceedings of the 6th EWA/JSWA/WEF Joint Conference "The Resilience of the Water Sector", Munich, Germany, 15–18 May 2018.
20. Lin, K.H.E.; Chang, Y.C.; Liu, G.Y.; Chan, C.H.; Lin, T.H.; Yeh, C.H. An interdisciplinary perspective on social and physical determinants of seismic risk. *Nat. Hazards Earth Syst. Sci.* **2015**, *15*, 2173–2182. [CrossRef]
21. Clark, C.S. Health effects associated with wastewater treatment and disposal. *J. Water Pollut. Control Fed.* **1986**, *27*, 566.
22. Michael, I.; Rizzo, L.; McArdell, C.S.; Manaia, C.M.; Merlin, C.; Schwartz, T.; Dagot, C.; Fatta-Kassinos, D. Urban wastewater treatment plants as hotspots for the release of antibiotics in the environment: A review. *Water Res.* **2013**, *47*, 957–995. [CrossRef]
23. Qi, W.K.; Sanuba, T.; Norton, M.; Li, Y.Y. Effect of the Great East Japan Earthquake and Tsunami on sewage facilities and subsequent recovery measures. *J. Water Sustain.* **2014**, *4*, 27–40.
24. Panico, A.; Basco, A.; Lanzano, G.; Pirozzi, F.; Santucci de Magistris, F.; Fabbrocino, G.; Salzano, E. Evaluating the structural priorities for the seismic vulnerability of civilian and industrial wastewater treatment plants. *Saf. Sci.* **2017**, *97*, 51–57. [CrossRef]

25. Panico, A.; Lanzano, G.; Salzano, E.; Santucci De Magistris, F.; Fabbrocino, G. Seismic vulnerability of wastewater treatment plants. *Chem. Eng. Trans.* **2013**, *32*, 13–18.
26. Zare, M.R.; Wilkinson, S.; Potangaroa, R. Vulnerability of Wastewater Treatment Plants and wastewater pumping stations to earthquakes. *Int. J. Strateg. Prop. Manag.* **2010**, *14*, 408–420. [CrossRef]
27. Schiff, A.J. *The Loma Prieta, California Earthquake of October 17, 1989-Lifelines*; USGS: Washington, DC, USA, 1998.
28. Knudson, M.; Ballantyne, D.; Stuhr, M.; Damewood, M. The Oregon Resilience plan for water and wastewater Systems. In *Proceedings of the Pipelines 2014*; American Society of Civil Engineers: Reston, VA, USA, 2014; pp. 2211–2220.
29. Tierney, K. Disaster governance: Social, political, and economic dimensions. *Annu. Rev. Environ. Resour.* **2012**, *37*, 341–363. [CrossRef]
30. Schwab, J.C. *Hazard Mitigation: Integrating Best Practices into Planning*; APA Planning Advisory Service: Chicago, IL, USA, 2010.
31. JRC EN Eurocode Parts. Available online: https://eurocodes.jrc.ec.europa.eu/showpage.php?id=13 (accessed on 22 January 2021).
32. *Homeland Security Water and Wastewater Systems Sector-Specific Plan*; 2015. Available online: https://www.cisa.gov/publication/nipp-ssp-water-2015 (accessed on 1 April 2020).
33. Prathivadi, K. Wastewater Resilience Planning. In Proceedings of the 6th EWA/JSWA/WEF Joint Conference "The resilience of the water sector", Munich, Germany, 15–18 May 2018; European Water Association: Hennef, Germany; pp. 304–323.
34. Kent, R.; Klostrman, R. GIS and mapping. *J. Am. Plan. Assoc.* **2000**, *66*, 189–198. [CrossRef]
35. FEMA Project "Hazus". Available online: https://www.fema.gov/hazus (accessed on 28 April 2020).
36. Giannopoulos, G.; Filippini, R.; Schimmer, M. *Risk Assessment Methodologies for Critical Infrastructure Protection. Part I: A State of the Art*; Publications Office of the European Union: Luxembourg, 2012.
37. Eidinger, J.; Avila, E. *Guidelines for the Seismic Evaluation and Upgrade of Water Transmission Facilities*; ASCE: Reston, VA, USA, 1999.
38. Pitilakis, K.; Franchin, P.; Khazai, B.; Wenzel, H. *SYNER-G: Systemic Seismic Vulnerability and Risk Assessment of Complex Urban, Utility, Lifeline Systems and Critical Facilities: Methodology and Applications*; Springer: Dordrecht, The Netherlands, 2014.
39. EPPO Structural Vulnerability Checking. Available online: https://www.oasp.gr/node/76 (accessed on 30 April 2020). (In Greek)
40. Al-Nimry, H.; Resheidat, M.; Qeran, S. Rapid assessment for seismic vulnerability of low and medium rise infilled RC frame buildings. *Earthq. Eng. Eng. Vib.* **2015**, *14*, 275–293. [CrossRef]
41. ASCE, A Comprehensive Assessment of America's Infrastructure. Available online: https://infrastructurereportcard.org/wp-content/uploads/2019/02/Full-2017-Report-Card-FINAL.pdf (accessed on 30 April 2020).
42. Juan-García, P.; Butler, D.; Comas, J.; Darch, G.; Sweetapple, C.; Thornton, A.; Corominas, L. Resilience theory incorporated into urban wastewater systems management. State of the art. *Water Res.* **2017**, *115*, 149–161. [CrossRef] [PubMed]
43. Wiemer, S.; Kraft, T.; Landtwing, D. *Seismic Risk*; ETH Research Collection: Zurich, Switzerland, 2015.
44. Kirchhoff, C.J.; Watson, P.L. Are wastewater systems adapting to climate change? *JAWRA J. Am. Water Resour. Assoc.* **2019**, *55*, 869–880. [CrossRef]
45. Nikolaou, A.D.; Golfinopoulos, S.K.; Kostopoulou, M.N.; Kolokythas, G.A.; Lekkas, T.D. Determination of volatile organic compounds in surface waters and treated wastewater in Greece. *Water Res.* **2002**, *36*, 2883–2890. [CrossRef]
46. Seifert, C.; Krannich, T.; Guenther, E. Gearing up sustainability thinking and reducing the bystander effect—A case study of wastewater treatment plants. *J. Environ. Manag.* **2019**, *231*, 155–165. [CrossRef]
47. Alexakis, D. Human health risk assessment associated with Co, Cr, Mn, Ni and V contents in agricultural soils from a Mediterranean site. *Arch. Agron. Soil Sci.* **2016**, *62*, 359–373. [CrossRef]
48. Golfinopoulos, S.K.; Nikolaou, A.D. Disinfection by-products and volatile organic compounds in the water supply system in Athens, Greece. *J. Environ. Sci. Health Part A Toxic Hazard. Subst. Environ. Eng.* **2001**, *36*, 483–499. [CrossRef]
49. Wycisk, P. 3D Geological and hydrogeological modelling—Integrated approaches in urban groundwater management. In *Management of Water, Energy and Bio-Resources in the Era of Climate Change: Emerging Issues and Challenges*; Springer: Cham, Switzerland, 2015; pp. 3–12.
50. Kassem, M.M.; Nazri, M.F.; Farsangi, N.E. The seismic vulnerability assessment methodologies: A state-of-the-art review. *Ain Shams Eng. J.* **2020**, *11*, 849–864. [CrossRef]
51. Duić, N.; Urbaniec, K.; Huisingh, D. Components and structures of the pillars of sustainability. *J. Clean. Prod.* **2015**, *88*, 1–12. [CrossRef]
52. Padrón-Páez, J.I.; Almaraz, S.D.L.; Román-Martínez, A. Sustainable wastewater treatment plants design through multiobjective optimization. *Comput. Chem. Eng.* **2020**, *140*, 16. [CrossRef]
53. Cossio, C.; Norrman, J.; McConville, J.; Mercado, A.; Rauch, S. Indicators for sustainability assessment of small-scale wastewater treatment plants in low and lower-middle income countries. *Environ. Sustain. Indic.* **2020**, *6*, 11. [CrossRef]
54. Müller, H.S.; Haist, M.; Vogel, M. Assessment of the sustainability potential of concrete and concrete structures considering their environmental impact, performance and lifetime. *Constr. Build. Mater.* **2014**, *67*, 321–337. [CrossRef]
55. Demis, S. Durability of reinforced concrete structures design—Problems and prospects. In Proceedings of the 22 Student Conf. Construction Repairs and Reinforcements, Patra, Greece, 16–17 February 2016; p. 80. (In Greek)
56. Song, H.W.; Saraswathy, V. Corrosion monitoring of reinforced concrete structures—A review. *Int. J. Electrochem. Sci.* **2007**, *2*, 28.
57. Bathrellos, G.D.; Gaki-Papanastassiou, K.; Skilodimou, H.D.; Papanastassiou, D.; Chousianitis, G.K. Potential suitability for urban planning and industry development using natural hazard maps and geological-geomorphological parameters. *Environ. Earth Sci* **2012**, *66*, 537–548. [CrossRef]

58. Kerpelis, P. Assessment of structural and non-structural vulnerability of sewage treatment plants, through a questionnaire. In Proceedings of the 6th International Conference Safe Corfu 2019, Corfu, Greece, 6–9 November 2019; pp. 133–136.
59. Kerpelis, P.; Golfinopoulos, S.; Alexakis, D. Proposing the critical structural characteristics of wastewater treatment plants (WTPs) for the estimation of their seismic vulnerability. In Proceedings of the International Conference VSU 2020, Sofia, Bulgary, 15–17 October 2020; pp. 825–830.
60. Keeney, R.L.; McDaniels, T.L.; Ridge-Cooney, V.L. Using values in planning wastewater facilities for Metropolitan Seattle. *J. Am. Water Resour. Assoc.* **1996**, *32*, 293–303. [CrossRef]
61. Allen, D.E.; Rainer, J.H. Guidelines for the seismic evaluation of existing buildings. *Can. J. Civ. Eng.* **1995**, *22*, 500–505. [CrossRef]
62. McFadden, L. An all-hazard approach to building resilience. In Proceedings of the 6th EWA/JSWA/WEF Joint Conference "The resilience of the Water sector", Munich, Germany, 15–18 May 2018; European Water Association: Hennef, Germany; pp. 34–58.
63. Qasim, S.; Zhu, G. *Wastewater Treatment and Reuse: Theory and Design Examples (Two-Volume Set)*; CRC Press: Boca Raton, FL, USA, 2018; ISBN 9781498762007.
64. Qasim, S. *Wastewater Treatment Plants: Planning, Design, and Operation*, 2nd ed.; CRC Press: Boca Raton, FL, USA, 1999.
65. Garcia, X.; Pargament, D. Reusing wastewater to cope with water scarcity: Economic, social and environmental considerations for decision-making. *Resour. Conserv. Recycl.* **2015**, *101*, 154–166. [CrossRef]
66. Vicente, R.; Lagomarsino, S.; Mendes Silva, R. Seismic vulnerability assessment, damage scenarios and loss estimation. Case study of the old city centre of Coimbra, Portugal. In Proceedings of the 14th World Conference on Earthquake Engineering, Beijing, China, 12–17 October 2008; p. 9.
67. Tang, M.; Pérez-Fernández, R.; De Baets, B. Fusing absolute and relative information for augmenting the method of nearest neighbors for ordinal classification. *Inf. Fusion* **2020**, *56*, 128–140. [CrossRef]
68. Hans, S.; Boutin, C.; Ibraim, E.; Roussillon, P. In situ experiments and seismic analysis of existing buildings. Part I: Experimental investigations. *Earthq. Eng. Struct. Dyn.* **2005**, *34*, 1513–1529. [CrossRef]
69. EPPO-Association of Civil Engineers of Greece. *Greek Code for Seismic Resistant Structures-GCSRS*, 2001st ed.; EPPO: Athens, Greece, 2001. (In Greek)
70. Jain, S.K. Earthquake safety in India: Achievements, challenges and opportunities. *Bull. Earthq. Eng.* **2016**, *14*, 1337–1436. [CrossRef]
71. Farokhnia, K.; Porter, K. Estimating the non-structural seismic vulnerability of building categories. In Proceedings of the 15th World Conference on Earthquake Engineering, Lisbon, Portugal, 24–28 September 2012; pp. 22083–22092.
72. Ballay, D.; Blais, J.F. Wastewater treatment. *Rev. Des Sci.* **1998**, *11*, 1–11.
73. Wang, Z. Understanding seismic hazard and risk assessments: An example in the new madrid seismic zone of the Central United States. In Proceedings of the 8th US National Conference on Earthquake Engineering, San Francisco, CA, USA, 18–22 April 2006; pp. 1–10.
74. Pavlou, K.; Kaviris, G.; Chousianitis, K.; Drakatos, G.; Kouskouna, V.; Makropoulos, K. Seismic hazard assessment in Polyphyto dam area (NW Greece) and its relation with the "unexpected" earthquake of 13 May 1995 (M s = 6.5, NW Greece). *Nat. Hazards Earth Syst. Sci.* **2013**, *13*, 141–149. [CrossRef]
75. Woessner, J.; Laurentiu, D.; Giardini, D.; Crowley, H.; Cotton, F.; Grünthal, G.; Valensise, G.; Arvidsson, R.; Basili, R.; Demircioglu, M.B.; et al. The 2013 European seismic hazard model: Key components and results. *Bull. Earthq. Eng.* **2015**, *13*, 3553–3596. [CrossRef]
76. Solomos, G.; Pinto, A.; Dimova, S. *A Review of the Seismic Hazard Zonation in National Building Codes in the Context of Eurocode 8. Support to the Implementation, Harmonization and Further Development of the Eurocodes*; Publications Office of the EU: Ispra, Italy, 2008.
77. Arslan, M.H.; Korkmaz, H.H. What is to be learned from damage and failure of reinforced concrete structures during recent earthquakes in Turkey? *Eng. Fail. Anal.* **2007**, *14*, 1–22. [CrossRef]
78. Akkar, S.; Sucuoğlu, H.; Yakut, A. Displacement-based fragility functions for low-and mid-rise ordinary concrete buildings. *Earthq. Spectra* **2005**, *21*, 901. [CrossRef]
79. Oliveira, C.F.; Varum, H.; Vargas, J. Earthen Constructio: Structural Vulnerabilities and Retrofit Solutions for Seismic Actions. In Proceedings of the 15th World Conference on Earthquake Engineering, Lisbon, Portugal, 24–28 September 2012; pp. 1–9.
80. Yao, Q.L. A Fuzzy Method for Evaluating the Influences of Some Geological Factors on Earthquake Disaster Risk. Available online: https://www.semanticscholar.org/paper/.../980ea730b149f70c847f9fcd46e53b817ceb0fed#paper-header (accessed on 30 January 2021).
81. Burland, J.B. Foundation engineering. *Struct. Eng.* **2008**, *86*, 45–51.
82. Baker, J.W.; Faber, M.H. Liquefaction risk assessment using geostatistics to account for soil spatial variability. *J. Geotech. Geoenvironmental Eng.* **2008**, *134*, 14–23. [CrossRef]
83. EPPO-Association of Civil Engineers of Greece. *Greek Code for Reinforced Concrete—GCRC*, 1st ed.; EPPO: Athens, Greece, 2001. (In Greek)
84. Liu, S.; Kuhn, R. Data loss prevention. *IT Prof.* **2010**, *12*, 10–13. [CrossRef]
85. Westenenk, B.; De La Llera, J.C.; Besa, J.J.; Jünemann, R.; Moehle, J.; Lüders, C.; Inaudi, J.A.; Elwood, K.J.; Hwang, S.J. Response of reinforced concrete buildings in concepción during the maule earthquake. *Earthq. Spectra* **2012**, *28*, 257–280. [CrossRef]

86. Iervolino, I.; Manfredi, G.; Polese, M.; Verderame, G.M.; Fabbrocino, G. Seismic risk of R.C. building classes. *Eng. Struct.* **2007**, *29*, 813–820. [CrossRef]
87. American Concrete Institute. Building Code Requirements for Structural Concrete (ACI 318-14). Available online: http://aghababaie.usc.ac.ir/files/1506505203365.pdf (accessed on 28 December 2020).
88. Korkmaz, K.A.; Sari, A.; Carhoglu, A.I. Seismic risk assessment of storage tanks in Turkish industrial facilities. *J. Loss Prev. Process Ind.* **2011**, *24*, 314–320. [CrossRef]
89. Anagnostopoulos, S.A.; Spiliopoulos, K.V. An investigation of earthquake induced pounding between adjacent buildings. *Earthq. Eng. Struct. Dyn.* **1992**, *21*, 289–302. [CrossRef]
90. NIST GCR 12-917-21 Soil-Structure Interaction for Building Structures. Available online: https://www.nehrp.gov/pdf/nistgcr12-917-21.pdf (accessed on 28 December 2020).
91. Behnamfar, F.; Banizadeh, M. Effects of soil-structure interaction on distribution of seismic vulnerability in RC structures. *Soil Dyn. Earthq. Eng.* **2016**, *80*, 73–86. [CrossRef]
92. Tokimatsu, K.; Mizuno, H.; Kakurai, M. Building damage associated with geotechnical problems. *Soils Found.* **1996**, *36*, 219–234. [CrossRef]
93. Issa, C.A.; Debs, P. Experimental study of epoxy repairing of cracks in concrete. *Constr. Build. Mater.* **2007**, *22*, 459–466. [CrossRef]
94. Chau Khun, M.; Nazirah Mohd, A.; Chin Siew Yung, S.; Jen Hau, N.; Wen Haur, L.; Abdullah Zawawi, A.; Wahid, O. Repair and rehabilitation of concrete structures using confinement: A review. *Constr. Build. Mater.* **2017**, *133*, 502–515.
95. Millard, A.; Pimienta, P. Modelling of Concrete Behaviour at High Temperature. In *RILEM State-of-the-Art Reports*; Springer: Berlin/Heidelberg, Germany, 2019.
96. Yépez, F.; Yépez, O. Role of construction materials in the collapse of R/C buildings after Mw 7.8 Pedernales—Ecuador earthquake, April 2016. *Case Stud. Struct. Eng.* **2017**, *7*, 24–31. [CrossRef]
97. Dritsos, S. The Application of the Rules of Interventions and the Eurocodes in the interventions in buildings made of Concrete. In Proceedings of the Information Meeting Day: "Restoration and Strengthening of Buildings", TCG/RDWG, Aigio, Greece, 4 June 2015; p. 38. (In Greek)
98. Park, Y.; Ang, A.H.-S.; Wen, Y.K. Seismic damage analysis of reinforced concrete buildings. *J. Struct. Eng.* **1985**, *111*, 740–757. [CrossRef]
99. Kroworz, A.; Katunin, A. Non-destructive testing of structures using optical and other methods: A review. *SDHM* **2018**, *12*, 1–17.
100. Naeim, F. Book Review: Seismic design of reinforced concrete buildings. *Earthq. Spectra* **2015**, *31*, 615–616. [CrossRef]
101. Randal, F.A. Waterstops. Available online: https://www.concreteconstruction.net/how-to/materials/waterstops-1_o (accessed on 30 January 2021).
102. Valente, M. *Collection of Guidelines on Concrete Raccolta di Linee Guida su Calcestruzzo*; Presidenza del Consiglio Superiore: Milano, Italy, 2003.
103. Cladera, A.; Weber, B.; Leinenbach, C.; Czaderski, C.; Shahverdi, M.; Motavalli, M. Iron-based shape memory alloys for civil engineering structures: An overview. *Constr. Build. Mater.* **2014**, *63*, 281–293. [CrossRef]
104. Mahamid, M.; Gaylord, E.H.; Gaylord, C.N. *Structural Engineering Handbook*, 5th ed.; McGraw-Hill: New York, NY, USA, 2020; ISBN 9780070231238.
105. Bungey, J.H.; Millard, S.G.; Grandham, M.G. *Testing of Concrete in Structures*, 4th ed.; Taylor & Francis: Abingdon, UK, 2006; ISBN 978-0-415-26301-6.
106. Wells, P.A.; Melchers, R.E. Findings of a 4 year study of concrete sewer pipe corrosion. In Proceedings of the Annual Conference of the Australasian Corrosion Association 2014: Corrosion and Prevention 2014, Darwin, Australia, 21–24 September 2014; pp. 182–193.
107. Rashidi, H.; Ghaffarianhoseini, A.; Ghaffarianhoseini, A.; Nik Sulaiman, N.M.; Tookey, J.; Hashim, N.A. Application of wastewater treatment in sustainable design of green built environments: A review. *Renew. Sustain. Energy Rev.* **2015**, *49*, 845–856. [CrossRef]
108. Bathrellos, G.D.; Skilodimou, H.D.; Chousianitis, K.; Youssef, A.M.; Pradhan, B. Suitability estimation for urban development using multi-hazard assessment map. *Sci. Total Environ.* **2017**, *575*, 119–134. [CrossRef]
109. Skilodimou, H.D.; Bathrellos, G.D.; Chousianitis, K.; Youssef, A.M.; Pradhan, B. Multi-hazard assessment modeling via multi-criteria analysis and GIS: A case study. *Environ. Earth Sci.* **2019**, *78*, 47. [CrossRef]

Article

The Formation Mechanism and Influence Factors of Highway Waterfall Ice: A Preliminary Study

Zhijun Zhou [1], Jiangtao Lei [1,*], Shanshan Zhu [1], Susu Qiao [1] and Hao Zhang [2]

[1] School of Highway, Chang'an University, Xi'an 710064, China
[2] Shaanxi Provincial Communication Construction Group, Xi'an 710075, China
* Correspondence: leijiangtao@chd.edu.cn

Received: 11 June 2019; Accepted: 25 July 2019; Published: 27 July 2019

Abstract: Highway waterfall ice hazards usually happen in cold regions. However, minimal research has addressed this so far due to its multidisciplinary nature. In this study, ground water monitoring tests were conducted for 2.5 years to study the relationship between ground water level changes and waterfall ice hazards. To explore the internal factors that lead to highway waterfall ice, gradation tests, penetration tests, and freezing tests were conducted which revealed that coarse-grained particles can enhance the permeability of aquifers. Further, volume expansion of free water freezing in a closed system is the main reason for pore pressure increasing aquifers in research areas. Furthermore, to understand the formation mechanism of highway waterfall ice further, a mathematical model of saturated coarse-grained soil at the state of phase transition equilibrium was obtained. This indicates that the essence of the aquifers' freezing (coarse-grained soil) in the waterfall ice area is the freezing of closed water. Finally, based on the abovementioned findings, the formation process of waterfall ice is defined as three stages: The drainage obstruction stage, the soil deformation stage, and the groundwater gushing stage, respectively. This definition can provide significant guidance on further research that focuses on prevention of highway waterfall hazards.

Keywords: highway waterfall ice; ground water; mathematical model; formation mechanism

1. Introduction

Underground aquifers are always disturbed during highway construction in cold regions. This disturbance is obvious during cutting excavation. More precisely, weak interfaces appear during excavating, and ground water squeezes out from these weak interfaces due to the pressure raise induced by the soil freeze. Then, it spreads over the road, and the freeze happens under negative air temperature. These ice blocks are called highway waterfall ice in the field of highway engineering. Highway waterfall ice, together with frost heave damage and snow blockage, are difficult issues when considering highways in cold regions [1–3]. High-grade highways in the northern part of China have always been damaged by waterfall ice [4,5]. The hazards have mainly presented traffic isolation and casualties induced by traffic accidents. To ensure the traffic flows, millions of RMB have been spent by the Chinese government for de-icing the highways [6,7]. Furthermore, the expansion of waterfall ice can last a considerable amount of time so that road maintenance workers have to work day and night.

Although there are research and engineering practices based on the study of waterfall ice, the research process is still slow [8,9]. The reason is that the study of highway waterfall ice is complex and multidisciplinary [10]. Considering the factors that affect the development of waterfall ice, only a few researchers have combined the indoor and outdoor experiments with theoretical analysis [11]. When referring to the formation mechanism of waterfall ice, the contraction of the underground cross section, and the increase of hydraulic pressure, which is caused by the compaction of subgrade soil, have been always considered as the main factors by scholars. There are also studies that considered frost depth increases that result from excavating during construction leading to highway waterfall ice hazards [12–14]. All of these viewpoints are reasonable, and the main reason for the difference is that the research purposes of these studies are different, and, in addition, the location of the research is different. However, there are insufficient systematic studies on the micro-formation mechanism of waterfall ice [15].

As the study of waterfall ice can also be defined as researching the mechanical properties and the interaction mechanism between groundwater and soil, the research methods of subgrade frost heaving can also be adopted to study waterfall ice [16–18]. According to Fourier's law, Darcy's law, and energy conservation law, many scholars have deduced the calculation model of the coupling effect of the temperature field, humidity field, and stress field in pavement structure systems [19–23]. Although these models explain the frost heaving mechanism of soil to some extent, they are not suitable for a theoretical analysis of waterfall ice. To explore the law of frost heaves, indoor and outdoor experiments have been conducted in recent years [24]. The general law of subgrade freezing damage has been discovered by many scholars based on in-situ investigations, thawing and freezing tests, and the law of capillary water rising [17,25]. Furthermore, some scholars have obtained many enlightening results by analyzing the microstructure differences of soils under freeze-thaw conditions by scanning electron microscopy (SEM) [26]. Nevertheless, these studies always focus on the freeze process under the function of capillary action. Therefore, with regard to waterfall ice which is caused by the flowing out of ground water, these test results are not applicable. In addition, with the development of numerical simulation technology, more and more scholars use numerical simulations to simulate the change of the soil temperature field, pore water migration and frost heave deformation [27,28]. Some significant macroscopic factors that lead to highway frost damage can be shown, but the generation mechanism and key parameters are far away from actual determination during simulations due to the complexity and variability of the internal structure of the soil. Moreover, based on the uncertainty analysis theory, some scholars also produced some interesting works [29].

This work chose Tongchuan-Huangling Expressway in China as a research site, and groundwater monitoring was undertaken at the place where waterfall ice hazards are prone to happen. Then, the relationship between ground water level changes and the scale of waterfall ice were analyzed. Following this, penetration tests and freezing tests were conducted in laboratories using in-situ soil, and a mathematical model of waterfall ice was established. Finally, the development process of the highway waterfall was discussed.

2. In-Situ and Indoor Experimental Investigation

Negative temperature, ground water, and the permeability of soil are basic factors that affect the scale of highway waterfall ice. The climate type of the research site belongs to the Monsoon Climate of Medium Latitudes. The negative temperature of the research site (Tongchuan-Huangling Expressway) can last several months during winter, and rainfall is seasonal. According to the site investigation, the soil composition of this area includes three layers (i.e., humus soil, crushed stone, and bed rock) from top to bottom. In addition, the topography along the expressway is mainly hills, and groundwater is abundant with its shallow buried depth. Therefore, highway waterfall ice hazards always happen. The soil composition and waterfall ice hazards are shown in Figure 1.

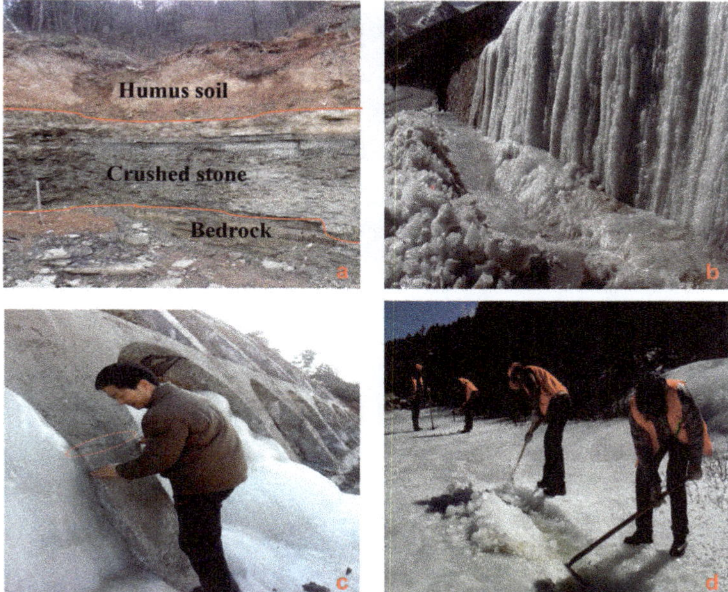

Figure 1. The soil composition and waterfall ice hazards of the research site (**a**) Soil composition (**b**) Large-scale highway waterfall ice (**c**) Facing wall dehiscence induced by waterfall ice (**d**) Highway waterfall ice extended to the road surface.

2.1. Ground Water Monitoring Results

The formation of highway waterfall ice is closely related to the change of the ground water level. In this study, eight monitoring sites at Tongchuan-Huangling Expressway were selected to conduct groundwater monitoring tests. The temperature and frozen depth change of Tongchuan-Huangling expressway is shown in Figure A1. To obtain ground water change statistics, two drill holes were assigned at each monitoring site. The information regarding drill holes is shown in Table 1. The ground water monitoring began when cutting started to excavate, and lasted 2.5 years. The Ground Water Level Monitoring System ZKGD2000-MB was adopted to conduct groundwater monitoring, and the details of the system are shown in Figure 2. The locations of monitoring sites are shown in Figure 3. The monitoring results are shown in Figure 4.

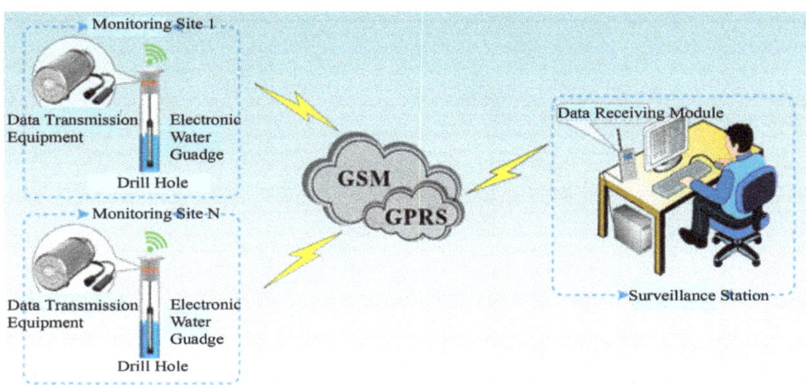

Figure 2. Ground Water Level Monitoring System [30].

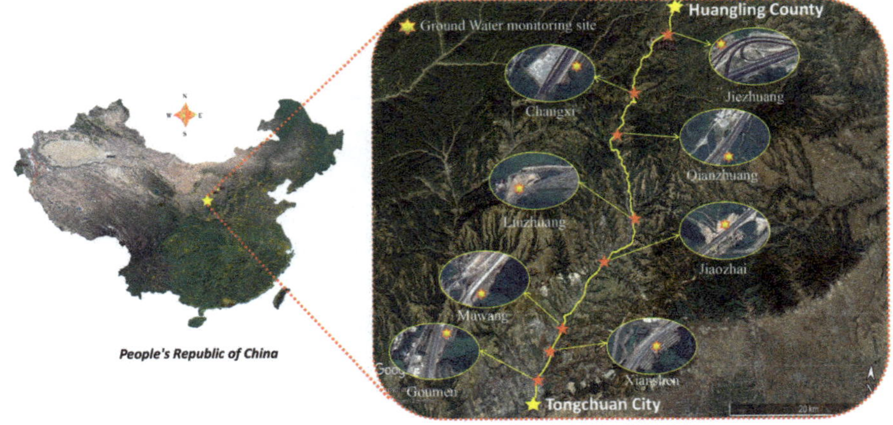

Figure 3. Location of monitoring sites [31].

Table 1. Information of drilled holes.

Monitoring Site	Code Name	Longitude	Latitude	Elevation (m)	Ground Water Burial Depth (m)	Lithological Description of Aquifer
Goumen	G-1	35°20′44.7″ N	108°58′07.7″ E	1321	3.10	Strong Weathered sandstone
	G-2			1319	2.46	Silty clay
Xianshen	X-1	35°21′56.9″ N	108°59′22.7″ E	1255	9.71	Sandy cobble layer
	X-2			1248	7.08	Strong weathered conglomerate
Mawang	M-1	35°22′35.0″ N	109°00′03.5″ E	1238	6.52	Strong weathered conglomerate
	M-2			1231	1.25	Strong weathered sandstone
Jiaozhai	J-1	35°24′36.8″ N	109°01′05.3″ E	1197	1.58	Strong weathered sandstone
	J-2			1204	5.23	Strong weathered sandstone
Liuzhuang	L-1	35°27′08.3″ N	109°02′54.7″ E	1125	3.29	Strong weathered sandstone
	L-2			1106	3.96	Strong weathered mudstone
Qianzhuang	Q-1	35°29′39.4″ N	109°04′18.9″ E	1094	1.14	Strong weathered mudstone
	Q-2			1087	1.52	Sandy cobble layer
Jiezhuang	J-3	35°32′18.9″ N	109°07′40.6″ E	889	5.08	Strong weathered sandstone
	J-4			881	2.30	Strong weathered sandstone
Changxi	C-1	35°33′54.2″ N	109°10′03.0″ E	993	4.46	Strong weathered sandstone
	C-2			985	5.71	Strong weathered mudstone

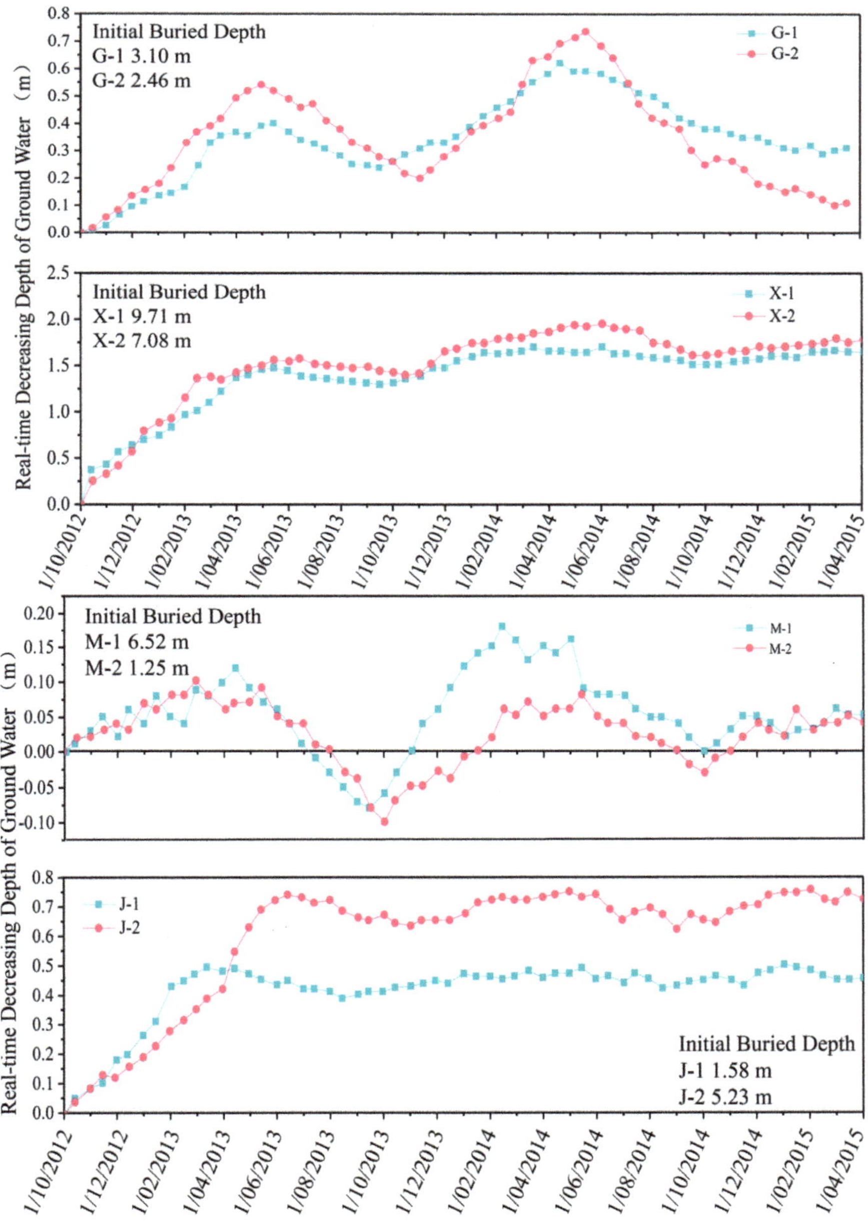

Figure 4. *Cont.*

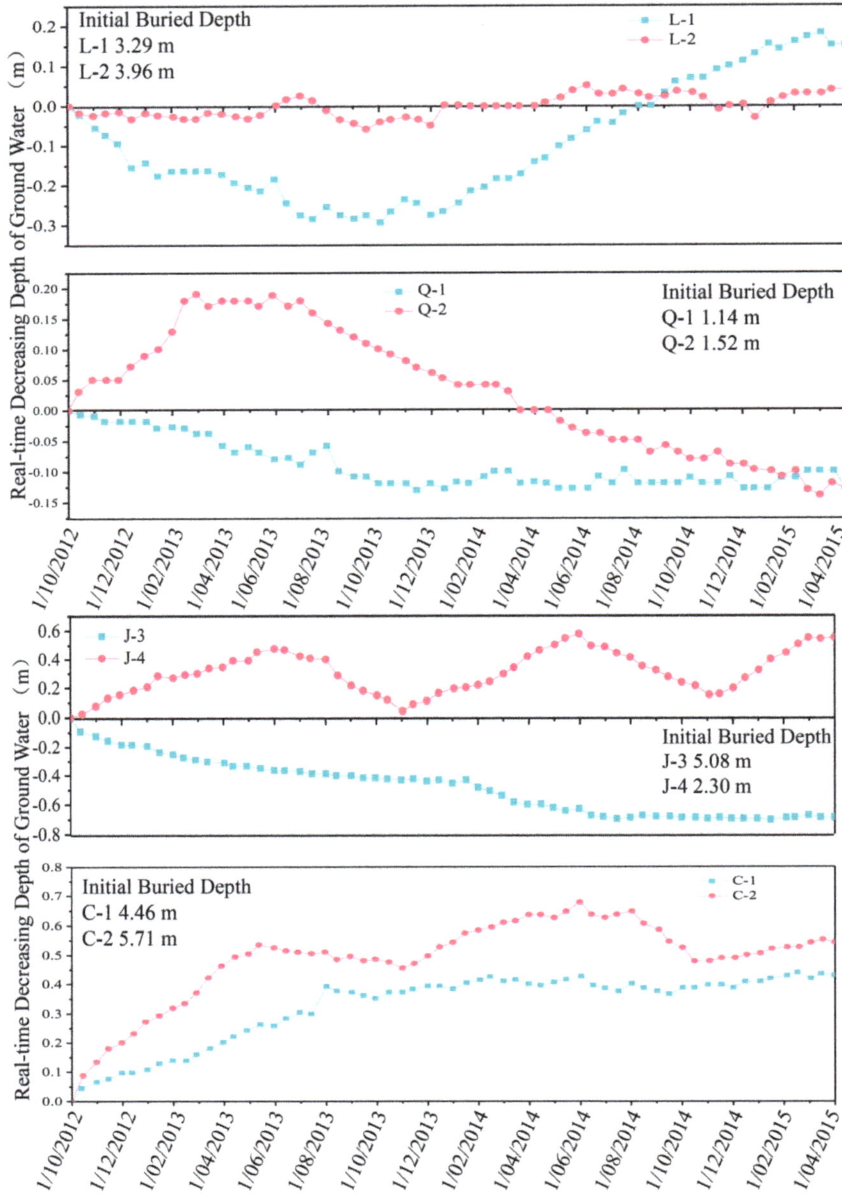

Figure 4. The groundwater monitoring results of drilled holes.

The groundwater monitoring results show that the trend of groundwater level changes of G-1, G-2, M-1, M-2, and J-4 are the same, and the decline and rise of groundwater shows a repeated trend. The reason for this phenomenon is that the ground water level of these test sites is sensitive to rainfall. Therefore, it decreased rapidly due to little rainfall in winter, then, it increased again after heavy rainfall in summer. This type of groundwater level change contributes considerably to the formation of highway waterfall ice in early winter, because the ground water level and its pressure are still high in this period, the scale of waterfall ice is expected to develop rapidly. In contrast, the scale of waterfall ice is expected to remain steady due to little rainfall in deep winter. The ground water level of X-1, X-2, Q-2, J-1, J-2, C-1, and C-2 decreased rapidly after the excavation of cuttings, and then it remained steady. The reason for this phenomenon is that the ground water level of these test sites is sensitive to excavation, and cannot be recharged effectively after ground water levels decrease. Therefore, the scale of highway waterfall ice is always small in these areas. The ground water level of Q-1, J-3 kept increasing during excavation, and kept steady afterwards. The reason is that the groundwater balance was broken after excavation, and perched ground water in the upper part of the mountain continued to flow to the hill foot. Highway waterfall ice in these areas is always difficult to eliminate owing to the continuous supplement of perched ground water. The ground water level of L-1 also increased during excavation, but it decreased after a year because the perched water supplement was insufficient. The ground water level of L-2 did not change during monitoring, hence, waterfall ice hazards did not happen in this area.

From monitoring the results and analysis, it can be seen that the ground water change that results from cutting excavation is the main reason for highway waterfall ice hazards, and the area that can easily get a seasonal rainfall supplement and perched ground water supplement suffers severe waterfall ice hazards. The purpose of monitoring is to explore the relationship between the ground water level change and highway waterfall ice hazards. In this way, an overall understanding of the formation mechanism of highway waterfall ice can be obtained. Furthermore, the monitoring results can give a significant reference on the route design of the expressway in the Tongchuan-Huangling area. More specifically, the route design can avoid the area where the highway waterfall ice hazards are prone to happen according to the ground water level change statistics. Moreover, it is also a new method to evaluate the occurrence possibility of highway water fall ice hazards by using ground water monitoring technology in groundwater-rich areas.

2.2. *Laboratory Tests of In-Situ Soil*

2.2.1. Gradation Tests and Penetration Tests

According to the above analysis, this study found that the occurrence of waterfall ice was closely related to ground water level change. However, the scale of waterfall ice is directly determined by the permeability of aquifers. Therefore, the gradation test and penetration test were conducted using the soil of aquifers, which were taken from eight monitor sites at Tongchuan-Huangling Expressway. The test equipment for the penetration tests is shown in Figure 5. The penetration test specimens were taken from eight monitoring sites by using the core drilling method [32]. Most of the specimens had difficulty keeping the cylindrical shape because of the high moisture content and the strong weathered property. Furthermore, the soil was disturbed and filled the container before the penetration tests because the diameter of the container (30 cm) was greater than the soil specimens (10 cm). According to the Test Methods of Soils for Highway Engineering (JTG E40-2007) [33], the constant head permeability test was adopted in this research. Before the penetration tests, the soil was put into the container in layers (3 cm/layer), and compacted slightly to control the porosity ratio. After filling one layer, the water-inflow switch opened to saturate the soil layer, and closed when the water level reached the soil layer's surface. Then, filling the layers stopped when the height of the soil was above the piezometric hole 3–4 cm. After that, the air free water under the temperature of 20 °C was poured into the container from the top to submerge the soil, and until it reached the position of the spilled water hole. After that, the soil was saturated for 24 h before the tests. Then, the penetration tests were conducted under the temperature of 20 °C. In this way, the hydraulic conductivity can be written as k_{20}. The tests results are shown in Table 2, among which the grain grade was divided into three groups (fine grained group, medium grained group, and coarse grained group) for convenient analysis.

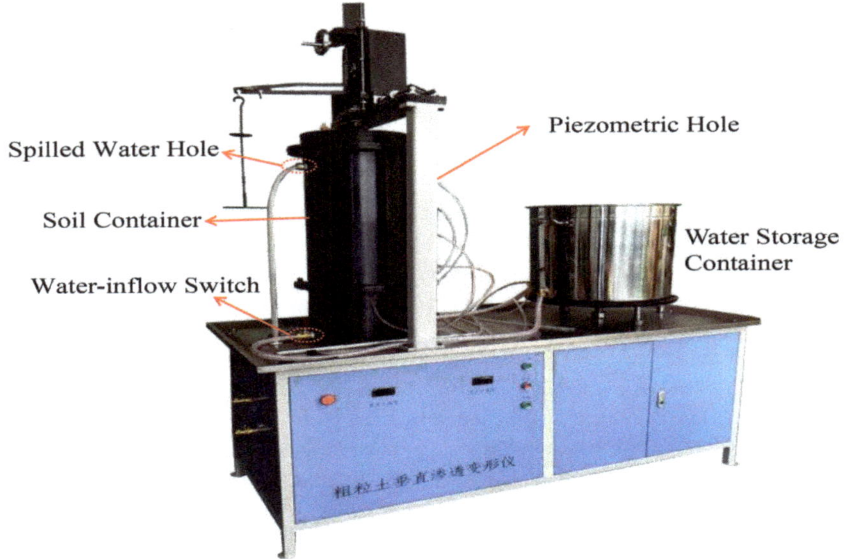

Figure 5. The equipment for penetration tests (Chang'an University).

Table 2. The test results of the gradation test and penetration test.

Sampling Sites		Mawang	Changxi	Jiaozhai	Goumen	Xianshen	Liuzhuang	Qianzhuang	Jiezhuang
Fine grained group/mm	<0.1	14%	15%	12%	5%	9%	5%	4%	2%
	0.1–0.5	15%	8%	5%	9%	2%	5%	3%	3%
Medium grained group/mm	0.5–2	11%	8%	11%	9%	14%	11%	16%	9%
	2–5	11%	9%	13%	13%	4%	12%	4%	8%
Coarse grained group/mm	5–10	27%	54%	50%	56%	66%	48%	52%	65%
	10–20	22%	6%	9%	8%	5%	19%	21%	13%
d_{10}/mm		0.083	0.079	0.091	0.25	0.105	0.23	0.81	1.27
d_{30}/mm		0.57	2.13	2.69	4.34	5.19	4.86	5.54	6.14
d_{60}/mm		6.61	6.81	6.74	7.11	7.34	7.51	8.15	8.70
Uniformity coefficient C_u		79.64	86.20	74.07	28.44	69.90	32.65	10.06	6.85
Curvature coefficient C_c		0.59	8.43	11.80	10.60	34.95	13.67	4.65	3.41
k_{20}/(cm/s)		0.037	0.094	0.187	0.214	0.279	0.335	0.54	0.688

The results show that the medium-sized group's weight percentage of eight sampling sites was fixed between 17% and 30%, and the variation range was relatively small from Table 2. Therefore, the effective grain diameter d_{10} is determined by the fine grained group, and the constrained diameter d_{60} is determined by the coarse grained group in these areas. Therefore, the effective grain diameter d_{10} and the constrained diameter d_{60} can be used as the measurement for the mass of fine grained soil and coarse grained soil, and then these two specific diameters can be used to study the relationship between them and hydraulic conductivity k_{20}. The results are shown in Figure 6. It can be seen that a proportional relationship exists between the specific diameters and hydraulic conductivity. According to the particle-size distribution of the aquifer's soil as shown in Table 2, d_{10} increases with the decrease of the fine grained group content, and d_{60} increases with the increase of the coarse grained group content. Combining this with the proportional relationship between specific diameters and hydraulic conductivity, it can be seen that the permeability of the aquifer can enhance with the decline of fine-grained particles and the increase of the coarse-grained particles, and the possibility of the occurrence of highway waterfall ice may increase.

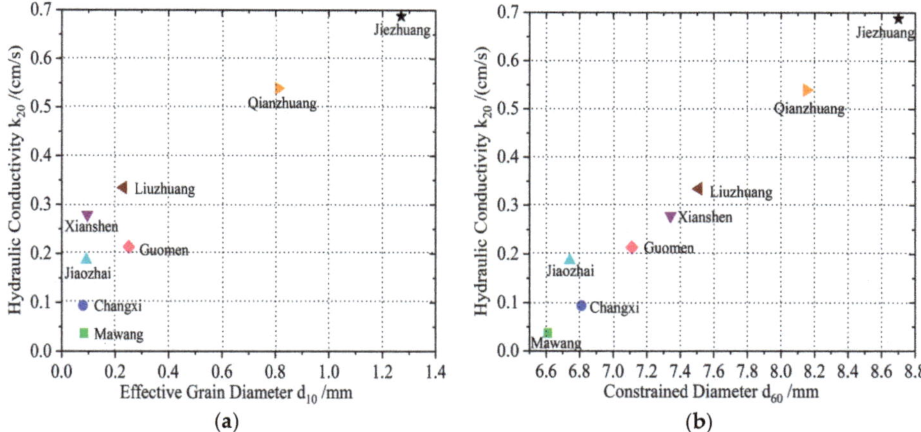

Figure 6. The relationship between specific diameters and hydraulic conductivity: (**a**) effective grain diameter (**b**) constrained diameter.

2.2.2. Freezing Tests

The above tests indicate that the coarse grain group can enhance the permeability of soil, which leads to highway waterfall ice hazards. In this section, freezing tests were carried out on saturated coarse-grained soil and fine-grained soil separately to explore the formation mechanism of waterfall ice further. The coarse-grained soil was taken from the area where waterfall ice hazards are the most serious, and the typical fine silt of the research site was used as fine grained soil for the tests. The physical characteristics of coarse-grained soil are as follows: Initial density $\rho = 1.965\ \mathrm{g/cm^3}$, maximum dry density $\rho_d = 2.05\ \mathrm{g/cm^3}$, and optimum water content $\omega = 6.8\%$. The physical characteristics of fine grained soil are as follows according to laboratory tests: Initial density $\rho = 1.489\ \mathrm{g/cm^3}$, maximum dry density $\rho_d = 1.734\ \mathrm{g/cm^3}$, and optimum water content $\omega = 18.9\%$. The grain size composition of coarse grained soil and fine grained soil are shown as Tables 3 and 4.

Table 3. The grain size composition of coarse-grained soil.

Grain size (mm)	10–20	5–10	1–5	0.1–1	0.05–0.1	<0.05
Weight percent (%)	5.8	30.5	45.3	16.4	1.3	0.7

Table 4. The grain size composition of fine-grained soil.

Grain size (mm)	0.05–0.1	0.01–0.05	0.005–0.01	<0.005
Weight percent (%)	14.8	45.7	32.2	7.3

The samples were saturated and consolidated under the pressure of 0.5 MPa before the tests, and after that, the water content of fine grained soil and coarse grained soil were detected, which were 39% and 20%, respectively. Then, the cylindrical specimens with the diameter of 10 cm and the height of 15 cm were made, and the soil was compacted in layers (3 cm/layer). The vibrating wire piezometers were installed into the samples (Figure 7) to test pore pressure. The samples were covered by transparent insulating material without the bottom, and a water layer was at the top to simulate ground water. The details of the experimental design is shown in Figure 7.

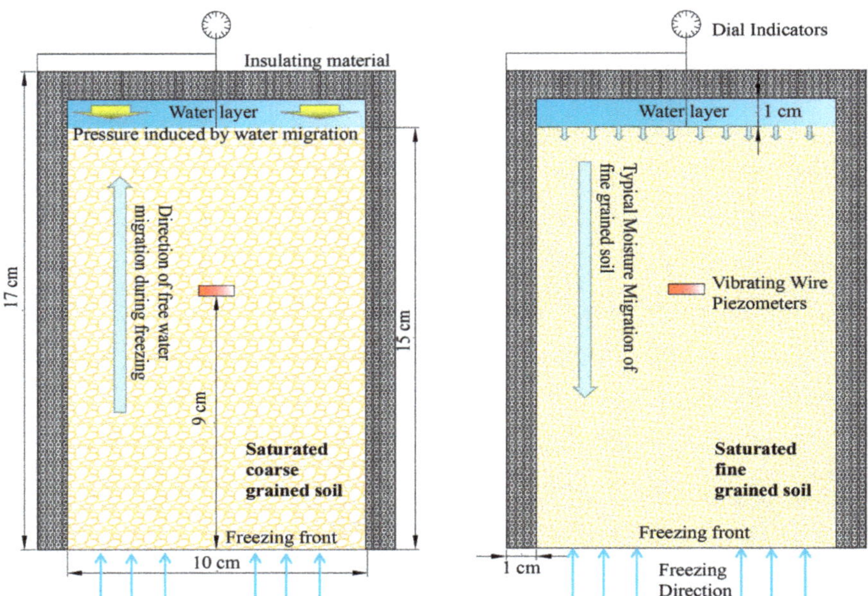

Figure 7. Freezing test set up design.

The experiment procedure includes two stages as follows:

1. The samples were put into an incubator at the temperature of 1 °C, and lasts for 24 h.
2. The temperature of incubator was set to −10 °C, and began to freeze. The freezing direction was from the bottom to the top. The waterproof thin film with good thermal conductivity was covered at the bottom of the samples to prevent the water seeping from the bottom during freezing. The incubator was opened and the frost-heaving ratio, pore pressure, and water inflow volume were measured (by testing the volume change of water layer) quickly every 2 h. The frost-heaving ratio was measured by dial indicators. The initial volume of the water layer was measured by using vernier caliper to measure the height, and calculated the volume afterwards. The volume change of the water layer was also measured by vernier caliper. As the soil samples were completely saturated during the tests, the downward water migration due to gravity was neglected.

The test results are shown in Figure 8. From test results, it can be seen that the pore pressure of saturated coarse-grained soil increased during freezing conditions, and decreased sharply after reaching the maximum value (84 KPa in this test). The frost heaving ratio and the inflow volume of it were almost zero during the freezing tests. The reason for the phenomena is that the process of free water freeze happened in saturated coarse-grained soil during the freezing test.

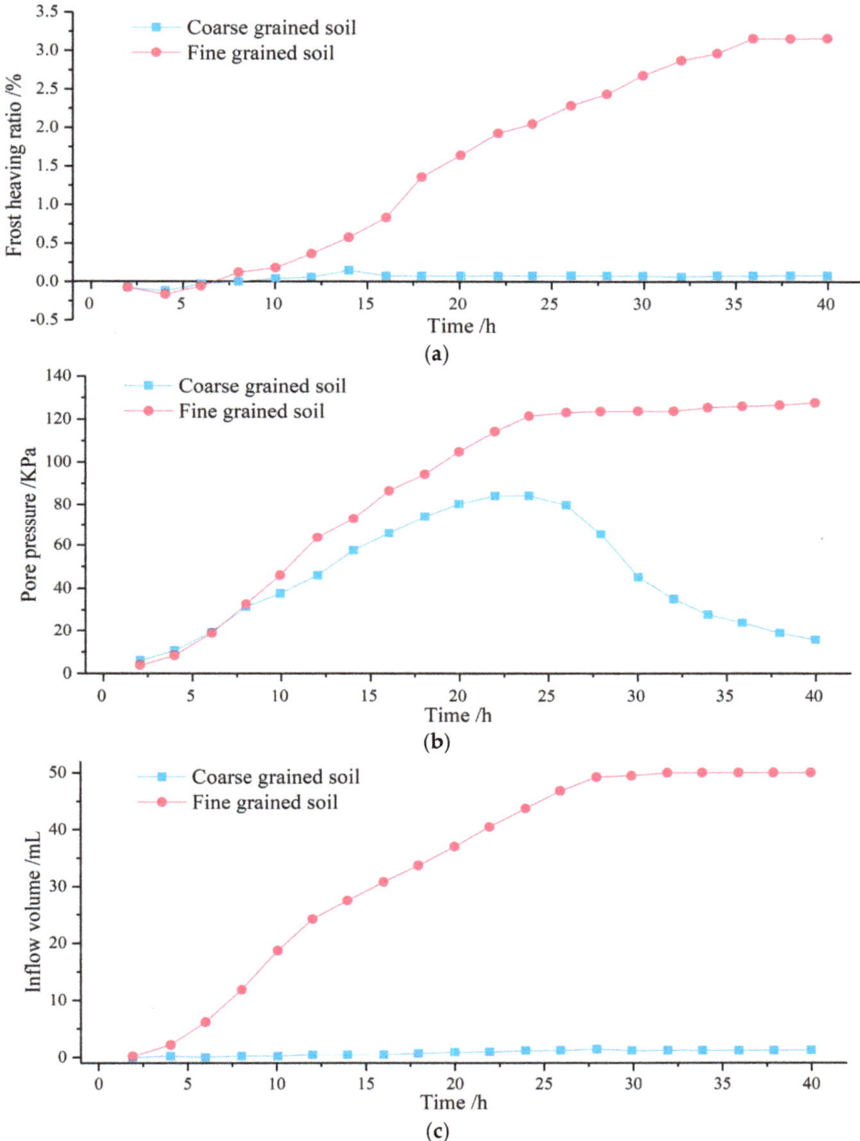

Figure 8. Freezing tests results (**a**) frost heaving ration (**b**) pore pressure (**c**) inflow volume.

As the volume of free water increases after freezing, the expansive force induced by the volume change can appear in a closed system, and the pore pressure of the part that do not freeze is expected to increase. When considering water migration, during the freezing process from the bottom to the top, the volume change of frozen free-water in coarse grained soil leads to increasing the free water level above the freezing surface. Therefore, the downward reaction force emerges at the top of the experimental container, then, the pore pressure of the test point increases. With the increase of the frozen depth, the position of the vibrating wire piezometer gradually freezes. Therefore, the downward reaction force cannot transfer to the position of the vibrating wire piezometer completely. Then, the pore pressure decreases. The microscopic freezing process of saturated coarse grained soil is shown in Figure 9. Moreover, this frozen mode of free water in coarse-grained soil made the inflow volume almost zero.

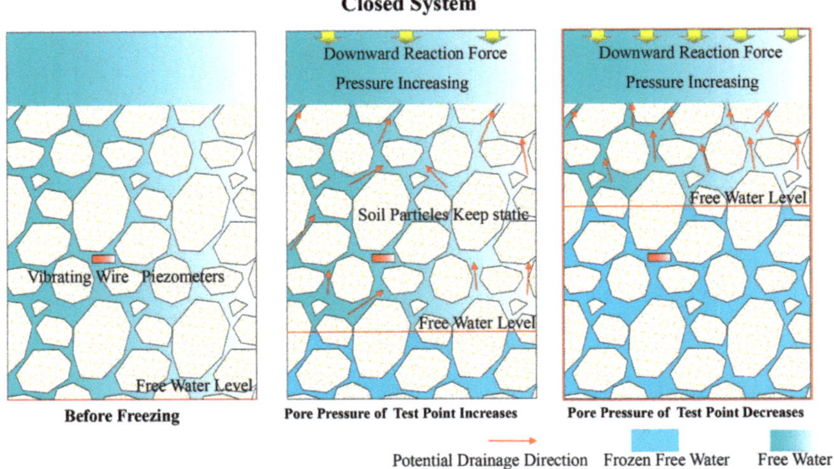

Figure 9. Microscopic freezing process of saturated coarse grained soil.

As mentioned above, the process of free water freeze happened in saturated coarse-grained soil during the freezing test. Therefore, the freezing process of the water in the coarse-grained soil is independent of soil particles, and this process leads to the upward drainage of the part that do not freeze. Although the whole tests were conducted in a closed system, and the drainage did not happen, the water pressure did not apply to the soil particles. In contrast, the pressure applied to the top part of the insulating material due to the potential drainage trend as shown in Figure 9. Therefore, if the soil is kept static, the frost heave is not expected to happen. The frost heaving ratio, the pore pressure, and the inflow volume of saturated fine-grained soil increased when it was freezing, and then kept steady after reaching maximum values (3.2%, 127.8 Kpa, and 50 mL, respectively). The reason for the phenomena is that the pore water in fine-grained soil belongs to weak-bound water, and moisture migration from the top to the freezing front happened when it was freezing. This migration led to the water content increasing near the freezing front. Therefore, the frost heaving happened. Furthermore, the moisture migration also induced the water inflow and increased pore pressure. After the position of the vibrating wire piezometer froze, the pore pressure induced by the weak-bound water was not be influenced by the moisture migration, therefore, it kept steady. The restriction effect of the experimental container may result in the appearance of steady stages of frost heaving and inflow volume.

Moreover, for saturated coarse grained soil, it can be seen from the test results that the decreasing speed of the pore pressure was becoming slow when the freezing process conducted 30 h, and kept stable at the end of the test. This means that at the end of the test, the position of the vibrating wire piezometer was completely frozen, and no reaction force was transferred to the position of the vibrating wire piezometer. Therefore, when the pore water around the vibrating wire piezometers was frozen, the pore pressure of saturated coarse grained soil was approximately 13 KPa (40 h) as shown in Figure 8. For saturated fine grained soil, the pore pressure kept stable when the freezing process was conducted at 23 h. Further, after the position of the vibrating wire piezometer froze, the pore pressure induced by the weak-bound water was not influenced by the moisture migration, therefore, it kept steady. Hence, when the pore water around the vibrating wire piezometers was frozen, the pore pressure of saturated fine grained soil was approximately 124 KPa. However, it can be seen that the sharp decrease of the pore pressure of saturated coarse grained soil and the pore pressure stable stage of saturated fine grained soil happened at the same time from the pore pressure curves. This phenomenon means that the freezing speed of saturated coarse grained soil and fine grained soil is almost the same. More specifically, the freezing of the position of the vibrating wire piezometer happened at 23 h for these two test specimens.

From the above tests, it can be seen that the coarse-grained soil has strong water permeability, and it is the major aquifer's soil in the area of waterfall ice hazards. Furthermore, if the drainage is blocked, the pore pressure of coarse-grained soil increases rapidly. Considering actual engineering, a frozen cutting surface can block the drainage of ground water, and then, the water pressure increases and squeezes out from the weak surface, and waterfall ice forms. Therefore, the aquifer with saturated coarse-grained soil is the internal condition that leads to highway waterfall ice.

3. Mathematical Model in Investigating Waterfall Ice

The percolation theory has the advantage of analyzing the frozen procedure of soil only focusing on temperature and pressure [34]. This theory is usually adopted for research on saturated fine-grained soil. Therefore, some theorems are reliable when establishing a mathematical model of saturated fine-grained soil. In this study, the frozen model of saturated fine-grained soil was established first. Following this, the freezing model of saturated coarse-grained soil was established based on the fine-grained soil model.

3.1. Freezing Model of Saturated Fine-Grained Soil

The moisture migration happens in fine-grained soil during freezing conditions. The microcosmic process of moisture migration is described only considering two-dimensional cases in Figure 10.

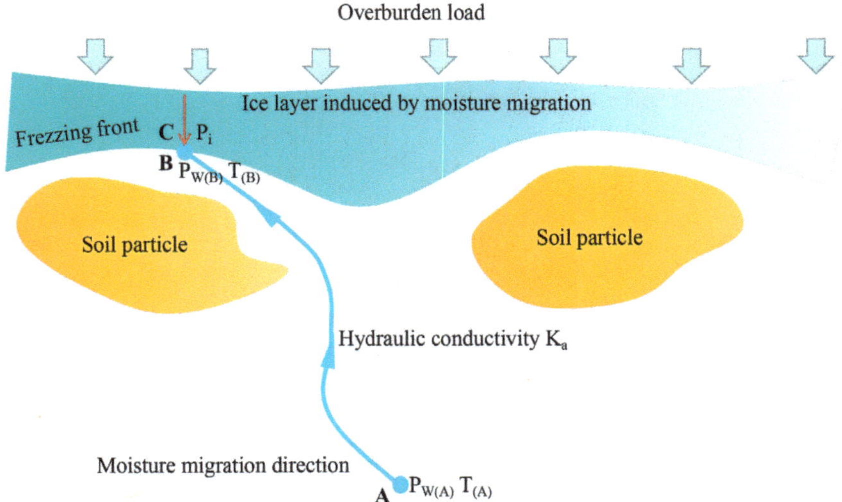

Figure 10. The microcosmic process of moisture migration in fine-grained soil.

As shown in Figure 10, the water transfers from point A to Point B during freezing conditions. The pressure and temperature of point A are described as $P_{W(A)}$ and $T_{(A)}$, respectively. The pressure and temperature of point B are described as $P_{W(B)}$ and $T_{(B)}$, respectively. Generally, the hydraulic conductivity K_a depends on the composition of soil particles and the temperature in soil. To simplify the calculation, the hydraulic conductivity between point A and point B is assumed to be the same. Point C at the freezing front, and the pressure is expressed as P_i, which is related to the overburden load. To illustrate the effect of osmotic pressure on water migration, a simplified theoretical freezing model of fine-grained soil was established as shown in Figure 11. The theoretical model consists of 4 parts: Part A, part B, part C, and part D. The parts are separated by three semipermeable membranes, which are named S1, S2, and S3, respectively. Part A and Part B contains pure water, and the temperatures of them are $T_{(A)}$ and $T_{(B)}$, respectively. The pressure of Part A and Part B are $P_{W(A)}$ and $P_{W(B)}$, respectively. Part C contains the solution, and the concentration of it increases with the volume of the ice layer increasing. Part D is fine-grained soil. To apply the thermodynamic equilibrium theory, the pressure of the solution in Part C is assumed to keep balance with pure water in Part B. In this way, the osmotic pressure $\Psi_{(B)}$ can be defined as Equation (1).

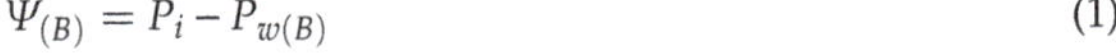

$$\Psi_{(B)} = P_i - P_{w(B)} \tag{1}$$

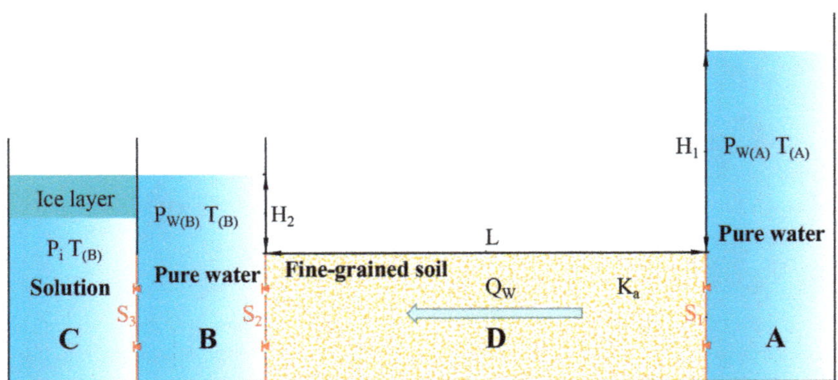

Figure 11. A simplified theoretical model of frozen saturated fine-grained soil.

The moisture migration of soil is induced by the pressure difference between Part A and Part B. The volume of flow (Q_w) can be described as Equation (2) based on Darcy's Law:

$$Q_w = -K_a A\frac{H_2 - H_1}{L} = -K_a AI = -K_a A\frac{\rho g H_2 - \rho g H_1}{\rho g L} = -K_a A\frac{P_{W(B)} - P_{W(A)}}{\rho g L} \tag{2}$$

where ρ represents density, I represents the hydraulic gradient, and A represents inflow area.

Therefore, the rate of flow (q_w) can be obtained as Equation (3):

$$q_w = -K_a\frac{P_{W(B)} - P_{W(A)}}{\rho g L} \tag{3}$$

According to Equations (1) and (3), the osmotic pressure can be described as Equation (4):

$$\Psi_{(B)} = P_i - P_{W(A)} + \frac{q_w \rho g l}{K_a} \tag{4}$$

As the simplified theoretical model belongs to a closed system, when the ice layer and the solution of Part-C are at the state of the phase transition equilibrium, the chemical potential of ice (μ_i) is equal to the chemical potential of the solution (μ_s) according to the first law of thermodynamics. To keep this balance, the change of μ_i is always equal to the change of μ_s as shown in Equation (5):

$$d\mu_s(P, \psi, T) = d\mu_i(P, T) \tag{5}$$

where P represents the pressure, ψ represents osmotic pressure, and T represents temperature.

According to the second law of thermodynamics, the expressions of $d\mu_w$ and $d\mu_i$ are shown in Equations (6) and (7):

$$d\mu_i = V_i dP - S_i dT \tag{6}$$

$$d\mu_s = V_s dP - V_s d\psi - S_s dT \tag{7}$$

where V represents the specific volume and S represents the specific entropy (ratio of heat absorbed or released by the system to corresponding temperature). V_i represents the specific volume of ice, V_s represents the specific volume of solution, S_i represents the specific entropy of ice, and S_s represents the specific entropy of solution.

The integrating Equation (5) from the initial state ($P = P_0$, $\psi_0 = 0$, and $T = T_0$) to the end state (P_i, $\psi_{(B)}$, and $T_{(B)}$), the expression is shown in Equation (8):

$$\iiint_{P0,\psi_0,T0}^{Pi,\Psi_{(B)},T(B)} V_i dP - S_i dT\, dpd\psi dT = \iiint_{P0,\psi_0,T0}^{Pi,\Psi_{(B)},T(B)} V_s dP - V_s d\psi - S_s dT\, dpd\psi dT \tag{8}$$

where P_0 represents atmospheric pressure, $T_0 = 276k$;

The calculation result of Equation (8) is obtained as shown in Equation (9):

$$(V_S - V_i)(P_i - P_0) - (S_S - S_i)(T_{(B)} - T_0) = V_S\left(\Psi_{(B)} - \Psi_0\right) \tag{9}$$

As mentioned above, $\psi_0 = 0$, and the specific entropy change can be described as the ratio of the internal energy change (Q) to the temperature of the phase transition as shown in Equation (10):

$$S_S - S_i = Q/T_0 \tag{10}$$

Therefore, by combining Equations (9) and (10), the expression of osmotic pressure $\Psi_{(B)}$ can be obtained as shown in Equation (11):

$$\Psi_{(B)} = \frac{(V_s - V_i)}{V_s}(P_i - P_0) - \frac{Q}{V_s T_0}(T_{(B)} - T_0) \tag{11}$$

when the phase transition keeps equilibrium, which means that the ice layers do not change, the rate of flow $q_w = 0$, and Equation (12) can be obtained by Equation (3):

$$P_{W(A)} = P_{W(B)} \tag{12}$$

Then, by combing Equations (1), (11) and (12), the osmotic pressure at the state of phase transition equilibrium ($\overline{\psi}_{(B)}$) can be obtained as shown in Equation (13):

$$\overline{\psi}_{(B)} = \frac{(V_S - V_i)}{V_i}\left(P_{W(A)} - P_0\right) - \frac{Q}{V_i T_0}\left((T_{(B)} - T_0\right) \tag{13}$$

According to above analysis, the mathematical frozen model of saturated fine-grained soil is obtained as shown in Equations (4), (11) and (13).

3.2. Freezing Model of Saturated Coarse-Grained Soil

There is no weak-bound water exists in saturated coarse-grained soil, therefore, the moisture migration cannot happen during freezing conditions, which means that there is no osmotic pressure ($\overline{\psi}_{(B)} = 0$) at the state of the phase transition equilibrium of saturated coarse-grained soil. At the same time, although the percolation theory cannot be fully be applied to the saturated coarse grained soil, the freezing process of saturated coarse grained soil and saturated fine grained soil both obey the first law of thermodynamics and the second law of thermodynamics. Therefore, considering saturated coarse grained soil or fine grained soil, the chemical potential of ice (μ_i) is always equal to the chemical potential of the solution (μ_s) when the ice layer and the water are at the state of the phase transition equilibrium in a closed system (This equation has nothing to do with the percolation theory). Therefore, Equations (5)–(7) can also be applied to saturated coarse grained soil. However, to give physical meaning to equations (there is no mathematical expression to describe the phase transition equilibrium of saturated coarse-grained soil), the frozen model of saturated fine-grained soil was established first as shown in Equation (13). Although the model includes osmotic pressure $\overline{\psi}_{(B)}$, it is a independent variable during mathematical integral. Therefore, the mathematical model of Equation (13) can extend to saturated coarse-grained soil when $\overline{\psi}_{(B)} = 0$ as shown in Equation (14).

$$\frac{P_{W(A)} - P_0}{T_{(B)} - T_0} = \frac{Q}{[T_0(V_S - V_i)]} \tag{14}$$

It can be seen that the formula does not include the parameters of the soil (K_a, L), and the osmotic pressure $\overline{\psi}_{(B)} = 0$. Therefore, the freezing model of saturated coarse-grained soil in a closed system cannot be influenced by the properties of the soil. Furthermore, it can be seen that Equation (14) is highly similar to the Clapeyron equation as shown in Equation (15). The Clapeyron equation can be applied to any pure substances at the state of the phase transition equilibrium.

$$\frac{dP}{dT} = \frac{Q}{T\Delta V} \tag{15}$$

In this way, the freezing model of saturated coarse-grained soil in a closed system is highly similar to pure water in a closed system. When considering the formation process of highway waterfall ice, from above analysis, it can be ensured that the essence of the aquifers' freezing in the waterfall ice area is the freezing of closed water. Then, the volume expansion due to freezing conditions increases the pressure of closed water, which squeezes out from the weak interfaces and forms the waterfall ice. This conclusion can give a primary reference for the discussion of the formation mechanism of highway waterfall ice. That is, pure water freezing helps to analyze the formation progress of highway waterfall ice.

4. Discussion of the Formation Mechanism of Highway Waterfall Ice

According to in-situ and indoor experimental investigations and the establishment of a mathematical model, this study found that the saturated coarse-grained soil in a closed system is the main factor that results in waterfall ice hazards. Consequently, to describe the formation mechanism of highway waterfall ice clearly, the soil layer model was established, which includes three layers, i.e., humus soil layer, saturated coarse-grained soil layer, and bedrock layer from top to bottom. Based on this model, when excavation occurs during highway construction, the frozen depth of the soil increases although aquifers are not exposed. This leads to the frozen depth of the aquifer (saturated coarse-grained soil) deeper than before. Therefore, the pressure of the groundwater increases, which has already been proved by the freezing tests mentioned in Section 3. According to the pressure increasing procedure during freezing conditions, the formation process of waterfall ice is defined as three stages (i.e., the drainage obstruction stage, the soil deformation stage, and the groundwater gushing stage) as shown in Figure 12.

At the drainage obstruction stage, the drainage path is partly obstructed with the increasing frozen depth. Hence, the pressure of the aquifer increases. The compaction process of the aquifer, which results from construction of machines, is also a main factor that leads to pressure increases when considering actual engineering. The groundwater flows to the weak area under this pressure.

At the soil deformation stage, with the increasing frozen depth, the drainage path of the aquifer is completely blocked, and a closed system is formed. The pressure of the aquifer increases rapidly, and the groundwater at weak area breaks through the humus soil layer and freezing under negative temperature, which results in top soil deformation. When the volume of ice reaches its maximum volume, the deformation of top soil also reaches its limit. Then, the ice stretches to the deformation soil as shown in Figure 12. At the same time, water separates from the coarse-grained soil and moves to the weak area constantly because of the increasing pressure.

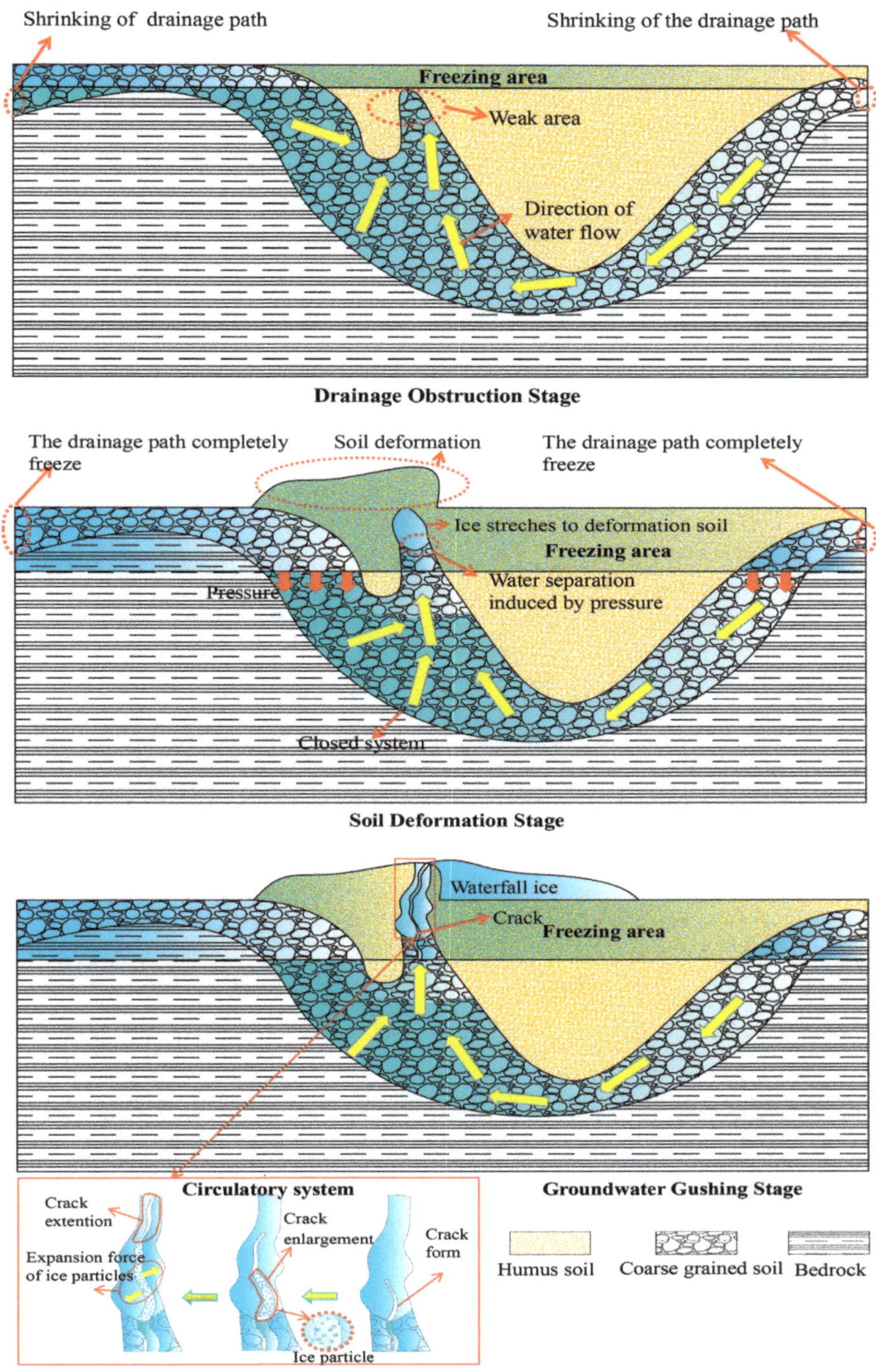

Figure 12. The formation process of waterfall ice.

At the groundwater gushing stage, the ice breaks through the deformation soil. Then, the huge expansive force generated by groundwater in a closed system fractures the ice block. After that, the groundwater flows into the cracks, and becomes ice particles under negative temperature. The cracks get larger due to the ice particle forming and the constant water inflow. With the increasing groundwater flow volume and the crack development process in ice blocks, the cracks cross the ice block completely, and groundwater flows out and becomes ice rapidly under negative temperature. After that, a closed system is formed again, and the groundwater gushing stage happens again. Therefore, a circulatory system is formed, and the volume of waterfall ice increases rapidly during this circulation.

5. Conclusions

In this paper, areas that suffer serious highway waterfall ice hazards (Tongchuan-Huangling expressway) were chosen to conduct ground water monitoring for 2.5 years. The relationship between the ground water level change and waterfall ice hazards is revealed. After that, the gradation test and penetration test of the aquifer soil, and the freezing test of typical saturated fine-grained soil and coarse-grained soil were conducted. Following this, the mathematical model of investigating waterfall ice was established. Finally, the formation mechanism of highway waterfall ice was discussed. The following primary conclusions can be drawn:

(1) Ground water level change that results from excavation is the main reason for highway waterfall ice hazards. Furthermore, the area that can easily get seasonal rainfall supplements, and the mountainous regions where ground water is replenished by perched water after cutting excavation suffers serious highway waterfall hazards. The monitoring results also provide significant references for the route design of expressways in related area.

(2) A proportional relationship between the content of coarse-grained particles and the hydraulic conductivity was obtained based on the gradation test and the penetration test using in-situ aquifer soil of eight research sites. The results indicate that the content of coarse-grained particles can significantly enhance the permeability of the aquifer, which is the internal factor that leads to highway waterfall ice.

(3) According to the freezing tests of saturated fine-grained soil and coarse-grained soil, this study found that, contrary to saturated fine-grained soil which showed obvious water migration during freezing, the pore pressure of coarse-grained soil increased rapidly first, and decreased after reaching the maximum value (84 KPa in this test). The frost heaving ratio and the inflow volume were almost zero during the test. The test results indicate that the volume expansion of free water freezing in a closed system is the main reason for pore pressure increasing the saturated coarse-grained soil.

(4) The freezing model of fine-grained soil was established based on the percolation theory. After that, the freezing model of saturated coarse-grained soil at the state of the phase transition equilibrium was obtained using the freezing model of fine-grained soil. This indicates that the essence of the aquifers' freezing (coarse-grained soil) in the waterfall ice area is the freezing of closed water.

(5) The formation mechanism of waterfall ice was discussed based on the above-mentioned results. The formation process of waterfall ice is defined as three stages (i.e., the drainage obstruction stage, the soil deformation stage, and the groundwater gushing stage). This definition can give significant guidance for further research that focuses on the prevention of highway waterfall hazards.

Author Contributions: Conceptualization, Z.Z., J.L., S.Z., and H.Z.; data curation, J.L., S.Z., S.Q., and H.Z.; funding acquisition, Z.Z.; investigation, J.L.; methodology, J.L., S.Z.; writing-original draft, J.L.

Funding: This work is financially supported by the National Key R&D Program of China (no. 2016YFC0802203-8).

Conflicts of Interest: The authors declare no conflicts of interest.

Appendix A

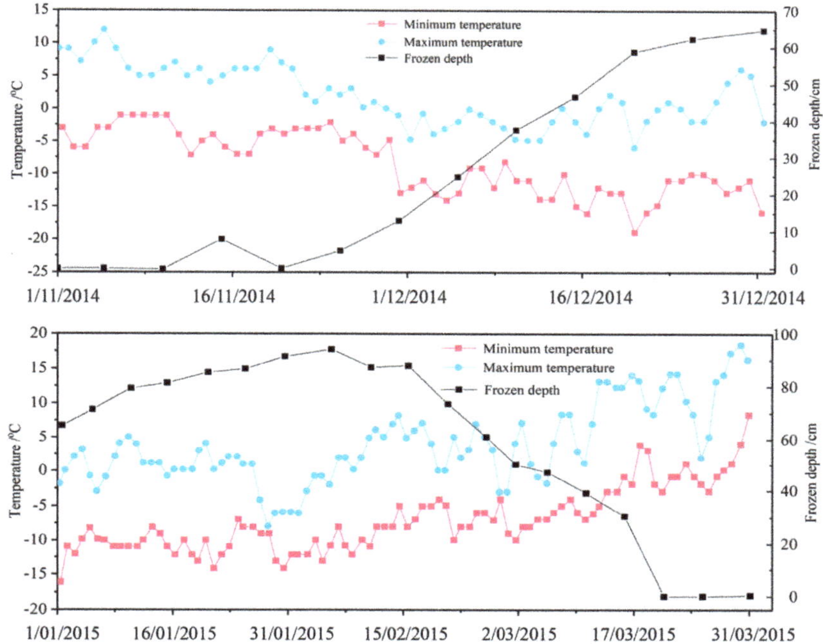

Figure A1. Temperature and frozen depth change of Tongchuan-Huangling expressway during winter 2014–2015.

References

1. Weiss, J.; Montagnat, M.; Cinquin-Lapierre, B.; Labory, P.; Moreau, L.; Damilano, F.; Lavigne, D. Waterfall ice: Mechanical stability of vertical structures. *J. Glaciol.* **2011**, *57*, 407–415. [CrossRef]
2. Gauthier, F. Statistical analysis of climate conditions conducive to falling ice blocks along road corridors in northern Gaspe. *Can. Geotech. J.* **2012**, *49*, 1408–1426. [CrossRef]
3. Lai, Y.; Zhang, S.; Yu, W. A new structure to control frost boiling and frost heave of embankments in cold regions. *Cold Reg. Sci. Technol.* **2012**, *79*, 53–66. [CrossRef]
4. Zhou, Z.; Zhan, H.; Hu, J.; Ren, C. Characteristics of Unloading Creep of Tuffaceous Sandstone in East Tianshan Tunnel under Freeze-Thaw Cycles. *Adv. Mater. Sci. Eng.* **2019**, 1–17. [CrossRef]
5. Lin, Z.; Niu, F.; Li, X.; Li, A.; Liu, M.; Luo, J.; Shao, Z. Characteristics and controlling factors of frost heave in high-speed railway subgrade, Northwest China. *Cold Reg. Sci. Technol.* **2018**, *153*, 33–34. [CrossRef]
6. Lai, J.; Wang, X.; Qiu, J.; Zhang, G.; Chen, J.; Xie, Y.; Luo, Y. A state-of-the-art review of sustainable energy based freeze proof technology for cold-region tunnels in China. *Renew. Sustain. Energy Rev.* **2018**, *82*, 3554–3569. [CrossRef]
7. Zhou, Z.; Zhu, S.; Kong, X.; Lei, J.; Liu, T. Optimization Analysis of Settlement Parameters for Postgrouting Piles in Loess Area of Shaanxi, China. *Adv. Civ. Eng.* **2019**, *12*, 1–16. [CrossRef]
8. Gauthier, F.; Hetu, B.; Allard, M. Forecasting method of ice blocks fall using logistic model and melting degree-days calculation: A case study in northern Gaspésie, Québec, Canada. *Nat. Hazards* **2015**, *79*, 855–880. [CrossRef]
9. Silvia, F.; Andrea, B.; Ian, J.F.; Frank, M. Calcite fabrics, growth mechanisms, and environments of formation in speleothems from the Italian Alps and southwestern Ireland. *J. Sediment. Res.* **2000**, *70*, 1183–1196. [CrossRef]

10. Raimo, S.; Pekka, H.; Ari, V. Effect of mild winter events on soil water content beneath snowpack. *Cold Reg. Sci. Technol.* **2008**, *51*, 56–67. [CrossRef]
11. Zhang, S.; Sheng, D.; Zhao, G.; Niu, F.; He, Z. Analysis of frost heave mechanisms in a high-speed railway embankment. *Can. Geotech. J.* **2016**, *53*, 520–529. [CrossRef]
12. Guthrie, W.; Shoop, S.; Ulring, J. Frost Damage in Pavement: Causes and Cures. *J. Cold Reg. Eng.* **2016**, *30*, 1–15. [CrossRef]
13. Montagnat, M.; Weiss, J.; Cinquin-Lapierre, B.; Labory, P.; Moreau, L.; Damilano, F.; Lavigne, D. Waterfall ice: Formation, structure and evolution. *J. Glaciol.* **2010**, *56*, 225–234. [CrossRef]
14. Yang, Z.; Zhang, L.; Ling, X.; Li, G.; Tu, Z.; Shi, W. Experimental study on the dynamic behavior of expansive soil in slopes under freeze-thaw cycles. *Cold. Reg. Sci. Technol.* **2019**, *163*, 27–33. [CrossRef]
15. Batenipour, H.; Alfaro, M.; Kurz, D.; Graham, J. Deformations and ground temperatures at a road embankment in northern Canada. *Can. Geotech. J.* **2014**, *51*, 260–271. [CrossRef]
16. Mahedi, M.; Cetin, B.; Cetin, K. Freeze-thaw performance of phase change material (PCM) incorporated pavement subgrade soil. *Constr. Build. Mater.* **2019**, *202*, 449–464. [CrossRef]
17. Lai, Y.; Xu, X.; Dong, Y.; Li, S. Present situation and prospect of mechanical research on frozen soils in China. *Cold Reg. Sci. Technol.* **2013**, *87*, 6–18. [CrossRef]
18. Mill, T.; Ellmann, A.; Aavik, A.; Horemuz, M.; Sillamaee, S. Determining ranges and spatial distribution of road frost heave by terrestrial laser scanning. *Balt. J. Road Bridge Eng.* **2014**, *9*, 225–234. [CrossRef]
19. Qin, Z.; Lai, Y.; Tian, Y.; Yu, F. Frost-heaving mechanical model for concrete face slabs of earthen dams in cold regions. *Cold Reg. Sci. Technol.* **2019**, *161*, 91–98. [CrossRef]
20. Lai, Y.; Pei, W.; Zhang, M.; Zhou, J. Study on theory model of hydro-thermal–mechanical interaction process in saturated freezing silty soil. *Int. J. Heat Mass Transf.* **2014**, *78*, 805–819. [CrossRef]
21. Tretyakova, O.; Yushkov, B. Inverted-Cone Piles for Transport Constructions in Seasonally Freezing Soils. *Soil Mech. Found. Eng.* **2017**, *54*, 173–176. [CrossRef]
22. Lee, J.; Kim, Y.; Chae, D.; Cho, W. Loading rate effects on strength and stiffness of frozen sands. *KSCE J. Civ. Eng.* **2016**, *20*, 208–215. [CrossRef]
23. Zhou, Z.; Ren, C.; Xu, G.; Zhan, H.; Liu, T. Dynamic Failure Mode and Dynamic Response of High Slope Using Shaking Table Test. *Shock. Vib.* **2019**, *8*, 1–19. [CrossRef]
24. Larson, L.; Kiemnec, G.; Johnson, E. Influence of Freeze-Thaw Cycle on Silt Loam Soil in Sagebrush Steppe of Northeastern Oregon. *Rangel. Ecol. Manag.* **2019**, *72*, 69–72. [CrossRef]
25. Gullu, H.; Khudir, A. Effect of freeze-thaw cycles on unconfined compressive strength of fine-grained soil treated with jute fiber, steel fiber and lime. *Cold Reg. Sci. Technol.* **2014**, *106*, 55–65. [CrossRef]
26. Ye, W.; Li, C. The consequences of changes in the structure of loess as a result of cyclic freezing and thawing. *Bull. Eng. Geol. Environ.* **2019**, *78*, 2125–2138. [CrossRef]
27. Yin, X.; Liu, E.; Song, B.; Zhang, D. Numerical analysis of coupled liquid water, vapor, stress and heat transport in unsaturated freezing soil. *Cold. Reg. Sci. Technol.* **2018**, *155*, 20–28. [CrossRef]
28. Ji, Y.; Zhou, G.; Zhou, Y.; Vandeginste, V. Frost heave in freezing soils: A quasi-static model for ice lens growth. *Cold Reg. Sci. Technol.* **2019**, *158*, 10–17. [CrossRef]
29. Rizzo, F.; Caracoglia, L. Examining wind tunnel errors in Scanlan derivatives and flutter speed of a closed-box. *Wind Struct.* **2018**, *26*, 231–251.
30. Zhou, Z.; Yang, T.; Fan, H. A Full-Scale Field Study on Bearing Characteristics of Cast-in-Place Piles with Different Hole-Forming Methods in Loess Area. *Adv. Civ. Eng.* **2019**, *4*, 1–16. [CrossRef]
31. Zhou, Z.; Lei, J.; Shi, S.; Liu, T. Seismic Response of Aeolian Sand High Embankment Slopes in Shaking Table Tests. *Appl. Sci.* **2019**, *9*, 1677. [CrossRef]
32. Liu, X.; Cai, G.; Liu, L.; Liu, S.; Anand, J. Thermo-hydro-mechanical properties of bentonite-sand-graphite-polypropylene fiber mixtures as buffer materials for a high-level radioactive waste repository. *Int. J. Heat Mass Transf.* **2019**, *141*, 981–994. [CrossRef]

33. Profession Standard of the People's Republic of China. *Test Methods of Soils for Highway Engineering (JTG E40-2007)*; Ministry of Communications of the PRC: Beijing, China, 2007.
34. Zhou, Z.; Dong, Y.; Jiang, P.; Han, D.; Liu, T. Calculation of Pile Side Friction by Multiparameter Statistical Analysis. *Adv. Civ. Eng.* **2019**, *6*, 1–16. [CrossRef]

Article

Level of Contamination Assessment of Potentially Toxic Elements in the Urban Soils of Volos City (Central Greece)

Evangelia E. Golia [1,*], Sotiria G. Papadimou [1,2], Christos Cavalaris [3] and Nikolaos G. Tsiropoulos [2]

1 Department of Agriculture Crop Production and Agricultural Environment, Laboratory of Soil Science, University of Thessaly, N. Ionia Volos, 38 446 Magnesia, Greece; spapadimou@uth.gr

2 Department of Agriculture Crop Production and Agricultural Environment, Analytical Chemistry and Pesticides Laboratory, University of Thessaly, N. Ionia Volos, 38 446 Magnesia, Greece; ntsirop@uth.gr

3 Department of Agriculture Crop Production and Agricultural Environment, Laboratory of Agricultural Engineering, University of Thessaly, N. Ionia Volos, 38 446 Magnesia, Greece; chkaval@uth.gr

* Correspondence: egol@uth.gr; Tel.: +30-24210-93290

Abstract: A three-year study, designed to record the level of potentially toxic elements within the urban complex in the city of Volos, Greece, was carried out between 2018 and 2020. For the needs of the aforementioned study, 62 surface (0–15 cm) soil samples were collected each year (i.e., 186 samples in total) from an urban area of 3.65 km^2, and the average value of pseudo-total metal concentration was measured. Soil pollution indices, such as the contamination factor (CF) and the geo-accumulation index (Igeo), were estimated regarding each of the metals of interest. The respective thematic maps were constructed, and the spatial variability of the contamination degree was displayed. Higher values of the CF and Igeo were obtained near the heavy traffic roads and beside the railway station, the bus stations, and the commercial port. The maps based on the pollution indices, along with the database that was constructed using the appropriate mathematical tools of geostatistical analysis, may be a useful tool for monitoring, prediction, and continuous verification of contamination in the urban soils of Volos city.

Keywords: contamination indices; geo-accumulation factor; heavy metals; GIS

Citation: Golia, E.E.; Papadimou, S.G.; Cavalaris, C.; Tsiropoulos, N.G. Level of Contamination Assessment of Potentially Toxic Elements in the Urban Soils of Volos City (Central Greece). *Sustainability* **2021**, *13*, 2029. https://doi.org/10.3390/su13042029

Academic Editor: George D. Bathrellos

Received: 18 January 2021
Accepted: 10 February 2021
Published: 13 February 2021

1. Introduction

Important functions vital for agricultural, environmental, land preservation, perspective architecture, and urban activities usually take place in the soil [1,2]. Urban soils are inside or nearby cities, i.e., in urban and suburban areas that are significantly and persistently disrupted by human activity [3]. They include: (a) soils consisting of a mixture of materials of different origins, with inorganic or organic compositions, and from agricultural or forest areas, which are largely transformed by human intervention [4]; (b) soils in parks and green gardens that offer different compositions, uses, and management of agricultural land [5,6]; and (c) soils or mixtures resulting from various construction or metallurgical activities within cities [7].

Urban soils are created when human intervention takes place for many years, or after an immediate and abrupt concentration of human activities in a specific area and in a specific environmental background [8–10]. Compared to natural and agricultural soils, urban soil inherits certain characteristics of the parent rock, but it is also influenced by the local microenvironment as well as the process of its formation. Changes in the physical and chemical properties and in the biological cycles of nutrients are often observed.

Urban areas are often densely populated, so there is an over-consumption of natural resources and high production of pollutants. In recent decades, there has been an increase in urbanization and industrialization, as short-term and intensive human activities lead to large numbers of organic pollutants (such as polyhalogenated compounds), along with inorganic pollutants (such as potentially toxic metals) [11,12].

It is well known that Potentially Toxic Elements (PTEs) accumulation in the soil leads to soil degradation, which can be a serious problem concerning human health, as metals can easily accumulate in the human body through food or ingestion of water as well as through respiration pathways [13]. Ecological risk assessments and their potential impact on human health are of great importance because they can point to unbearable dangers related to man and the environment [14]. In addition, they also require the identification of the specific driving factors that contribute to these risks [15]. The accumulation of metals in the upper soil can threaten health through consumption, digestion, and skin contact [3], and soil toxicity by heavy metals is highly correlated with metal levels and nature. In many cases, damage to human health is inevitable: headaches, insomnia, insanity, joint pain, and cancer. They can have a stronger effect on human health, plants, and soil organisms as well as in water filtration [16]. Problems due to urban soil systems are distinguished by other strongly affected soils, such as those found in quarries, mines, and airports far from cities [17]. There are different routes of exposure, such as direct ingestion through the food chain, inhalation through the mouth and nose, and direct skin contact. Therefore, it is important to assess the contamination and risk to human health of soil metals to improve the environment and protect human health [18,19].

Identifying potential sources of metals in soils is important for controlling priority pollutants. Recently, various mathematical models and methods of statistical analysis have been applied, such as the various indices usually called environmental factors (EFs) such as, the contamination factor (CF), the geo-accumulation index (Igeo), the availability factor (AF), principal component analysis (PCA), and factor analysis (FA) [20,21]. These tools are highly used to distinguish the natural and human origins of metals in soil systems [17] On the other hand, GIS in environmental pollution is a valuable tool for mapping out the contaminants in soil. Spatial display and pollutant distribution allow for a more efficient perspective to monitor soil contamination [22,23]. Spatial data management has a large range of applications along with great advantages [24,25] Methodologies incorporating GIS can be applied in soil contamination projects to facilitate the interpretation, estimation, and evaluation of information between stakeholders, which will improve stakeholder communication in the decision making process [26].The potential burden on the urban environment due to the presence of heavy metals in urban soils, along with the particles suspended in the air and deposited on the soil surface, has gained intense research interest in recent years [27,28]. However, limited studies are focused on risk assessment in urban and suburban areas, on densely populated areas, or on the industrial cities of Greece [29–31].

The objectives of the present study were (1) to evaluate heavy metal contamination levels in the urban soils of Volos, (2) to identify which PTEs are of most concern using soil contamination indices, and (3) to highlight the areas (sites) that are most at risk of pollution by constructing proper thematic maps to illustrate the spatial distribution in the study area. The novelty of the present research lies in the fact that, for the first time, a database has been created that enables the continuous monitoring of contamination assessment after recording the changes of the contamination index values in the area studied.

2. Materials and Methods

2.1. Site Description

Volos is a seaside port of Thessaly located in the middle of Greece. It was built inside the Pagasiticos Gulf and spreads along the plain beneath Pelion. It has a population of about 150,000 inhabitants. The area has a typical Mediterranean climate with wet and cold winters, dry and hot summers, a 14.4 °C average yearly temperature, and a mean yearly precipitation of 450 mm. At the west side, about 12–14 km from the city center, resides an industrial area that also includes a steel plant. Another large cement industry, producing seven types of cement, clinker, solid fuels, and aggregates, is located a distance of 4 km to the east part of the city (Figure 1). The present study is focused on the downtown city center, covering an area of 3.65 km^2.

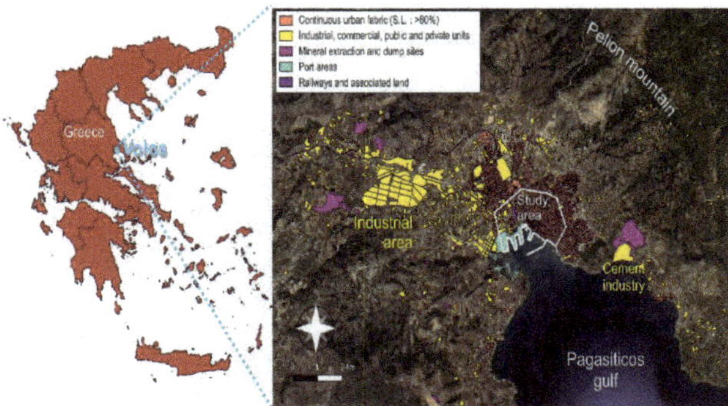

Figure 1. The city of Volos in Thessaly, Central Greece and the study area.

2.2. Soil Sampling

Sixty-two soil samples per year, consisting of three sub-soils each [32], were collected from diverse green spaces, including parks, playgrounds, squares close to the harbor, flower beds located close to main streets, and a neighborhood adjacent to a cement factory. The investigation began in June 2018. In the following two years (2019–2020), soil samples were collected from the same sampling points (i.e., 186 soil samples in total) of Volos.

2.3. Chemical Analysis of Soil Samples

The physicochemical analyses of the soil samples were conducted using the methods described by Page et al. [33]. The analyses were preceded by air-drying the soil samples for three days and sieving them using a 2 mm sieve. Clay, sand, and silt percentage, as well as cation exchange capacity (CEC), pH values, electrical conductivity, organic matter content (Table 1), and pseudo-total content (Table 2), were measured using the aqua regia method [34,35]. Metals were determined using an atomic absorption spectrophotometer (flame and/or graphite furnace) with the following detection limits: 0.1, 0.09, 0.09, 0.1, 0.08, 0.2, 0.1, 0.15 (mg L^{-1}) for Cu, Zn, Cd, Mn, Pb, Ni, Cr, and Co, respectively. A series of 11 standard solutions (Perkin Elmer®) for each metal were used after conducting the proper solutions with 5% c. HNO3 [35]. For the verification of the accuracy of the analyses, a certified reference material (CRM) (No 141R, calcareous loam soil) from the Community Bureau of Reference (BCR) was analyzed with the soil samples. The results of the metal determination in the CRM are presented in the Supplementary Materials Section. The European Council Directive 86/278/EEC [36] regulated the permissible limits for Cd, Cu, Co, Ni, Pb, and Zn. The Cr and Mn maximum values were reported in Kabata–Pendias [37] as the maximum allowable concentrations.

Table 1. Physicochemical parameters of the soil samples (mean values of the three-year study, n = 186).

Descriptive Statistics	pH (1:1)	EC (μS cm^{-1})	OM (%)	Clay (%)	Sand (%)	$CaCO_3$ (%)
Minimum Value	6.57	1123.00	0.30	2	22	9.61
Maximum Value	8.92	6957.46	4.40	52	78	20.54
Mean Value	7.44	3235.60	2.44	19	57	14.57
Relative Standard Deviation	0.41	12.76	0.91	7.50	6.99	1.54
Kurtosis Coefficient	1.415	−0.495	−0.676	−0.006	−0.071	−0.680
Skewness Coefficient	0.710	0.747	−0.128	1.0096	−0.932	0.231

Table 2. Pseudo-total concentrations of potentially toxic elements (mean values of the three years of the study, n = 186).

.	Cu	Zn	Pb	Ni	Cd	Co	Cr	Mn
.				mg kg^{-1}				
Minimum Value	28.90	88.42	5.43	24.98	0.57	4.62	26.07	254.33
10th-perc [a]	33.91	106.17	13.05	32.96	0.71	10.81	47.72	308.64
50th-perc [b]	51.48	126.40	41.31	63.41	0.96	22.38	110.86	714.13
Average	53.91	131.02	35.94	67.28	0.94	22.56	93.24	663.71
90th-perc [c]	80.26	166.07	55.12	114.86	1.14	35.40	122.33	942.59
Maximum Value	89.27	218.11	58.14	117.89	1.25	38.14	123.25	951.18
EU Limits [d]	140	300	300	75	3	-	200	-
BG (mg kg^{-1}) [e]	24.57	64.35	29.69	22.92	0.49	9.62	23.84	540.13

[a] 10th percentile, [b] 50th percentile, [c] 90th percentile; [d] For Cr: maximum allowable concentrations ([37], p. 24). For Cd, Cu, Ni, Pb, and Zn: 86/278/EEC Directive [36]; [e] Average of all reported background reference values in Kabata–Pendias [37].

2.4. Contamination Risk Assessment Indices of PTEs

The following PTE contamination indices were estimated. [17,38].

The contamination factor (CF) is as follows:

$$CF = C_{AR}/C_{AR\,ref}, \tag{1}$$

where C_{AR} is aqua regia extracted metal (mean values of the 3 years of the study) (mg kg^{-1} soil). $C_{AR\,ref}$ is the background reference element concentration in uncontaminated areas (mg kg^{-1}, PTE concentrations in pristine soils, "world soil average") [37].

The CF follows the classification: class I: CF < 1 (pristine soil), class II: CF = 1–3 (contamination is rated "moderate"), class III: CF = 3–6 ("considerable"), class IV: CF > 6 ("very high").

The geo-accumulation Index (I_{geo}) is as follows:

$$I_{geo} = \log_2(C_{AR}\ /1.5\ C_{AR\,ref}) \tag{2}$$

Igeo [39] is designated using Latin numbering [40]: class I: Igeo < 0, class II: Igeo = 0–1, class III: Igeo = 1–2, class IV: Igeo = 2–3, class V: Igeo = 3–4, class VI: Igeo = 4–5, class VII: Igeo > 5.

2.5. Statistical and Geostatistical Procedures

The 3-year data taken from the 62 sample sites were subjected to a one-way ANOVA at a $p < 0.05$ level using the SPSS-26 package, in order to identify potential timescale changes [23,41]. The results showed no significant differences for the mean three-year values, although an upward trend from year to year was detected (ANOVA results are listed in the Supplementary Materials Section). The mean values were used for further analysis and the construction of thematic maps. Initially, the mean values were checked for normality and homogeneity by means of the Kolmogorov–Smirnov and Shapiro–Wilks tests [42]. The results showed an abnormal distribution. Based on that distribution, a spatial interpolation process using the ordinary kriging along with data logarithmic transformation and a grid of 5 m spatial resolution, was performed in the SAGA 7.9.0 GIS (SAGA User Group Association, 2020) [43]. Prior to kriging, the datasets were also checked for anisotropy in two perpendicular directions. Showing a similar spatial autocorrelation pattern, an omnidirectional variogram was computed for every variable, and a suitable model was fitted to that. The geospatial statistics are presented in Tables 3 and 4. The ordinary kriging process resulted in raster images of pollution for each element. These rasters, along with the sampling point vector information and local points of interests (POIs), were integrated on the thematic maps using the QGIS 3.16.1 (QGIS Development Team, 2020) [44] and utilizing the web map service OpenStreetMap as a base layer (Figure 3).

Table 3. Descriptive statistics for the contamination factor (CF) of the potentially toxic elements and geostatistical parameters, for the creation and validation of the contour maps.

CF	Cu	Zn	Pb	Ni	Cd	Co	Cr	Mn
			Descriptive statistics					
Mean Value	1.38	1.85	1.30	2.32	2.27	3.99	2.21	1.34
Standard Deviation	0.46	0.40	0.62	0.99	0.43	1.72	1.21	0.50
Minimum Value	0.74	1.15	0.09	0.86	1.20	0.15	0.19	0.48
Maximum Value	2.30	3.12	2.15	4.07	3.06	6.75	3.91	1.95
Skewness Coefficient	0.401	1.018	−0.381	0.368	−0.405	−0.234	0.120	−0.360
Kurtosis Coefficient	−1.016	1.249	−1.177	−0.935	−0.650	−0.796	−1.583	−1.373
			Geostatistical parameters					
Model	Cubic	Exp.	Exp.	Exp.	Spher.	Spher.	Spher.	Spher.
Nugget	0.017	0.004	0.109	0.143	0.064	0.016	0.027	0.026
Range	985	985	973	1209	981	984	996	985
Sill	0.231	0.176	0.426	1.191	0.198	0.833	0.488	0.207
Slope	-	-	-	-	-	-	-	-
Nugget sill ratio	0.08	0.02	0.26	0.12	0.32	0.02	0.05	0.13
R^2	0.53	0.68	0.40	0.47	0.60	0.64	0.46	0.30

Table 4. Descriptive statistics for the Igeo of the potentially toxic elements and geostatistical parameters, for the creation and validation of the contour maps.

Igeo	Cu	Zn	Pb	Ni	Cd	Co	Cr	Mn
			Descriptive statistics					
Mean Value	−0.20	0.27	−0.47	0.49	0.57	1.20	6.18	−0.28
Standard Deviation	0.48	0.29	1.02	0.66	0.29	0.96	0.98	0.64
Minimum Value	−1.01	−0.38	−4.14	−0.80	−0.33	−3.31	2.91	−1.65
Maximum Value	0.61	1.05	0.52	1.44	1.03	2.17	7.28	0.38
Skewness Coefficient	0.007	0.400	−1.379	−0.285	−0.819	−2.218	−0.777	−0.712
Kurtosis Coefficient	−1.262	0.225	1.644	−0.893	0.253	7.210	0.416	−0.942
			Geostatistical parameters					
Model	Exp.	Exp.	Cubic	Cubic	Exp.	Spher.	Cubic	Exp.
Nugget	0.055	0.004	0.031	0.006	0.001	0.031	0.006	0.082
Range	985	984	979	971	983	995	846	983
Sill	0.272	0.093	1.140	0.490	0.095	0.655	0.386	0.443
Slope	-	-	-	-	-	-	-	-
Nugget sill ratio	0.20	0.04	0.03	0.01	0.01	0.05	0.01	0.19
R^2	0.49	0.66	0.50	0.52	0.60	0.65	0.46	0.34

3. Results

3.1. Physicochemical Properties of Soil Samples

The soils ranged from sandy loam to clayey, with soil reaction values between 6.6 and 8.9 (Table 1). High clay content was found in 19% of the soil samples, while 57% had a sandy texture. According to the soil pH values, 82.3% of the soil samples were alkaline, while 17.7% had a slightly acid soil reaction (<7). Organic matter (OM) values ranged between 0.3% and 4.4%. Soil electrical conductivity (EC) varied between 1122 and 6958 μS cm^{-1}, with the mean value at 3204 μS cm^{-1}.

3.2. Levels of Potentially Toxic Elements

The mean values of metal concentrations during the three years' research along with their statistical characteristics are presented in Table 2. The mean values were found to be lower than maximum permitted values. The mean Cu concentration was 53.9, while the CEC directive value [36] was 140 mg kg^{-1}. The other values were as follows: for Zn, the mean was 131.0, while the CEC limit was 300; for Pb, the mean was 35.9, while the CEC limit was 300; for Cd, the mean was 0.94, while the CEC limit was 3; for Cr, the mean was

93.2, while the Kabata–Pendias [37] value was 200; and for Ni, the average was 67.3, while the CEC limit was 75mg kg^{-1}). Average Ni concentration was lower than the EU limit, while the Co and Mn mean concentrations were higher than the background levels.

3.3. Geostatistical analysis, Construction of Thematic Maps of Availability Indices of PTEs

The thematic maps, presented in Figures 2 and 3, were created for the optimal depiction of spatial variability of the metal pollution assessment. The contamination factor (CF) and geo-accumulation index (Igeo) were used as useful tools to assess the pollution of the area. Descriptive statistics for the contamination factor (CF) and geo-accumulation index (Igeo) calculations along with the geostatistical parameters, for the creation and validation of the contour maps, are presented in Tables 3 and 4, respectively.

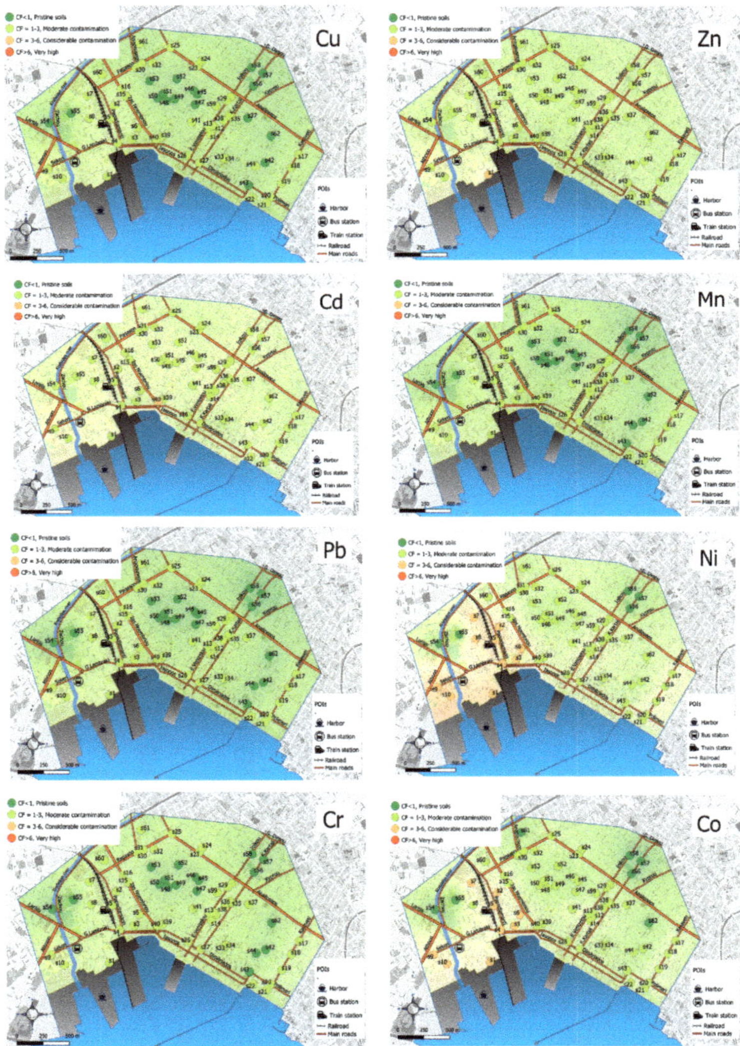

Figure 2. Thematic maps of potentially toxic metals based on contamination factor (CF) values in Volos (mean values of the three years of the study).

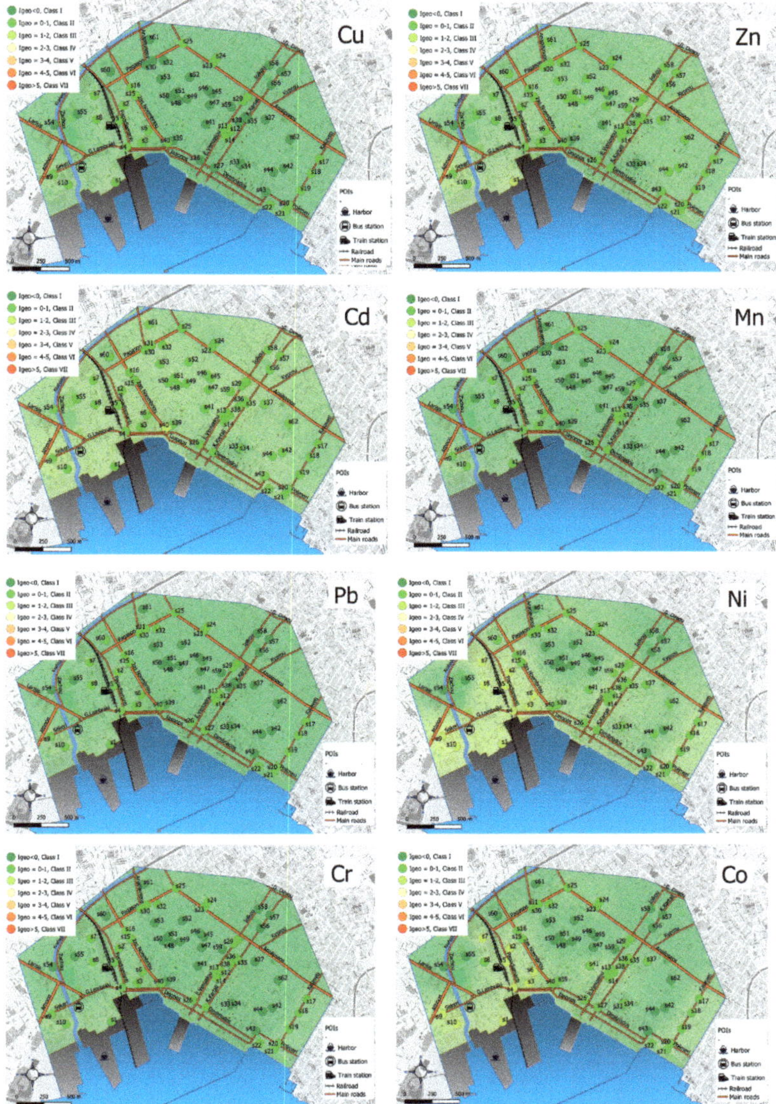

Figure 3. Thematic maps of potentially toxic metals based on the geo-accumulation index (Igeo) values in Volos city (mean values of the three-year study).

The largest number of soil samples had CF values belonging to class II (1–3) and were characterized as "moderate contamination". In class II, the order of CFs was as follows: Cd = Zn > Ni > Co > Cu = Pb = Cr = Mn. In class III (CF: 3–6), characterized as "considerable contamination", the order changed as follows: Ni > Co > Zn = Cd. There were no soil samples with CF > 6, i.e., belonging to class IV with "very high contamination".

The Igeo values were also calculated. In class II (0–1), the order of the metal elements was as follows: Cd > Zn > Cr = Ni > Pb > Co > Mn > Cu. In class III (1–2), Ni was the most enriched element in the soil, followed by Co. Zn and Cd exhibited lower Igeo values in the decreasing sequence: Ni > Co > Zn = Cd. None of the metals studied had Igeo values > 2.

4. Discussion

4.1. Physicochemical Properties of Soil Samples

The soil samples were neutral to slightly alkaline. The presence of alkaline soils in the urban environment is rather common [7]. High clay content has usually been found in urban and peri-urban locations [28]. On the other hand, 19% of the samples had a sandy texture, which is probably due to the proximity of the samples to the coastal zone [45] or to the area near the river, which flows into the Pagasiticos Gulf. The organic matter content was rather low, but usually observed in Mediterranean soils with the characteristic climate, relatively high temperatures, and low humidity [46]. Anthropogenic activities create conditions of impaired salinity in soils, and especially in the capitals of the prefectures. A survey conducted in a Mediterranean site also revealed that soil salinity is controlled mainly by seawater intrusion and the usual application of salt-rich water [12]. The lower part of the study area was adjacent to the edges of the Pagasiticos Gulf, which naturally contributes to the increase in salinity [47].

4.2. Levels of Potentially Toxic Elements and Contamination Indices

The average CF values of the metals were >1, which indicates the strong anthropogenic influence in the area studied. The PTE contents were lower than those in soil samples in Athens, as described by Massas et al. [18] and Kelepertzis and Argyraki [5]. The soil samples seemed to be less contaminated than soils in areas with high anthropogenic activities [26,30,31,48]. The previous studies of Antoniadis et al. [48] and Kelepertzis et al. [30] made extensive reference to the impact of industrial activity on the peri-urban environment of Volos. The soil contamination indices (CF and Igeo) indicated low or moderate contamination. The elevated Ni Igeo may have been influenced by activities leading to enhanced Ni deposition [49]. Moreover, high Ni levels were previously reported in the nearby site, probably due to the geochemical composition of the parental material [48].

4.3. Construction of Thematic Maps and Geostatistical Analysis

The constructed maps based on the contamination indices [50] seemed to be a useful tool for the investigation of spatial distribution of metals in the city of Volos. To avoid misleading interpretations and alerts from visual observations, a uniform full-scale of the CF and Igeo values was used for the present maps' color classification. Despite reducing the spatial discrimination, this choice resulted in the maps depicting the potential toxicity of the studied metals [23]. In the present study, the cubic, exponential, and spherical models were used for the construction of the thematic maps in Figures 2 and 3, of the CF and Igeo, respectively. The cubic model seemed to be the proper one for the construction of maps based on the contamination factor of Cu along with the Pb, Ni, and Cr maps based on the Igeo indices [24]. The exponential and spherical models were preferred for the spatial distribution of Zn and Co in the CF and Igeo indices, respectively. Similar techniques and mathematical models have been used by other researchers [51,52] as they provide the best predictions. The low nugget/sill ratio (<0.25) obtained for most of the parameters indicated the spatial dependence of the variables [53].

4.3.1. Maps of Contamination Factors

The Cu, Pb, Mn, and Cr contamination factor maps (Figure 2) revealed that the highest concentrations of these metals occurred near the heaviest traffic roads (Analipseos, K., Kartali, Demetriados), as well as in the areas adjacent to the railway station, the bus stations, and the commercial port, as mentioned in many studies [13,30,54]. The long-time and increased vehicle traffic using lead-based fuels, before the banning of lead-containing gasoline, may have led to increased PTE contents in urban soils [54,55]. Nevertheless, the CF values were high, and the ranking led to the rating of "moderate contamination" conditions. Cd and Zn revealed "moderate contamination" in all sampling points, with the exception of S40. The higher Cd concentrations were probably because of road construction works carried out during the sampling period (second year, 2019). The high CF value of Zn in S1 near the commercial port led to "considerable contamination". Prolonged use and storage of metal objects and scrubs may have led to high Zn levels locally [29]. Ni and Co concentrations led to "considerable contamination" around the railway, the bus stations, and the port [56]. However, in the main area of Volos, the CF of Ni and Co indicated a "moderate contamination".

4.3.2. Maps of Geo-Accumulation Index

The maps (Figure 3), based on the Igeo distribution, revealed moderate contamination of Cu and Mn at the main roads and railway, and beside the bus stations and the city port (Igeo, 0–1), and the values suggested lower contamination level compared to other researches [28,29,37]. The Igeo values of Cd and Zn also belonged to class II (0–1), but in the S1 and S40 soil samples, an exaggeration was observed as higher values were recorded. High levels of metals beside ports are usually observed [56]. It is well known that metal concentration in aquatic ecosystems, such as ports, is measured in water and in sediment. Usually, metals exist in the lowest content in water and in higher concentrations in sea sediments [6,12]. The permanent presence of metals near ports is a problem of utmost concern as their constant contact with salts and the particularly high percentage of moisture creates oxidizing conditions. The metals therefore corrode and their concentrations in the soils accumulate [45]. In 16.1% of the soil samples, the Igeo values of nickel belonged to class III (1–2). Nickel has a large number of industrial uses and applications [57]. The occurrence of unexpectedly higher concentrations of nickel at the railway station and on the west side of the study area may be related to its geochemical origin, as recorded in a previous study [48]. The Igeo values of class II (0–1) were found for Pb, Co, and Cr near the main city roads, possibly affected by heavy traffic [57–60].

The factories located around the city of Volos may contribute to the pollution of the area, especially when the weather conditions are favorable for the spread of pollutants [30,48]. It is also worth mentioning that Volos is a rural area surrounded by the Pelion mountain to the north and the Pagasiticos Gulf to the south, creating a microenvironment favorable to the frequent entrapment of atmospheric pollution [29,30]. The burning of materials in the fireplaces of the houses to the right of the study area may also contribute to the increase in PTE pollution [61]. There is also evidence that the coal or crude oil used for household heating, gasoline, and diesel vehicle exhaust is of the major problems in urban soil pollution [29]. Concluding, the contamination seems to be significantly increasing near the commercial port of the city of Volos, close to the railway station, and in the intercity bus station along with the big and heaviest traffic roads of Volos [62]. The present study could give rise to further and more thorough research in the future. The identification of contamination sources at the local and national level could control or even reduce the levels of PTEs in the area of interest.

5. Conclusions

The level of PTEs in the soil samples of Volos were found to be lower than the maximum permitted values set by the European Union. Soil contamination indices (CF and Igeo) also confirmed the low to moderate level of contamination with PTEs. The formation of a database with three-year records of the PTE level is an innovation, as it has not been described before. Map formation, based on the values of the contamination indices, is a valuable, useful, and easy-to-handle tool for recording, monitoring, managing, and predicting contamination levels in the area of interest. Nevertheless, continuous long-term monitoring for the assessment of PTE levels is of paramount importance, as high PTE levels could ultimately have negative effects on the environment, food, and consequently, on the health of the citizens of Volos and the adjacent communities. Spatial environmental monitoring could be very useful for the assessment of environmental conditions and trends in the cities of Greece, and even worldwide, as it leads to an evidence-based approach for reporting to national policy makers. After all, maintaining good urban soil environmental quality is crucial for several socio-economic reasons along with human health and well-being.

Supplementary Materials: The following are available online at https://www.mdpi.com/2071-1050/13/4/2029/s1, (Table S1(SM) Pseudo-total concentrations of heavy metals of the reference material BCR CRM 141 R, Table S2(SM) Results of the analysis of variance (ANOVA) for the three-year data.

Author Contributions: Conceptualization, E.E.G. and N.G.T.; methodology, E.E.G. and C.C.; software, C.C.; validation, E.E.G. and N.G.T.; formal analfiguysis, E.E.G., S.G.P. and C.C.; investigation, E.E.G. and S.G.P.; resources, E.E.G., C.C. and N.G.T.; data curation, E.E.G., C.C. and S.G.P.; writing—original draft preparation, E.E.G., S.G.P. and C.C.; writing—review and editing, E.E.G., C.C. and N.G.T.; visualization, C.C. and S.G.P.; supervision, E.E.G. and N.G.T.; All authors have read and agreed to the published version of the manuscript.

Funding: This research received no external funding.

Institutional Review Board Statement: Not applicable.

Informed Consent Statement: Not applicable.

Data Availability Statement: Data that support the findings of this study are available from the corresponding author upon reasonable request.

Acknowledgments: The authors would like to express their gratitude to the Postgraduate Program of the University of Thessaly, entitled: Environmental Changes Management and Circular Economy. Part of the results come from the post-graduate dissertation of student Sotiria G. Papadimou.

Conflicts of Interest: The authors declare no conflict of interest.

References

1. Pan, L.; Wang, Y.; Ma, J.; Hu, Y.; Su, B.; Fang, G.; Wang, L.; Xiang, B. A review of heavy metal pollution levels and health risk assessment of urban soils in Chinese cities. *Environ. Sci. Pollut. Res.* **2018**, *25*, 1055–1069. [CrossRef] [PubMed]
2. Adimalla, N.; Chen, J.; Qian, H. Spatial characteristics of heavy metal contamination and potential human health risk assessment of urban soils: A case study from an urban region of South India. *Ecotoxicol. Environ. Saf.* **2020**, *194*. [CrossRef] [PubMed]
3. Mehmood, K.; Ahmad, H.R.; Abbas, R.; Saifullah, R.; Murtaza, G. Heavy metals in urban and peri-urban soils of a heavily-populated and industrialized city: Assessment of ecological risks and human health repercussions. *Hum. Ecol. Risk Assess.* **2020**, *26*, 1705–1722. [CrossRef]
4. Ok, Y.S.; Lee, S.S.; Jeon, W.T.; Oh, S.E.; Usman, A.R.A.; Moon, D.H. Application of eggshell waste for the immobilization of cadmium and lead in a contaminated soil. *Environ. Geochem. Health* **2011**, *33*, 31–39. [CrossRef]
5. Kelepertzis, E.; Argyraki, A. Geochemical associations for evaluating the availability of potentially harmful elements in urban soils: Lessons learnt from Athens, Greece. *Appl. Geochem.* **2015**, *59*, 63–73. [CrossRef]
6. Papadopoulou-Vrynioti, K.; Alexakis, D.; Bathrellos, G.D.; Skilodimou, H.D.; Vryniotis, D.; Vassiliades, E.; Gamvroula, D. Distribution of trace elements in stream sediments of Arta plain (western Hellas): The influence of geomorphological parameters. *J. Geochem. Explor.* **2013**, *134*, 17–26. [CrossRef]
7. Hanfi, M.Y.; Mostafa, M.Y.A.; Zhukovsky, M. V Heavy metal contamination in urban surface sediments: Sources, distribution, contamination control, and remediation. *Environ. Monit. Assess.* **2020**, *192*, 1–21. [CrossRef]
8. Adimalla, N. Heavy metals pollution assessment and its associated human health risk evaluation of urban soils from Indian cities: A review. *Environ. Geochem. Health* **2020**, *42*, 173–190. [CrossRef] [PubMed]
9. Bathrellos, G.D.; Skilodimou, H.D.; Kelepertsis, A.; Alexakis, D.; Chrisanthaki, I.; Archonti, D. Environmental research of groundwater in the urban and suburban areas of Attica region, Greece. *Environ. Geol.* **2008**, *56*, 11–18. [CrossRef]
10. Gamvroula, D.; Alexakis, D.; Stamatis, G. Diagnosis of groundwater quality and assessment of contamination sources in the Megara basin (Attica, Greece). *Arab. J. Geosci.* **2013**, *6*, 2367–2381. [CrossRef]
11. Papadopoulou-Vrynioti, K.; Alexakis, D.; Bathrellos, G.D.; Skilodimou, H.D.; Vryniotis, D.; Vassiliades, E. Environmental research and evaluation of agricultural soil of the Arta plain, western Hellas. *J. Geochemical Explor.* **2014**, *136*, 84–92. [CrossRef]
12. Alexakis, D.; Gotsis, D.; Giakoumakis, S. Evaluation of soil salinization in a Mediterranean site (Agoulinitsa district—West Greece). *Arab. J. Geosci.* **2015**, *8*, 1373–1383. [CrossRef]
13. Han, Q.; Wang, M.; Cao, J.; Gui, C.; Liu, Y.; He, X.; He, Y.; Liu, Y. Health risk assessment and bioaccessibilities of heavy metals for children in soil and dust from urban parks and schools of Jiaozuo, China. *Ecotoxicol. Environ. Saf.* **2020**, *191*. [CrossRef] [PubMed]
14. Wei, B.; Yang, L. A review of heavy metal contaminations in urban soils, urban road dusts and agricultural soils from China. *Microchem. J.* **2010**, *94*, 99–107. [CrossRef]
15. Trujillo-González, J.M.; Torres-Mora, M.A.; Jiménez-Ballesta, R.; Zhang, J. Land-use-dependent spatial variation and exposure risk of heavy metals in road-deposited sediment in Villavicencio, Colombia. *Environ. Geochem. Health* **2019**, *41*, 667–679. [CrossRef]

16. Gu, Y.G.; Gao, Y.P.; Lin, Q. Contamination, bioaccessibility and human health risk of heavy metals in exposed-lawn soils from 28 urban parks in southern China's largest city, Guangzhou. *Appl. Geochem.* **2016**, *67*, 52–58. [CrossRef]
17. Kowalska, J.B.; Mazurek, R.; Gąsiorek Michałand Zaleski, T. Pollution indices as useful tools for the comprehensive evaluation of the degree of soil contamination–A review. *Environ. Geochem. Health* **2018**, *40*, 2395–2420. [CrossRef]
18. Massas, I.; Ehaliotis, C.; Kalivas, D.; Panagopoulou, G. Concentrations and availability indicators of soil heavy metals; The case of children's playgrounds in the city of Athens (Greece). *Water. Air. Soil Pollut.* **2010**, *212*, 51–63. [CrossRef]
19. Wijesiri, B.; Egodawatta, P.; McGree, J.; Goonetilleke, A. Process variability of pollutant build-up on urban road surfaces. *Sci. Total Environ.* **2015**, *518–519*, 434–440. [CrossRef]
20. Seleznev, A.A.; Yarmoshenko, I.V.; Malinovsky, G.P. Assessment of total amount of surface sediment in urban environment using data on solid matter content in snow-dirt sludge. *Environ. Process.* **2019**, *6*, 581–595. [CrossRef]
21. Alexakis, D.; Gamvroula, D.; Theofili, E. Environmental availability of potentially toxic elements in an agricultural Mediterranean site. *Environ. Eng. Geosci.* **2019**, *25*, 169–178. [CrossRef]
22. Mehmood, K.; Chang, S.; Yu, S.; Wang, L.; Li, P.; Li, Z.; Liu, W.; Rosenfeld, D.; Seinfeld, J.H. Spatial and temporal distributions of air pollutant emissions from open crop straw and biomass burnings in China from 2002 to 2016. *Environ. Chem. Lett.* **2018**, *16*, 301–309. [CrossRef]
23. Tong, S.; Li, H.; Li, W.; Tudi, M.; Yang, L. Concentration, spatial distribution, contamination degree and human health risk assessment of heavy metals in urban soils across china between 2003 and 2019—A systematic review. *Int. J. Environ. Res. Public Health* **2020**, *17*, 3099. [CrossRef]
24. Guo, G.; Wu, F.; Xie, F.; Zhang, R. Spatial distribution and pollution assessment of heavy metals in urban soils from southwest China. *J. Environ. Sci.* **2012**, *24*, 410–418. [CrossRef]
25. Zhang, X.; Wei, S.; Sun, Q.; Wadood, S.A.; Guo, B. Source identification and spatial distribution of arsenic and heavy metals in agricultural soil around Hunan industrial estate by positive matrix factorization model, principle components analysis and geo statistical analysis. *Ecotoxicol. Environ. Saf.* **2018**, *159*, 354–362. [CrossRef] [PubMed]
26. Saha, N.; Rahman, M.S.; Jolly, Y.N.; Rahman, A.; Sattar, M.A.; Hai, M.A. Spatial distribution and contamination assessment of six heavy metals in soils and their transfer into mature tobacco plants in Kushtia District, Bangladesh. *Environ. Sci. Pollut. Res.* **2016**, *23*, 3414–3426. [CrossRef]
27. Khoder, M.; Al Ghamdi, M.; Shiboob, M. Heavy Metal Distribution in Street Dust of Urban and Industrial Areas in Jeddah, Saudi Arabia. *J. King Abdulaziz Univ. Environ. Arid L. Agric. Sci.* **2012**, *23*, 55–75. [CrossRef]
28. Logiewa, A.; Miazgowicz, A.; Krennhuber, K.; Lanzerstorfer, C. Variation in the concentration of metals in road dust size fractions between 2 μm and 2 mm: Results from three metallurgical centres in Poland. *Arch. Environ. Contam. Toxicol.* **2020**, *78*, 46–59. [CrossRef] [PubMed]
29. Botsou, F.; Moutafis, I.; Dalaina, S.; Kelepertzis, E. Settled bus dust as a proxy of traffic-related emissions and health implications of exposures to potentially harmful elements. *Atmos. Pollut. Res.* **2020**, *11*, 1776–1784. [CrossRef]
30. Kelepertzis, E.; Argyraki, A.; Chrastný, V.; Botsou, F.; Skordas, K.; Komárek, M.; Fouskas, A. Metal(loid) and isotopic tracing of Pb in soils, road and house dusts from the industrial area of Volos (central Greece). *Sci. Total Environ.* **2020**, *725*. [CrossRef]
31. Christoforidis, A.; Stamatis, N. Heavy metal contamination in street dust and roadside soil along the major national road in Kavala's region, Greece. *Geoderma* **2009**, *151*, 257–263. [CrossRef]
32. McLaughlin, M.J.; Zarcinas, B.A.; Stevens, D.P.; Cook, N. Soil testing for heavy metals. *Commun. Soil Sci. Plant Anal.* **2000**, *31*, 1661–1700. [CrossRef]
33. Page, A.L. Methods of soil analysis-Part 2: Chemical and microbiological properties. In *American Sociaty of Agronomy*, 2nd ed.; Phosphurus Inc.: Madison, WI, USA, 1982; Volume 9, pp. 421–422.
34. ISO. *Environment Soil Quality*; ISO/DIS 11466; ISO Standards Compendium: Switzerland, Geneva, 1994.
35. Golia, E.E.; Tsiropoulos, N.G.; Dimirkou, A.; Mitsios, I. Distribution of heavy metals of agricultural soils of central Greece using the modified BCR sequential extraction method. *Int. J. Environ. Anal. Chem.* **2007**, *87*, 1053–1063. [CrossRef]
36. Council of the European Communities. The protection of the environment, and in particular of the soil, when sewage sludge is used in agriculture; Council Directive of 12 June 1986. *Off. J. Eur. Commun.* **1986**, *181*, 6.
37. Meharg, A.A. *Trace Elements in Soils and Plants*, 4th ed.; Kabata-Pendias, A., Ed.; CRC Press/Taylor & Francis Group: Boca Raton, FL, USA, 2010; ISBN 9781420093704.
38. Antoniadis, V.; Golia, E.E.; Shaheen, S.M.; Rinklebe, J. Bioavailability and health risk assessment of potentially toxic elements in Thriasio Plain, near Athens, Greece. *Environ. Geochem. Health* **2017**, *39*, 319–330. [CrossRef]
39. Mueller, G. Schwermetalle in Den Sedimenten Des Rheins—Veranderungen Seit 1971. *Umsch. Wissensch. Techn.* **1979**, *79*, 778–783.
40. Kasa, E.; Felix-Henningsen, P.; Duering, R.A.; Gjoka, F. The occurrence of heavy metals in irrigated and non-irrigated arable soils, NW Albania. *Environ. Monit. Assess.* **2014**, *186*, 3595–3603. [CrossRef]
41. Ding, Q.; Shi, X.; Zhuang, D.; Wang, Y. Temporal and spatial distributions of ecological vulnerability under the influence of natural and anthropogenic factors in an eco-province under construction in China. *Sustainability* **2018**, *10*, 87. [CrossRef]
42. Johnson, R.A.; Wichern, D.W. *Applied Multivariate Statistical Analysis*, 6th ed.; Prentice-Hall: Hoboken, NJ, USA, 2007.
43. SAGA User Group Association. SAGA 7.9.0—System for Automated Geoscientific Analyses. Available online: http://www.saga-gis.org (accessed on 29 December 2020).

44. QGIS Development Team. QGIS Geographic Information System. Available online: http://qgis.osgeo.org (accessed on 10 December 2020).
45. Shirani, M.; Afzali, K.N.; Jahan, S.; Strezov, V.; Soleimani-Sardo, M. Pollution and contamination assessment of heavy metals in the sediments of Jazmurian playa in southeast Iran. *Sci. Rep.* **2020**, *10*. [CrossRef] [PubMed]
46. Antoniadis, V.; Golia, E.E. Sorption of Cu and Zn in low organic matter-soils as influenced by soil properties and by the degree of soil weathering. *Chemosphere* **2015**, *138*, 364–369. [CrossRef] [PubMed]
47. Wu, Y.; Lu, X.; Zhuang, S.; Han, X.; Zhou, Y. Contamination characteristics and assessment of manganese, Zinc, Chrome, Lead, Copper and Nickel in Bus Station Dusts of Xifeng, Northwest China. *Preprints* **2016**, 1–13. [CrossRef]
48. Antoniadis, V.; Golia, E.E.; Liu, Y.T.; Wang, S.L.; Shaheen, S.M.; Rinklebe, J. Soil and maize contamination by trace elements and associated health risk assessment in the industrial area of Volos, Greece. *Environ. Int.* **2019**, *124*, 79–88. [CrossRef]
49. Ivankovic, N.; Kasanin-Grubin, M.; Brceski, I.; Vukelic, N. Possible sources of heavy metals in urban soils: Example from Belgrade, Serbia. *J. Environ. Prot. Ecol.* **2010**, *11*, 455–464.
50. Briggs, D.J. The use of GIS to evaluate traffic-related pollution. *Occup. Environ. Med.* **2007**, *64*, 1–2. [CrossRef] [PubMed]
51. Isaaks, E.H.; Srivastava, R.M. *An Introduction to Applied Geostatistics*; Oxford University Press: New York, NY, USA, 1989.
52. Mihailović, A.; Budinski-Petković, L.; Popov, S.; Ninkov, J.; Vasin, J.; Ralević, N.M.; Vasić, M.V. Spatial distribution of metals in urban soil of Novi Sad, Serbia: GIS based approach. *J. Geochemical Explor.* **2015**, *150*, 104–114. [CrossRef]
53. Kerry, R.; Oliver, M.A. Determining nugget:sill ratios of standardized variograms from aerial photographs to krige sparse soil data. *Precis. Agric.* **2008**, *9*, 33–56. [CrossRef]
54. Zhang, Y.; Lu, H.; Qu, W. Geographical detection of traffic accidents spatial stratified heterogeneity and influence factors. *Int. J. Environ. Res. Public Health* **2020**, *17*, 572. [CrossRef] [PubMed]
55. Wu, Y.; Lu, X. Physicochemical properties and toxic elements in bus stop dusts from Qingyang, NW China. *Sci. Rep.* **2018**, *8*. [CrossRef] [PubMed]
56. Jahan, S.; Strezov, V. Assessment of trace elements pollution in the sea ports of New South Wales (NSW), Australia using oysters as bioindicators. *Sci. Rep.* **2019**, *9*. [CrossRef]
57. Rizo, O.D.; Hernández, I.C.; López, J.A.; Arado, O.D.; Pino, N.L.; Rodríguez, K.D. Chromium, cobalt and nickel contents in urban soils of Moa, northeastern Cuba. *Bull. Environ. Contam. Toxicol.* **2011**, *86*, 189–193. [CrossRef]
58. Maeaba, W.; Prasad, S.; Chandra, S. First assessment of metals contamination in road dust and roadside soil of Suva City, Fiji. *Arch. Environ. Contam. Toxicol.* **2019**, *77*, 249–262. [CrossRef] [PubMed]
59. Różański, S.; Jaworska, H.; Matuszczak, K.; Nowak, J.; Hardy, A. Impact of highway traffic and the acoustic screen on the content and spatial distribution of heavy metals in soils. *Environ. Sci. Pollut. Res.* **2017**, *24*, 12778–12786. [CrossRef] [PubMed]
60. Szwalec, A.; Mundała, P.; Kędzior, R.; Pawlik, J. Monitoring and assessment of cadmium, lead, zinc and copper concentrations in arable roadside soils in terms of different traffic conditions. *Environ. Monit. Assess.* **2020**, *192*. [CrossRef] [PubMed]
61. Navarro, M.; D'Agostino, A.; Neri, L. The effect of urbanization on subjective well-being: Explaining cross-regional differences. *Socioecon. Plann. Sci.* **2020**, *71*. [CrossRef]
62. Vaiškūnaitė, R.; Jasiūnienė, V. The analysis of heavy metal pollutants emitted by railway transport. *Transport* **2020**, *35*, 213–223. [CrossRef]

Article

Spatial Distribution and Evaluation of Arsenic and Zinc Content in the Soil of a Karst Landscape

Dimitrios E. Alexakis [1], George D. Bathrellos [2,*], Hariklia D. Skilodimou [3] and Dimitra E. Gamvroula [1]

[1] Laboratory of Geoenvironmental Science and Environmental Quality Assurance, Department of Civil Engineering, School of Engineering, University of West Attica, 250 Thivon & P.Ralli Str., 12241 Athens, Greece; d.alexakis@uniwa.gr (D.E.A.); d.gamvroula@uniwa.gr (D.E.G.)

[2] Division of General, Marine Geology and Geodynamics, Department of Geology, University of Patras, 26504 Rio Patras, Greece

[3] Department of Geography and Climatology, Faculty of Geology and Geoenvironment, National and Kapodistrian University of Athens, University Campus, 15784 Zografos Athens, Greece; hskilodimou@geol.uoa.gr

* Correspondence: gbathrellos@upatras.gr

Abstract: Karst features such as polje are highly vulnerable to natural and anthropogenic pollution. The main objectives of this study were to investigate the soil quality in the Ioannina polje (north-west Greece) concerning arsenic (As) and zinc (Zn), and delineate their origin as well as compare the As and Zn content in soil with criteria recorded in the literature. For this purpose, the geomorphological settings, the land use, and the soil physicochemical properties were mapped and evaluated, including soil texture and concentrations of aqua-regia extractable As and Zn. The concentration of elements was spatially correlated with the land use and the geology of the study area, while screening values were applied to assess land suitability. The results reveal that 72% of the total study area has a very gentle slope. This relief favors urban and agricultural activity. Thus, the urban and agricultural land used cover 92% of the total area. The spatial distribution for As and Zn in the soil of the study area is located on very gentle slopes and is strongly correlated with the geological parent materials and human-induced contamination sources. Arsenic and Zn can be considered enriched in the soil of the area studied. The median topsoil contents (in mg kg^{-1}) for As (agricultural soil 16.0; urban soil 17.8) and Zn (agricultural soil 92.0; urban soil 95.0) are higher compared to the corresponding median values of European topsoils. Land evaluation suitability concerning criteria given from the literature is discussed. The proposed work may be helpful in the project of land use planning and the protection of the natural environment.

Keywords: Ioannina polje; soil quality; agricultural soil; aircraft emissions; GIS

Citation: Alexakis, D.E.; Bathrellos, G.D.; Skilodimou, H.D.; Gamvroula, D.E. Spatial Distribution and Evaluation of Arsenic and Zinc Content in the Soil of a Karst Landscape. *Sustainability* **2021**, *13*, 6976. https://doi.org/10.3390/su13126976

Academic Editors: Giorgio Anfuso and Francesco Faccini

Received: 14 May 2021
Accepted: 17 June 2021
Published: 21 June 2021

1. Introduction

Karst terrains occupy about 7–10% of the planet and are mainly developed in carbonate rocks and evaporate [1]. It is estimated that carbonate rocks cover 12% of the Earth's continental area [1]. Karst formations consist of surface and subsurface forms such as poljes and caves. Polje represents an extraordinary element in a karst landscape. A polje is an extensive and closed basin with a flat bottom, karstic drainage, and a steep slope at least on one side [2].

Karst landscapes are areas of abundant resources, such as water supplies, limestone quarries, and minerals. Thus, karst aquifers constitute 25% of the Earth's groundwater resources. Worldwide, the vast majority of surface karstic formations have some form of urban and agricultural activity [1,3]. In Greece, carbonate rocks have a widespread outcrop and occupy about 40% of the surface area [3,4]. Karst regions are fragile landscapes and highly vulnerable to contamination [5–10]. Anthropogenic activities may lead to the degradation of a karst landscape and loss of karst landforms and can cause soil erosion and contamination [11–13]. Since water moves very rapidly in karstic systems and regions, pollutants can be transported for long distances in a short period [14–26].

The leaching of soluble contaminants from the soil in a karst landscape may deteriorate water resources. The chemical composition of water flowing through and out of the rocks and soils of a karst landscape reflects the element's content in the weathering material. Furthermore, karst systems are highly vulnerable compared to other systems because potential contaminants can easily reach groundwater [27,28]. Therefore, the proposal of protection measures to preserve soil quality in a karst region is a crucial issue.

In recent years, environmental studies revealed high contents of potentially toxic elements in soils, sediments, and groundwater [28–40]. Several research studies revealed that the total concentration of metals could not indicate deficiency and toxicity [41–43]. Furthermore, several soil properties, such as pH, Cation Exchange Capacity (CEC), and organic matter, should be studied in combination with trace element contents to better understand the trace elements' behavior [40–45].

Many researchers reported a high concentration of toxic elements such as arsenic (As) and zinc (Zn) in the soils of urban, suburban, industrial, traffic, and mining areas [30–34,40–46]. Therefore, the assessment of As and Zn contents in the soil of a karst area is a necessary step for proposing remedial measures and protecting the natural environment.

In the present study, an attempt was made to identify and evaluate the spatial distribution of As and Zn in the soil of the polje of the Ioannina karst basin (north-west Greece). The specific aims of this study were: (a) to determine the quality status of the soil in the Ioannina basin concerning As and Zn; (b) to investigate the origin of As and Zn in relation to land uses.

The findings of this study may contribute to the international database of investigations on As and Zn content in soil and provide a stable scientific basis for further investigating the specific sources of As and Zn accumulation and their geochemical mobility in the urban and suburban environment. The novelty of this study lies in the fact that in this research work, As and Zn content in the soil of the Ioannina basin are investigated and evaluated to assess the land suitability for residential and agricultural use, applying criteria provided by the literature. The outcome of this study may be helpful for stakeholders and policymakers monitoring the area. No other previous studies reported a potentially toxic elements content in the soil (or any other sampling material) of the Ioannina basin to the best of the authors' knowledge.

2. Materials and Methods

2.1. Primary Data

The data collected for this study include four topographic maps with a scale of 1:50,000, published by the Hellenic Army Geographical Service (H.A.G.S.), along with four geological maps of Greece on a scale of 1:50,000 (sheets Ioannina, Klimatia, Doliana, Tselepovo), published by the Institute of Geology and Mineral Exploration-IGME [47–50]. The land cover of the study area was obtained from the CORINE 2012 Land Cover (CLC) map of the Copernicus Program [51]. The soil map of Greece (scale 1:500,000) published by O.P.E.K.E.P.E. [52] was used for the soil classification of the study area. The soil groups were digitized and inserted as a polygon layer in the GIS database, using the Soil Map Of Greece [52]. GIS-based data have been used in this study for manipulating, analyzing, and presenting.

2.2. Study Area, Geological, Geomorphological, and Land Use Setting

The studied area is a part of the Ioannina karstic basin, and it is located in Epirus, in north-west Greece (Figure 1A). The altitude in the plain that is part of the study area varies between 460 and 760 m, while the eastern edge of the basin is bound by the Mitsikeli mountain (Figure 1B). The study area hosts several settlements and a part of the Ioannina city, and lies on the western shore of Lake Pamvotis (Figure 1B). The dominant wind in the Ioannina basin blows from the of West [53].

The geological formations of the study area are presented in Figure 1C. Rocks from the Ionian geotectonic zone and post-alpine sediments are part of the study area's geological structure [47–50]. The alpine rocks of this zone include (a) Sinion-Pantokrator limestone of the Late Triassic (Early Lias) age; (b) Jurassic shales and carbonate rocks; (c) Upper Jurassic Vigla limestones; (d) Senonian micro-breccia limestones; (e) Eocene limestones [49]; (f) Eocene-Oligocene lignite deposits [49,54]; and (g) Upper Eocene (Priabonian) to Early Miocene flysch [49,55]. The post-alpine sediments consist of Pliocene limnic sediments and Quaternary deposits such as terra-rossa and Fe-Mn oxides, old siliciclastic deposits, soil talus cones and scree, recent lacustrine and fluvial deposits, and some glacial scree deposits [56–60].

The morphology of an area is often related to the development of land uses, soils, and spatial distribution of the trace elements [5,25]. Geomorphologic features such as slope angle are considered. The slope of the study area was computed by using a digital elevation model (DEM). The slope values are categorized into five classes (Figure 2), as follows: (i) < 5°, (ii) 5–10°, (iii) 10–20°, (iv) 20–30°, and (v) > 30°. The slopes were correlated with the spatial distributions of land uses and soils, as well as the As and Zn content in the soil.

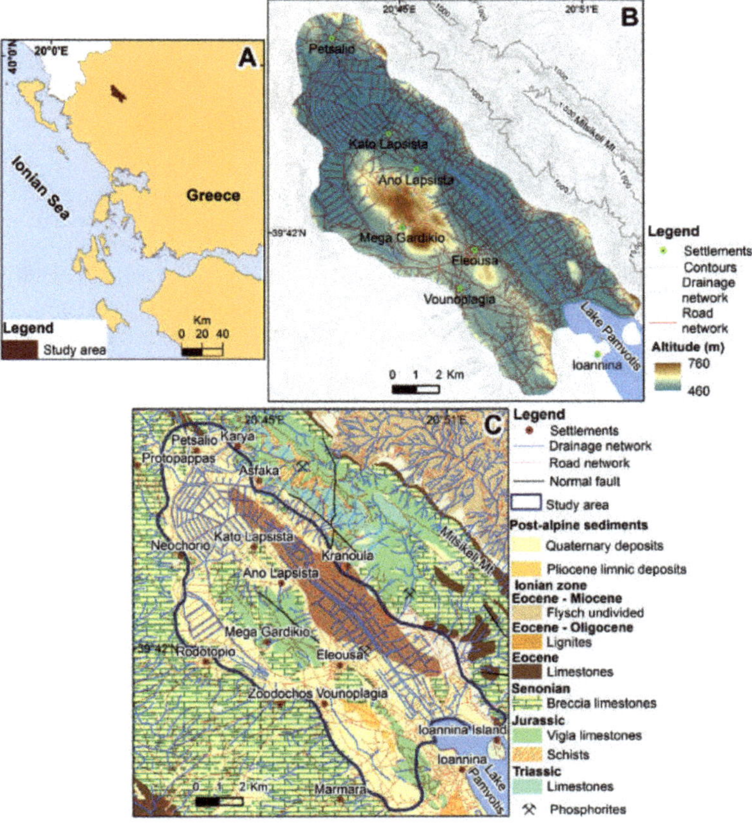

Figure 1. (**A**) Location map of the study area; (**B**) Road network, hydrographic network, and elevation of the study area; and (**C**) Geological map of the area studied.

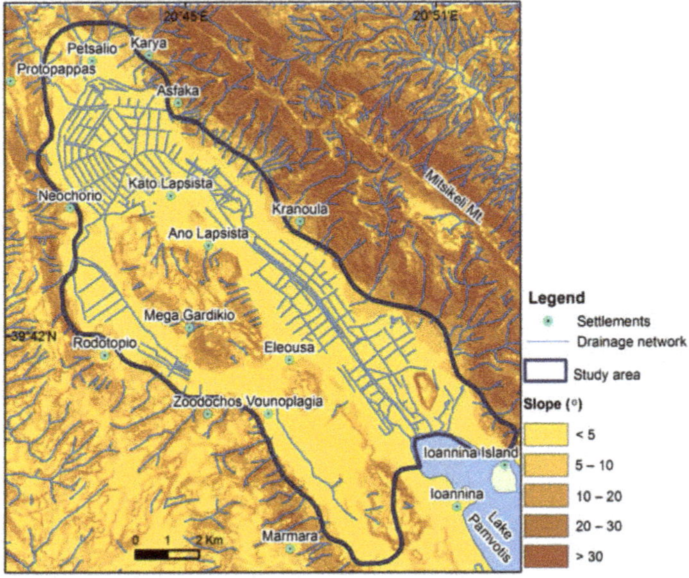

Figure 2. Map showing the morphologic slopes of the study area.

Land use was divided into urban areas, agricultural areas, shrub and sparsely vegetation areas, and wetlands (Figure 3). Urban fabric, roadway network, cottage industries, the Ioannina city's airport, and a vehicle recycling, crushing, and dismantling plant are anthropogenic activities and land-use types in the study area.

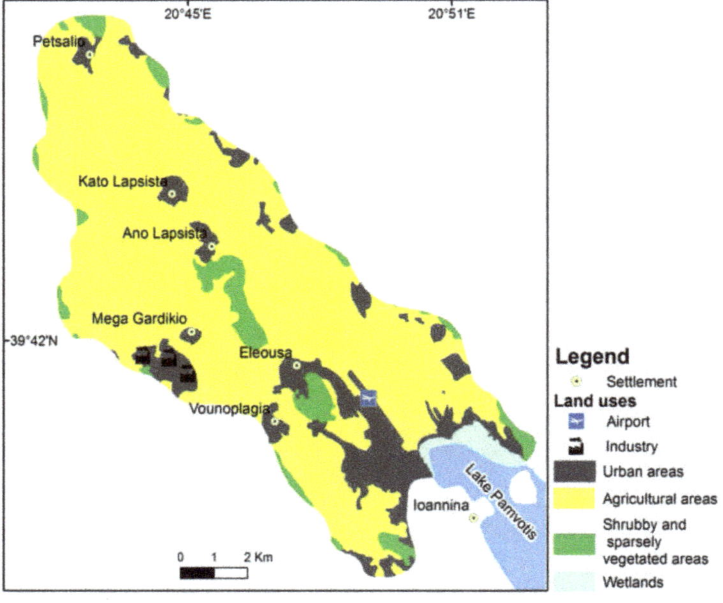

Figure 3. Map showing the land uses of the study area.

2.3. Soil Sampling and Laboratory Procedures

Topsoil samples (0–30 cm) were collected at 112 sampling sites from the Ioannina karst basin (Figure 4) following the soil sampling procedure of Papadopoulou-Vrynioti et al. [28]. Sampling sites were selected by applying the square grid procedure (1 km × 1 km), covering about 110 km^2. Composite soil samples for each sampling site consisted of five topsoil sub-samples which were taken from five points over a patch of land of 5 m^2, approximately. Soil samples were mixed thoroughly to obtain a composite sample (about 2 kg) for each sampling site.

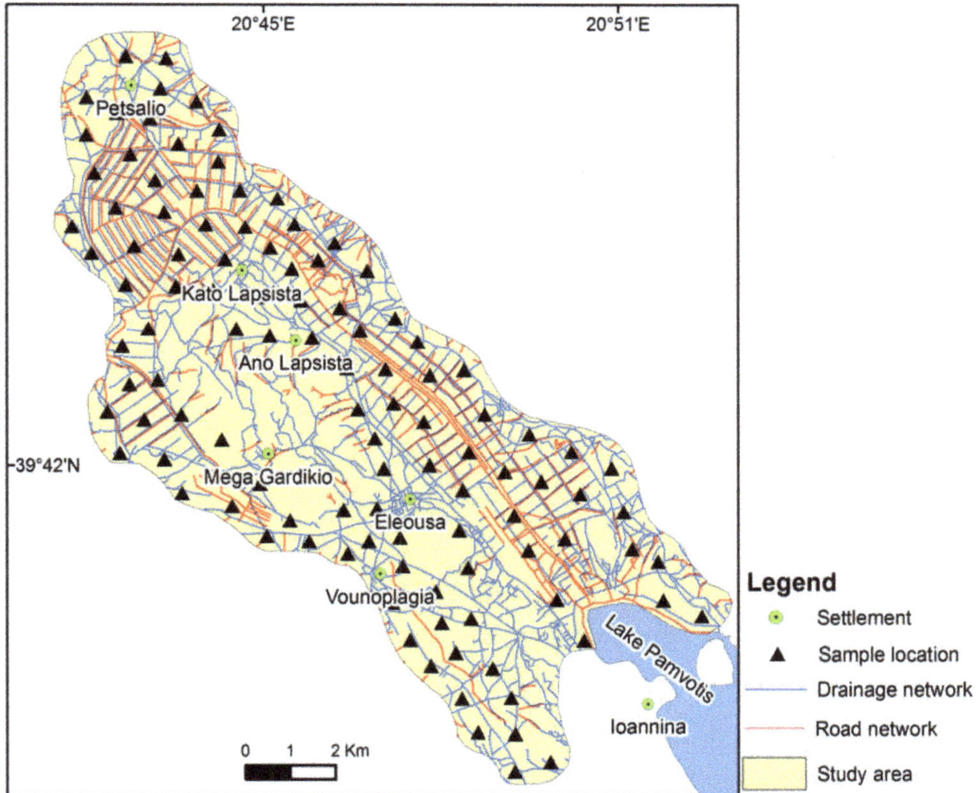

Figure 4. Map showing soil sampling locations in the studied area.

The soil samples were further treated using procedures as described by Papadopoulou-Vrynioti et al. [28]. All soil samples were dried at room temperature (< 25 °C) and divided into two soil subsample sets: (a) subsamples that have been stored in clean polythene bags for the determination of soil grain size and organic matter; and (b) subsamples that had passed through a 2.00 mm sieve and were stored in clean polyethene bags for the analysis of aqua-regia extractable elements including As and Zn.

The Bouyoucos hydrometer method [61] has been applied to determine the clay (<0.002 mm), silt (0.002–0.020 mm), and sand (0.02–2.00 mm) contents (in %). The organic matter content was determined by the dichromate oxidation method [62]. The dichromate oxidation method was also used by many researchers for the determination of the organic matter and organic carbon in soil [28,46,63,64]. To measure the aqua-regia extractable element concentration, 2.5 g of the fine-grained fraction of each sample was placed into glass tubes, and then aqua regia (HCl-HNO_3-H_2O) was added [28]. The glass tubes were heated until fuming on a hot plate. The solutions were left to cool down and made up a final volume of 50 mL with deionized double-distilled water [28]. The soil contents of As and Zn were measured by atomic absorption spectroscopy (AAS). The reference samples and method blanks were analyzed throughout the entire laboratory process to record the chemical analysis quality. Prepared blanks were always below the instrumental detection limits. The chemical analyses were repeated until an accuracy of 95–105%, and the precision of ± 5% was accomplished. The treatment of soil samples was performed at the laboratory of the Institute of Geological and Mineral Exploration (IGME).

2.4. Data Treatment

The software code Surfer 11.0 (Golden Software®, Golden, CO, USA) was used for the triangular diagram. The inverse distance weighted (IDW) technique was applied to interpolate the textured classes' values and the spatial distribution of soil properties. The numerical values were categorized applying the natural breaks classification method using the software code ArcMap 10.8 GIS (ESRI®) (Environmental Systems Research Institute; Redlands, CA, USA). The contents of the elements recorded in the soil of the study area were compared with the corresponding median values for European topsoil [65]. The software codes Microsoft® Excel (Redmond, Washington, DC, USA) and IBM® SPSS 26.0 (International Business Machines Corporation; Statistical Product and Service Solutions; Armonk, NY, USA) for Windows were applied for the data treatment. The Pearson correlation coefficient (r) of the examined variables was estimated applying Equation (1):

$$r_{i,m} = \frac{\sum_i (x_i - x_m)(y_i - y_m)}{\left\{\left[\sum_i (x_i - x_m)\right]^2 \left[\sum_i (y_i - y_m)\right]^2\right\}^{1/2}} \quad (1)$$

where x and y are the *i-th* values of the standardized variables x and y and x_m, y_m are their respective means.

The land suitability evaluation in the studied area was performed by adopting criteria for the soil content of As and Zn established by the Environmental Protection Agency (EPA) [66], the Canadian Council of Ministers of the Environment (CCME) [67], and the Department of Environment and Conservation (DEC) [68].

3. Results and Discussion

3.1. Geomorphologic and Land Use Characteristics

Carbonate rocks control the karst characteristics of the basin. According to Papadopoulou-Vrynioti et al. [69], alluvial sediments primarily derived from the flysch outcrops in the north-eastern part of the area. The transport capacity of the surface runoff increases from the steep slopes of the western flank of the Mitsikeli mountain [70–75].

The slopes of the Ioannina karst basin are shown in Figure 2. The percentage proportion of the area for each slope class in the study area was estimated. More specifically, 72% of the total study area has a very gentle slope (<5°), 15% a gentle slope (5–10°), 10% moderate slopes (10–20°), 5% a steep slope (20–30°) and 1% very steep slopes (>30°). The very gentle slopes appear in the Ioannina karst basin, while the steep slopes are presented mainly in the hilly area and the western edge of the Ioannina karst basin. The percentage proportion of the area for each land use in the study area was estimated. The urban land covers 15% of the total area, the agricultural land 77%, the shrubby and sparsely vegetation 6%, and the wetlands 2%. The urban and the agricultural areas are the most extensive land uses of the study area, covering 92% of the total area. Many researchers reported that the relief controls the land uses in an area [76–78]. The smooth morphology supports urban development and agricultural activity [79,80]. As much as 69% of the urban area is located on very gentle slopes, 22% on gentle slopes, and 9% on moderate slopes. Thus, the urban area includes an urban pattern, a dense roadway network, cottage industries, the Ioannina city's airport, and a vehicle recycling, crushing, and dismantling industry. Regarding the other land uses, 78% of the agricultural land is located on very gentle slopes, 14% on gentle slopes, and 8% on moderate slopes, and 6% of the shrubby and sparsely vegetation area is situated in very gentle slopes, 14% on gentle slopes, 8% on moderate slopes, and 5% on steep slopes, while 100% of the wetlands is located on very gentle slopes. The Ioannina Airport has served as a civil airport since 1965. It is a mid-sized airport with an annual mean of 2219 flights. During the period 1994–2016, 48,732 flights carried 2,615,336 passengers [53].

3.2. Physicochemical Characteristics of the Soil

There is a high variability of soil pH values across the studied area, ranging from 5.0 to 7.8 [69]. The study area's soil pH values suggest acid to alkaline conditions for all agricultural soil samples. The main geological factors controlling the distribution of pH in the studied area involves the weathering process of the calcareous material and the decomposition process of the organic matter of the lignite/peat occurrences. The CEC values ranged from 7.70 to 40.90 meq/100 g, with a median value of 17.50 meq/100 g [69]. The CEC values suggest that the mobility of As and Zn would be expected to be low in the agricultural soil of the study area. The calcium carbonate ($CaCO_3$) content in the soil of the study area ranged from 0.05 to 59%, with a median of 0.9% [69]. According to the O.P.E.K.E.P.E. [52], the following Reference Soil Groups are recorded in the Ioannina basin (Figure 5A): (a) Cambisols (CM), which are characterized by the absence of appreciable quantities of organic matter, illuviated clay, and compounds of Al or Fe; (b) Fluvisols (FL) accommodate genetically young soils in alluvial deposits. Fluvisols present a layering of the sediments rather than pedogenic horizons; (c) Histosols (HL) are shallow soils over hard rock and comprise very gravelly or highly calcareous material; (d) Leptosols (LP) are shallow over hard rock and comprise highly calcareous material; and (e) Luvisols (LV) are fertile soils suitable for a wide range of agricultural uses.

The spatial correlation of the soil groups and slopes proved that 68% of Cambisols are located in very gentle slopes, 23% in gentle slopes, and 9% in moderate slopes. As much as 91% of Fluvisols are situated on very gentle slopes, 6% on gentle slopes, and 3% on moderate slopes, while 93% of Histosols are located in very gentle slopes, 4% on gentle slopes, and 3% on moderate slopes. Regarding Leptosols, 52% are on very gentle slopes, 21% on gentle slopes, 23% on moderate slopes, and 4% on steep slopes. Finally, 80% of Luvisols are located in very gentle slopes, 21% on gentle slopes, and 5% on moderate slopes. Consequently, the vast majority of soil groups are located on very gentle slopes. The Fluvisols and Luvisols soils containing the highest proportion of clay are recorded mainly in the central part of the study area (Figure 5A). The Fluvisols are mainly derived from the erosion of flysch (Figure 1). The Histosols are observed in the eastern part of the polje, while the Leptosols are distributed primarily in the northern and western parts of the area (Figure 5A). These two soil groups include calcareous material and can be attributed to the karstification process of the carbonate rocks outcropping in the study area.

The mean, median, minimum, and maximum values of organic matter, clay, silt, and sand content in the soil of the studied area are tabulated in Table 1. The soil of the study area presents high proportions of silt and clay (Table 1). The spatial distribution of organic matter in the soil of the study area is shown in Figure 5B. The higher organic matter content is recorded mainly in the eastern part of the studied area, suggesting a higher capacity of immobilization for As and Zn (Figure 5B).

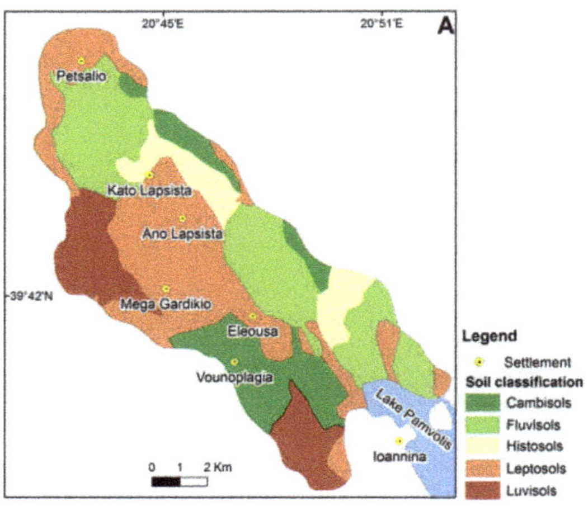

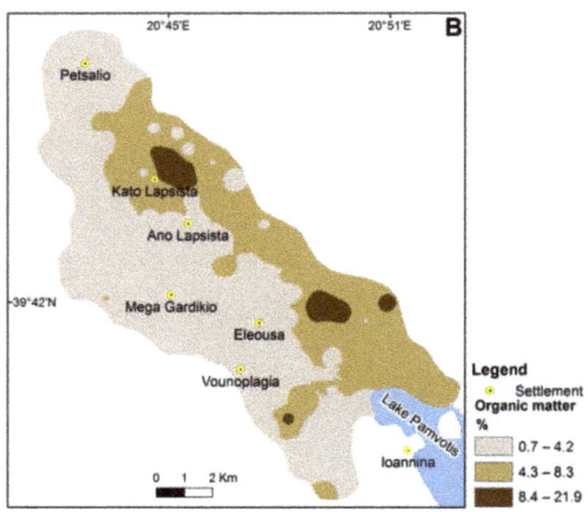

Figure 5. (**A**) Classification of soils in the studied area; (**B**) Spatial distribution of organic matter in the soil of the study area.

Table 1. Organic matter, clay, silt, and sand content (%) in the soil of the Ioannina basin.

	Organic Matter	Clay	Silt	Sand
		Agricultural land use (n = 102)		
Mean	4.2	32.3	38.1	29.8
Median	3.3	31.0	38.5	27.0
Minimum	0.7	5.0	13.0	4.0
Maximum	22.0	64.0	64.0	82.0
		Urban land use (n = 8)		
Mean	4.1	35.9	36.5	27.8
Median	3.6	31.5	37.5	23.5
Minimum	1.0	11.0	20.0	7.0
Maximum	6.9	65.0	56.0	69.0
		Wetlands (n = 2)		
Minimum	4.4	42.0	31.0	12.0
Maximum	6.8	57.0	35.0	23.0

The mean, median, minimum, and maximum values of aqua-regia extractable element contents at the Ioannina karst basin's soils and the analytical methods' detection limits are presented in Table 2. The study area's soil is highly enriched with the elements As and Zn, as shown by their median values, which are higher than that of the corresponding values given by Salminen et al. [65].

Table 3 tabulates the Pearson correlation coefficient (r) matrix for soil data of the studied area. It is shown that only some of the correlation coefficients are statistically significant (p-value < 0.05). Sand showed a significant negative correlation with the clay and silt contents (Table 3). In contrast, there are significant positive weak correlations for As-OM, Zn-OM, and Zn-Clay, indicating weak associations of these elements with organic matter and Zn with clay, respectively (Table 3). This finding suggested an additional anthropogenic or geological source, except for the known lignite deposits and clay minerals, which control the distribution of the As and Zn content in the soil of the study area.

Table 2. Detection limits and descriptive statistics of As and Zn concentration (in mg kg^{-1}) in the soil of the Ioannina basin compared to criteria provided from the literature.

	As	Zn
Detection limit	2	2
Agricultural land use (n = 102)		
Mean	19.8	94.6
Median	16.0	92.0
Minimum	3.6	17.0
Maximum	76.0	465.0
Urban land use (n = 8)		
Mean	18.7	94.9
Median	17.8	95.0
Minimum	13.0	60.0
Maximum	27.0	146.0
Wetlands (n = 2)		
Minimum	23.0	143.0
Maximum	33.0	173.0
European topsoil [65]	6.0	48.0
EPA Residential soil [66]	0.68	2300
EPA Plant-Avian-Mammalian [66]	5.7	6.62
CCME Agricultural land use [67]	12	200
CCME Residential land use [67]	12	200
DEC Ecological Investigation level [68]	20	200

Table 3. Pearson correlation coefficient (r) matrix for soil data (n = 112) of the study area. The *p*-value is calculated at the 95% confidence level (OM: Organic matter).

		As	Zn	OM	Clay	Sand	Silt
As	r	1.00	−0.10	0.27	−0.03	0.10	−0.13
	p-value		*0.317*	*0.004*	*0.774*	*0.309*	*0.166*
Zn	r		1.00	0.24	0.17	−0.18	0.09
	p-value			*0.010*	*0.082*	*0.065*	*0.323*
OM	r			1.00	−0.12	0.11	−0.04
	p-value				*0.197*	*0.249*	*0.710*
Clay	r				1.00	−0.82	0.16
	p-value					*0.000*	*0.104*
Sand	r					1.00	−0.70
	p-value						*0.000*
Silt	r						1.00

3.3. *Spatial Distribution of Elements in Soil*

3.3.1. Arsenic

The spatial distribution of As in the soil of the Ioannina basin is illustrated in Figures 6 and 7. The concentration of As in soil at the eastern and western part of the studied area is higher than the European topsoil's median value [65]. The elevated As (> 33.1 mg kg^{-1}) content in soil is recorded north of the Pamvotis lake and the areas lying between Petsalio and Ano Lapsista-Kato Lapsista (Figure 7A,B). The vast majority (93%) of elevated As concentration is found on very gentle slopes. The elevated As concentration is mainly observed in the Fluvisols and Leptosols soil classes (Figure 6A). The elevated As content in the Ioannina basin's soil is spatially related to the highest organic matter content (Figure 6B) and lignite deposits (Figure 7B), consequently reducing the mobility of As and leading to the high content of As in soil. A similar enrichment mechanism of trace elements in soil involving organic matter is also reported by Li et al. [81] in the soil of Tongguan

County (China). Furthermore, the elevated As concentration in soil is also well related to agricultural land use (Figure 7A), suggesting that an additional anthropogenic source of As in the soil of the study area is the application of agricultural chemicals, phosphate, and nitrogen fertilizers. Since there are no detailed data about the doses of the fertilizers applied in the study area and their As content, only relevant information from the literature can be reported and discussed. According to Kabata-Pendias and Pendias [82], the As concentration (mg kg^{-1}) in phosphate and nitrogen fertilizers range from 2 to 1200 and 22 to 120, respectively. Furthermore, according to Kabata-Pendias and Mukherjee [42], agricultural practices may be a significant source of As, as its concentration is often elevated in sludge, manure, fertilizer, and pesticides.

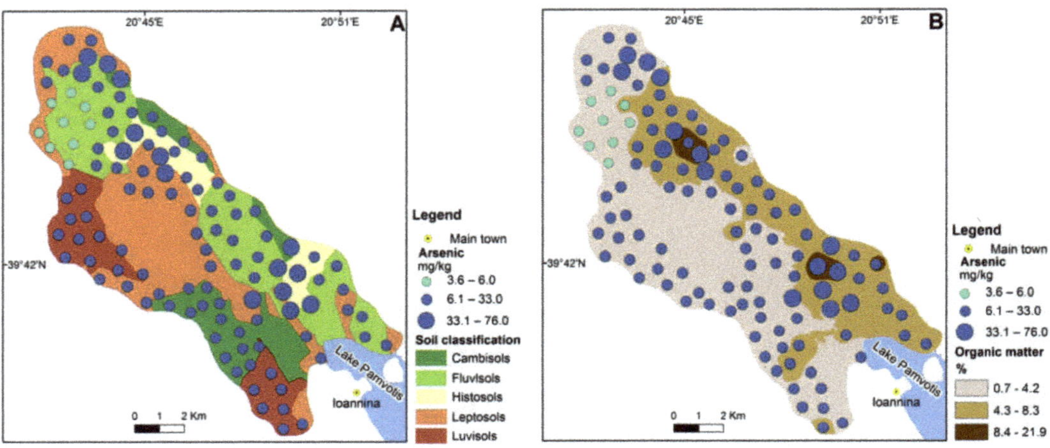

Figure 6. Graduated symbol plots of As content in the soil of the Ioannina basin in relation to the Reference Soil Groups (**A**); and organic matter (**B**).

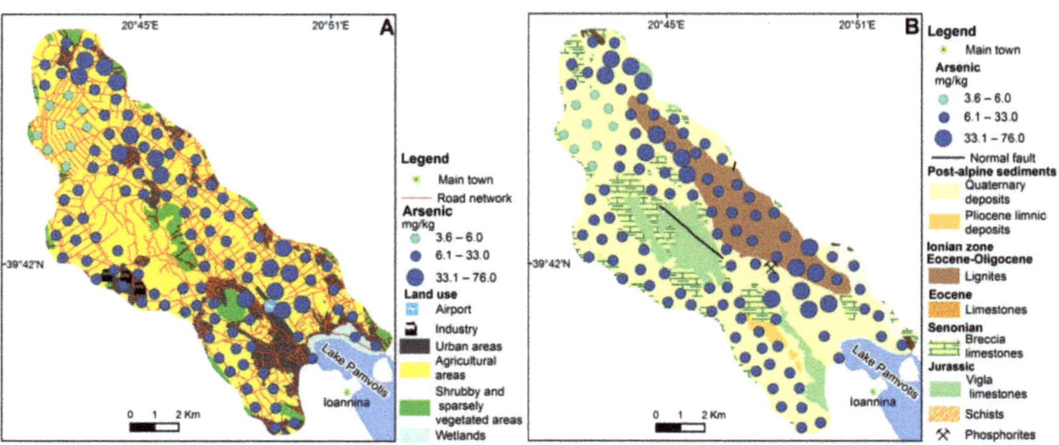

Figure 7. Graduated symbol plots of As content in the soil of the Ioannina basin in relation to land uses (**A**) and geology (**B**).

The arsenic content in the lignite deposits of the Ioannina basin varies between 1.93 and 46.40 mg kg^{-1} [54], suggesting that lignite deposits are a natural source of As in the soil of the Ioannina karst basin. Similar findings regarding the contribution of lignite deposits as a contamination source of As for soil and groundwater have also been reported by many researchers in the Oropos-Kalamos basin (Greece) [83,84]. According to Golfinopoulos et al. [85], the most important reason for the As occurrence in the environment of Greece is the geogenic sources. The airborne particulate matter, especially the lighter particles, attributable to aircraft operations, is another possible anthropogenic source of the As content in the soil of the Ioannina basin. Amato et al. [86] reported that aircraft operations at Barcelona airport (Spain) resulted in As contents of up to 6.3 ng m^{-3} in particulate matter samples. Agrawal et al. [87] studied emission indices of various elements for four planes as a function of engine load and reported that the As emission by jet engines is up to 0.00487 mg per 1 kg of combusted fuel. Moreover, the As content in PM_{10} collected at Barcelona airport is up to 6.3 ng m^{-3} [86]. The AsO_3 content in brake abrasion dust reaches up to 0.01% [88], suggesting that road traffic emissions are also an additional source of As in the Ioannina karst basin's soil. An increase of As concentration in roadside soil was also reported by Wiseman et al. [89] in Toronto (Canada).

An arsenic content in soil exceeding the plant-avian-mammalian screening level (P,A,M) established by EPA is recorded in the 99.9% (14.8 km^2), 91.8% (71.4 km^2), 91.6% (6.0 km^2), and 99.8% (1,6 km^2) of the urban agricultural, shrubby-vegetated, and wetland areas, respectively (Figures 8A and 9A). All the sampling sites at the Ioannina karst basin present an As content in soil higher than the EPA (RS) (Figure 9A).

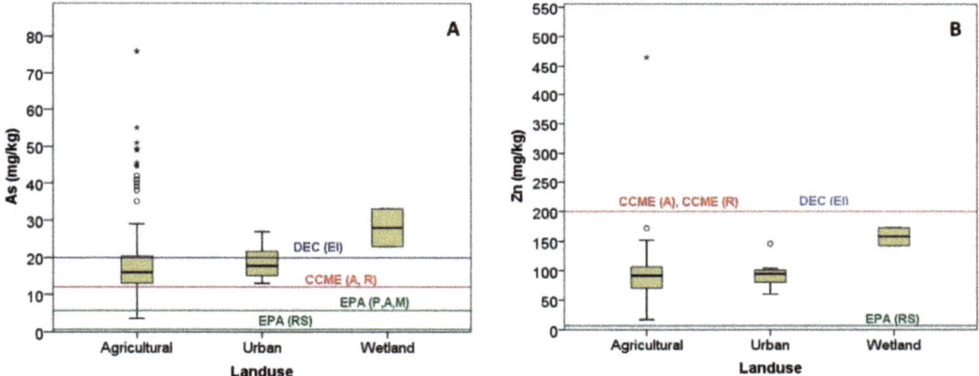

Figure 8. Boxplots for: (**A**) As and (**B**) Zn, comparing the elements' concentration to standards from the literature [65–67]. EPA (RS): Residential soil's regional screening level; EPA (P,A,M): Plant-Avian-Mammalian screening level; EPA (A,M): Avian-Mammalian screening level; CCME (A): Soil quality guideline for agricultural land use; CCME (R): Soil quality guideline for residential land use; DEC (EI): Ecological investigation level; asterisks and circles indicate outliers.

An arsenic concentration in soil that exceeds CCME (A) and CCME (R) is recorded in 83.6% (12.4 km^2), 74.7% (58.1 km^2), 75.8% (5 km^2), and 99.8% (1.6 km^2) of the urban, agricultural, shrubby-vegetated, and wetland area, respectively (Figure 9B). In the 37.5% (5.5 km^2), 28.2% (21.9 km^2), 25.5% (1.7 km^2), and 99.8% (1.6 km^2) of the urban agricultural, shrubby-vegetated, and wetland area, respectively, the As content in soil is higher than the DEC (EI) (Figure 9C). The median As values in the soil of agricultural, urban, and wetland land uses are higher than the CCME (A), CCME (R), EPA(P,A,M), and EPA (RS) values, suggesting a potential threat to residents, plants, and terrestrial ecological receptors of the study area. Only the median value for As in the soil of wetlands of the Ioannina karst basin exceeds the DEC (EI) (Figure 9A). A decrease in the growth of vegetables and chlorosis is among the expected impacts of As on plants [90], while As is known to deactivate enzymes in animals [91]. Adverse effects of As-contaminated vegetables on human health include, among others, abnormalities of the nervous system and endocrine disruptions [92].

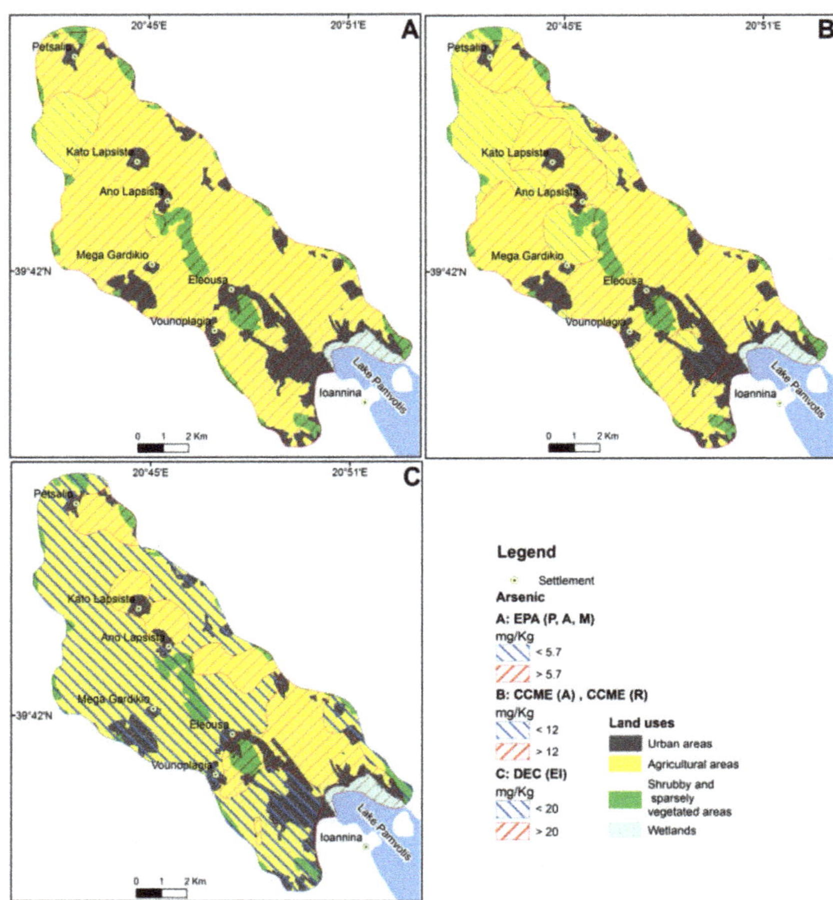

Figure 9. Suitability map comparing As content in the soil of the Ioannina basin in relation to EPA (P,A,M) (**A**); CCME (A) and CCME (R) (**B**); and DEC (EI) (**C**).

3.3.2. Zinc

Only 5% of the sampling sites contained a Zn concentration in soil lower than the European topsoil's median value [65]. The vast majority (94%) of the maximum Zn concentration is located on very gentle slopes. The highest Zn content is mainly recorded in the Fluvisols, Leptosols, and Luvisols soil classes (Figure 10A). The highest Zn concentration in the soil of the studied area is spatially related to the highest organic matter content, as depicted in the eastern edge of the Ioannina basin (Figure 10B). The spatial pattern of Zn is not very complex, probably because its sources are distributed all over the study area, suggesting that the Zn content in soil is attributed mainly to geogenic sources (Quaternary deposits and Fe-Mn oxides) (Figure 11A,B), as also reported by Vryniotis [93], who concluded that Fe-Mn oxides are the primary geological source of Zn in the soil of the Ioannina karstic basin. Automobile emissions, aircraft operations, usage of fertilizers, vehicles scrap, and related activities are also continuous anthropogenic sources of Zn in the soil of the Ioannina basin. Many researchers [94,95] measured Zn levels in the soil surrounding airports and have revealed a direct link between aircraft operations and Zn, thus supporting the results of this study. According to Kabata-Pendias and Mukherjee [42], agricultural practices are considered among the main Zn sources which significantly contribute to Zn level in soils, suggesting that agricultural activities in the studied area are an additional anthropogenic source of contamination. The Zn concentration in fertilizers varies between 61 and 625 mg kg^{-1} [42]. Furthermore, Zn content is a well-known key tracer of traffic emissions [96–99]. Alexakis [100] reported a high Zn content (up to 720 mg kg^{-1}) in a road sediment of the West Attica region (Greece), which can be explained by the wear of vehicle tires and traffic emissions. Agrawal et al. [87] observed up to 0.0441 mg Zn per kg of fuel combusted in aircraft engines, while the Zn content in PM_{10} samples collected at Barcelona airport (Spain) was up to 646 ng m^{-3} [86]. Zinc is primarily applied in galvanized iron and steel products, which are used in automobile door panels. Zinc is also used for grills, carburettors, and pumps, and is also required for paints [101]. Jaradat et al. [101] studied the heavy metal distribution at a scrapyard of discarded vehicles (Zarqa city, Jordan) and observed an elevated Zn concentration in the plants gathered inside (58.35 mg kg^{-1}) and outside (52.11 mg kg^{-1}) of the scrapyard. Jaradat et al. [101] suggested that crushing and dismantling activities are a source of contamination in the vicinity of vehicle scrapyards.

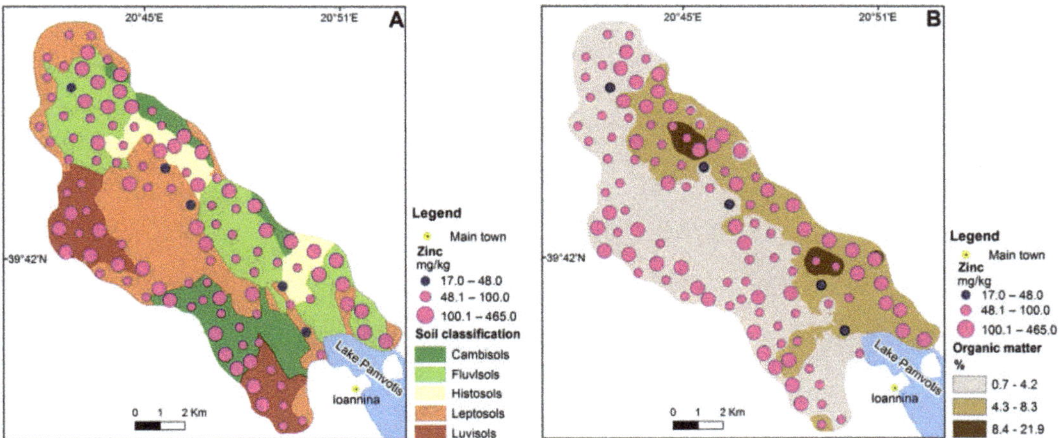

Figure 10. Graduated symbol plots of Zn content in the soil of the Ioannina basin in relation to the Reference Soil Groups (**A**); and organic matter (**B**).

All the sampling sites at the study area present a Zn content in soil [except only one sampling site for the screening values of CCME (A), CCMR (R), and DEC (EI)] higher than EPA (P,A,M) and lower than CCME (A), CCME(R), and DEC(E) screening values (Figure 11). Because Zn content in the soil does not exceed CCME (A), CCME(R), EPA (RS), and DEC (EI), there is no potential risk to human health and cultivated plants (Figure 12). Only 0.6% (0.1 km^2) and 1.1% (0.9 km^2) of the urban and agricultural land use areas, respectively, present Zn values higher than the screening values given by CCME (A), CCME (R), and DEC (EI) (Figure 12). Zinc causes alterations in plants' chloroplast structure [102], while Zn-rich vegetables' adverse impacts on human health mainly include kidney and liver failure, icterus, and bloody urine [103].

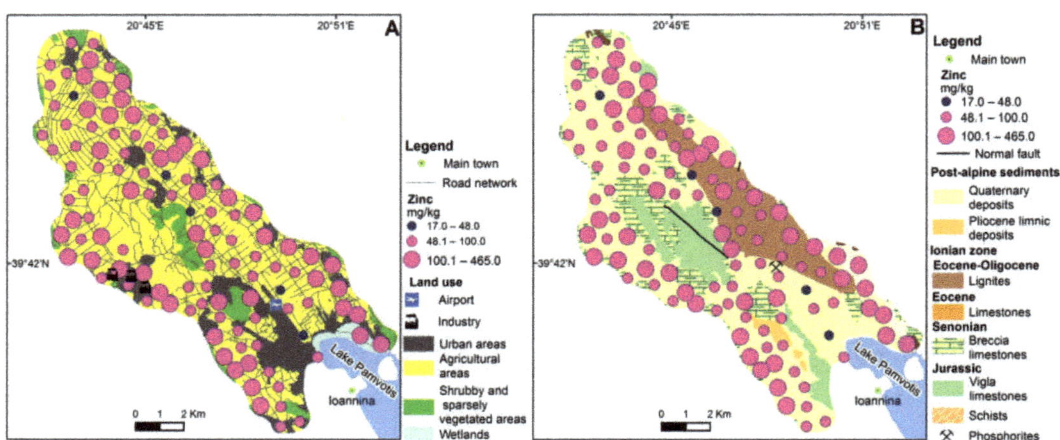

Figure 11. Graduated symbol plots of Zn content in the soil of the Ioannina karst basin in relation to land use (**A**) and geology (**B**).

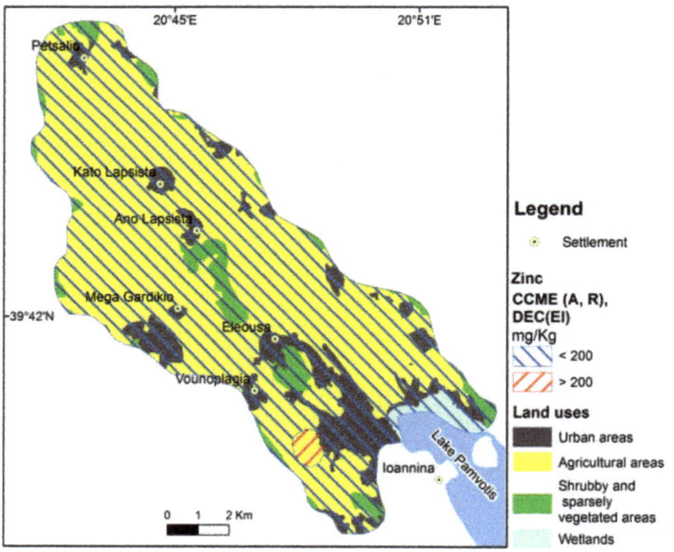

Figure 12. Suitability map comparing the Zn content in the soil of the study area to CCME (A), CCME (R) and DEC (EI).

4. Conclusions

This study delineates the spatial distribution of As and Zn content in the soil of a karst landform and assesses the land suitability for residential and agricultural use. Almost 72% of the total study area has a very gentle slope, favoring urban and agricultural land, which covers 92% of the total area. The median topsoil content for As and Zn is higher than the corresponding median values of European topsoils and was found to be in areas presenting very gentle slopes. The arsenic concentration in the topsoil of the karst landscape appeared to be influenced mainly by lignite deposits. The arsenic content in the soil of the Ioannina basin exceeds at least one or more of the criteria provided by the literature, and hence poses a potential risk to human and terrestrial ecological receptors. The comparison of Zn concentration in topsoil with the literature's criteria revealed that Zn does not pose a potential threat to human health and cultivated plants, except for 0.6% and 1.1% of the urban and agricultural land-use area, respectively. The evaluation of the land suitability of the study area revealed that the As content poses a potential risk to plants, avians and mammals. All the soil sampling sites in the study area present an As concentration higher than the screening value for residential soil. Moreover, the As content in soil is higher than the criteria given by CCME for 83.6% and 74.7% of the urban and agricultural areas, respectively. Scientists, engineers, stakeholders and planners may utilize the outcomes of this study in the environmental monitoring of the area and the forthcoming projects of spatial planning. Additionally, the local authorities may use this work for proposing remedial measures and policies to protect the natural environment.

Author Contributions: Conceptualization, D.E.A., G.D.B., H.D.S. and D.E.G.; methodology, D.E.A., G.D.B. and H.D.S.; software, H.D.S.; validation, D.E.A., G.D.B. and D.E.G.; formal analysis, D.E.A.; investigation, G.D.B.; resources, H.D.S.; data curation, D.E.G.; writing—original draft preparation, D.E.A., and H.D.S.; writing—review and editing, D.E.A.; visualization, D.E.G.; supervision, D.E.A., G.D.B., H.D.S. and D.E.G. All authors have read and agreed to the published version of the manuscript.

Funding: This research received no external funding.

Institutional Review Board Statement: Not applicable.

Informed Consent Statement: Not applicable.

Data Availability Statement: Data that support the findings of this study are available from the corresponding author upon reasonable request.

Acknowledgments: The Second Community Support Framework financially supported this work. The authors gratefully acknowledge the staff of IGME for their assistance during the sample collection and laboratory work.

Conflicts of Interest: The authors declare no conflict of interest.

References

1. Ford, D.C.; Williams, P.W. *Karst Hydrogeology and Geomorphology*; Wiley: Chichester, UK, 2007.
2. Gams, I. The polje: The problem of definition. *Z. Geomorph.* **1978**, *55*, 170–181.
3. Papadopoulou-Vrynioti, K. The role of epikarst in the morphogenesis of the karstic forms in Greece and especially of the karstic hollow forms. *Acta Carsolog.* **2004**, *33*, 219–235. [CrossRef]
4. Papadopoulou-Vrynioti, K.; Bathrellos, G.D.; Skilodimou, H.D.; Kaviris, G.; Makropoulos, K. Karst collapse susceptibility mapping considering peak ground acceleration in a rapidly growing urban area. *Eng. Geol.* **2013**, *158*, 77–88. [CrossRef]
5. Bathrellos, G.D.; Skilodimou, H.D. Land use planning for natural hazards. *Land* **2019**, *8*, 128. [CrossRef]
6. Tsolaki-Fiaka, S.; Bathrellos, G.D.; Skilodimou, H.D. Multi-criteria decision analysis for abandoned quarry restoration in Evros Region (NE Greece). *Land* **2018**, *7*, 43. [CrossRef]
7. Skilodimou, H.D.; Bathrellos, G.D.; Koskeridou, E.; Soukis, K.; Rozos, D. Physical and anthropogenic factors related to landslide activity in the Northern Peloponnese, Greece. *Land* **2018**, *7*, 85. [CrossRef]
8. Bathrellos, G.D.; Skilodimou, H.D.; Zygouri, V.; Koukouvelas, I.K. Landslide: A recurrent phenomenon? Landslide hazard assessment in mountainous areas of central Greece. *Z. Geomorphol.* **2021**, *63*, 95–114. [CrossRef]
9. Skilodimou, H.D.; Bathrellos, G.D.; Alexakis, D.E. Flood hazard assessment mapping in burned and urban areas. *Sustainability* **2021**, *13*, 4455. [CrossRef]
10. Novel, G.P.; Dimadi, A.; Zervopoulou, M.; Bakalowicz, M. The Aggitis karst system, Eastern Macedonia, Greece: Hydrologic functioning and development of the karst structure. *J. Hydrol.* **2007**, *334*, 477–492. [CrossRef]
11. Gutiérrez, F.; Parise, M.; De Waele, J.; Jourde, H. A review on natural and human-induced geohazards and impacts in karst. *Earth Sci. Rev.* **2014**, *138*, 61–88. [CrossRef]
12. Parise, M.; Gabrovsek, F.; Kaufmann, G.; Ravbar, N. *Advances in Karst Research: Theory, Fieldwork and Applications*; Geological Society of London, Special Publications: London, UK, 2018; p. 466. [CrossRef]
13. Veni, G.; DuChene, H.; Crawford, C.N.; Groves, G.C.; Huppert, N.G.; Kastning, H.E.; Olson, R.; Wheeler, B. *Living with Karst—A Fragile Foundation*; American Geological Institute: Alexandria, VA, USA, 2001; p. 66.
14. Prohic, E.; Peh, Z.; Miko, S. Geochemical characterization of a karst polje—An example from Sinjsko polje, Croatia. *Environ. Geol.* **1998**, *33*, 263–273. [CrossRef]
15. Bathrellos, G.D.; Skilodimou, H.D.; Soukis, K.; Koskeridou, E. Temporal and spatial analysis of flood occurrences in drainage basin of Pinios River (Thessaly, central Greece). *Land* **2018**, *7*, 106. [CrossRef]

16. Bathrellos, G.D.; Karymbalis, E.; Skilodimou, H.D.; Gaki-Papanastassiou, K.; Baltas, E.A. Urban flood hazard assessment in the basin of Athens Metropolitan city, Greece. *Environ. Earth Sci.* **2016**, *75*, 319. [CrossRef]
17. Bathrellos, G.D.; Gaki-Papanastassiou, K.; Skilodimou, H.D.; Papanastassiou, D.; Chousianitis, K.G. Potential suitability for urban planning and industry development using natural hazard maps and geological–geomorphological parameters. *Environ. Earth Sci.* **2012**, *66*, 537–548. [CrossRef]
18. Bathrellos, G.D.; Gaki-Papanastassiou, K.; Skilodimou, H.D.; Skianis, G.A.; Chousianitis, K.G. Assessment of rural community and agricultural development using geomorphological-geological factors and GIS in the Trikala prefecture (Central Greece). *Stoch. Environ. Res. Risk A* **2013**, *27*, 573–588. [CrossRef]
19. Rozos, D.; Bathrellos, G.D.; Skilodimou, H.D. Comparison of the implementation of Rock Engineering System (RES) and Analytic Hierarchy Process (AHP) methods, based on landslide susceptibility maps, compiled in GIS environment. A case study from the Eastern Achaia County of Peloponnesus, Greece. *Environ. Earth Sci.* **2011**, *63*, 49–63. [CrossRef]
20. Panagopoulos, G.P.; Bathrellos, G.D.; Skilodimou, H.D.; Martsouka, F.A. Mapping urban water demands using multi-criteria analysis and GIS. *Water Resour. Manage.* **2012**, *26*, 1347–1363. [CrossRef]
21. Coxon, C. Agriculture and Karst. In *Karst Management*; Van Beynen, P., Ed.; Springer: Cham, Switzerland, 2011; pp. 103–113. [CrossRef]
22. Papadopoulou-Vrynioti, K.; Alexakis, D.; Bathrellos, G.; Skilodimou, H.; Vryniotis, D.; Vassiliades, E.; Gamvroula, D. Distribution of trace elements in stream sediments of Arta plain (western Hellas): The influence of geomorphological parameters. *J. Geochem. Explor.* **2013**, *134*, 17–26. [CrossRef]
23. Papadopoulou, K.; Vryniotis, D. Quality of soil and water in deltaic deposits of Louros and Arachthos rivers related to karstic rocks of the wider area. *Bull. Geol. Soc. Greece* **2007**, *40*, 1599–1608. [CrossRef]
24. Kazakis, N.; Chalikakis, K.; Mazzilli, N.; Ollivier, C.; Manakos, A.; Voudouris, K. Management and research strategies of karst aquifers in Greece: Literature overview and exemplification based on hydrodynamic modelling and vulnerability assessment of a strategic karst aquifer. *Sci. Total Environ.* **2017**, *643*, 592–609. [CrossRef] [PubMed]
25. Green, R.T.; Painter, S.L.; Sun, A.; Worthington, S.R. Groundwater contamination in karst terranes. *Water Air Soil Pollut. Focus* **2006**, *6*, 157–170. [CrossRef]
26. Alexakis, D.; Tsakiris, G. Drought impacts on karstic spring annual water potential. Application on Almyros (Heraklion Crete) brackish spring. *Desalin. Water Treat.* **2010**, *16*, 1–9. [CrossRef]
27. Tsakiris, G.; Alexakis, D. Karstic spring water quality: The effect of groundwater abstraction from the recharge area. *Desalin. Water Treat.* **2014**, *52*, 2494–2501. [CrossRef]
28. Papadopoulou-Vrynioti, K.; Alexakis, D.; Bathrellos, G.; Skilodimou, H.; Vryniotis, D.; Vassiliades, E. Environmental research and evaluation of agricultural soil of the Arta plain, western Hellas. *J. Geochem. Explor.* **2014**, *136*, 84–92. [CrossRef]
29. Alexakis, D. Human health risk assessment associated with Co, Cr, Mn, Ni and V contents in agricultural soils from a Mediterranean site. *Arch. Agron. Soil Sci.* **2016**, *62*, 359–373. [CrossRef]
30. Alexakis, D. Suburban areas in flames: Dispersion of potentially toxic elements from burned vegetation and buildings. Estimation of the associated ecological and human health risk. *Environ. Res.* **2020**, *183*, 109153. [CrossRef]
31. Alexakis, D. Contaminated land by wildfire effect on ultramafic soil and associated human health and ecological risk. *Land* **2020**, *9*, 409. [CrossRef]
32. Alexakis, D.; Gotsis, D.; Giakoumakis, S. Assessment of drainage water quality in pre- and post-irrigation seasons for supplemental irrigation use. *Environ. Monit. Assess.* **2012**, *184*, 5051–5063. [CrossRef] [PubMed]
33. Alexakis, D.; Gotsis, D.; Giakoumakis, S. Evaluation of soil salinization in a Mediterranean site (Agoulinitsa district—West Greece). *Arab. J. Geosci.* **2014**, *8*, 1373–1383. [CrossRef]
34. Kelepertsis, A.; Alexakis, D.; Kita, I. Environmental geochemistry of soils and waters of Susaki area, Korinthos, Greece. *Environ. Geochem. Health* **2001**, *23*, 117–135. [CrossRef]
35. Kelepertsis, A.; Alexakis, D. The impact of mining and metallurgical activity of the Lavrion sulfide deposits on the geochemistry of bottom sea sediments East of the Lavreotiki Peninsula, Greece. *Res. J. Chem. Environ.* **2004**, *8*, 40–46.
36. Bathrellos, G.D.; Skilodimou, H.D.; Kelepertsis, A.; Alexakis, D.; Chrisanthaki, I.; Archonti, D. Environmental research of groundwater in the urban and suburban areas of Attica region, Greece. *Environ. Geol.* **2008**, *56*, 11–18. [CrossRef]
37. Gamvroula, D.; Alexakis, D.; Stamatis, G. Diagnosis of groundwater quality and assessment of contamination sources in the Megara basin (Attica, Greece). *Arab. J. Geosci.* **2013**, *6*, 2367–2381. [CrossRef]
38. Makri, P.; Stathopoulou, E.; Hermides, D.; Kontakiotis, G.; Zarkogiannis, S.D.; Skilodimou, H.D.; Scoullos, M. The Environmental Impact of a Complex Hydrogeological System on Hydrocarbon-Pollutants' Natural Attenuation: The Case of the Coastal Aquifers in Eleusis, West Attica, Greece. *J. Mar. Sci. Eng.* **2020**, *8*, 1018. [CrossRef]
39. Alexakis, D. Diagnosis of stream sediment quality and assessment of toxic element contamination sources in East Attica, Greece. *Environ. Earth Sci.* **2011**, *63*, 1369–1383. [CrossRef]
40. Ivezić, V.; Singh, B.R.; Almås, A.R.; Lončarić, Z. Water extractable concentrations of Fe, Mn, Ni, Co, Mo, Pb and Cd under different land uses of Danube basin in Croatia. Acta Agriculturae Scandinavica. *Sect. B Soil Plant Sci.* **2011**, *61*, 747–759.
41. Abbaslou, H.; Martin, F.; Abtahi, A.; Moore, F. Trace element concentrations and background values in the arid soils of Hormozgan Province of southern Iran. *Arch. Agron. Soil Sci.* **2014**, *60*, 1125–1143. [CrossRef]
42. Kabata-Pendias, A.; Mukherjee, A.B. *Trace Elements from Soil to Human*; Springer: Berlin/Heidelberg, Germany, 2007.

43. Mendyk, Ł.; Hulisz, P.; Świtoniak, M.; Kalisz, B.; Spychalski, W. Human activity in the surroundings of a former mill pond (Turznice, N Poland): Implications for soil classification and environmental hazard assessment. *Soil Sci. Ann.* **2020**, *71*, 371–381. [CrossRef]
44. Uzarowicz, Ł.; Charzyński, P.; Greinert, A.; Hulisz, P.; Kabała, C.; Kusza, G.; Pędziwiatr, A. Studies of technogenic soils in Poland: Past, present, and future perspectives. *Soil Sci. Ann.* **2020**, *71*, 281–299. [CrossRef]
45. Gruszecka-Kosowska, A.; Kicińska, A. Long-Term Metal-Content Changes in Soils on the Olkusz Zn–Pb Ore-Bearing Area, Poland. *Int. J. Environ. Res.* **2017**, *11*, 359–376. [CrossRef]
46. Alexakis, D.; Gamvroula, D.; Theofili, E. Environmental availability of potentially toxic elements in an agricultural Mediterranean site. *Environ. Eng. Geosci.* **2019**, *25*, 169–178. [CrossRef]
47. IGME. *Geological Maps of Greece, Sheet Ioannina, Scale 1:50,000*; Institute of Geology and Mineral Exploration: Thessaloniki, Greece, 1967.
48. IGME. *Geological Maps of Greece, Sheet Klimatia, Scale 1:50,000*; Institute of Geology and Mineral Exploration: Thessaloniki, Greece, 1968.
49. IGME. *Geological Maps of Greece, Sheet Doliana, Scale 1:50,000*; Institute of Geology and Mineral Exploration: Thessaloniki, Greece, 1968.
50. IGME. *Geological Maps of Greece, Sheet Tselepovo, Scale 1:50,000*; Institute of Geology and Mineral Exploration: Thessaloniki, Greece, 1970.
51. Copernicus. Urban Atlas 2018. *Copernicus Land Monitoring Service*. 2021. Available online: https://land.copernicus.eu/local/urban-atlas/urban-atlas-2018 (accessed on 7 May 2021).
52. (Payment and Control Agency for Guidance and Guarantee Community Aid. Soil Map of Greece (Scale 1:500,000) Project Framework: Development of an Integrated System for Soil Geographic Data and Delineation of Agricultural Zones in Greece. *Scientific Coordinator N. Misopollinos*. Available online: https://esdac.jrc.ec.europa.eu/public_path/shared_folder/maps/eudasm_newmaps/GR/GRSoilMap_ENG.jpg (accessed on 7 May 2021).
53. Hellenic Civil Aviation Authority (HCAA). 2021. Available online: http://www.hcaa.gr/en/our-airports/kratikos-aerolimenas-iwanninwn-basileys-pyrros-kaiwp (accessed on 20 March 2021).
54. Gentzis, T.; Goodarzi, F.; Foscolos, A. Geochemistry and mineralogy of Greek lignites from the Ioannina basin. *Energy Sources* **1997**, *19*, 111–128. [CrossRef]
55. Karakitsios, V. Western Greece and Ionian Sea petroleum systems. *AAPG Bull.* **2013**, *97*, 1567–1595. [CrossRef]
56. Bathrellos, G.D.; Skilodimou, H.D.; Maroukian, H. The spatial distribution of Middle and Late Pleistocene cirques in Greece. *Geogr. Ann. A* **2014**, *96*, 323–338. [CrossRef]
57. Bathrellos, G.D.; Skilodimou, H.D.; Maroukian, H.; Gaki-Papanastassiou, K.; Kouli, K.; Tsourou, T.; Tsaparas, N. Pleistocene glacial and lacustrine activity in the southern part of Mount Olympus (central Greece). *Area* **2017**, *49*, 137–147. [CrossRef]
58. Bathrellos, G.D.; Skilodimou, H.D.; Maroukian, H. The significance of tectonism in the glaciations of Greece. *Geol. Soc. SP.* **2017**, *433*, 237–250. [CrossRef]
59. Roucoux, K.H.; Tzedakis, P.C.; Lawson, I.T.; Margari, V. Vegetation history of the penultimate glacial period (MIS 6) at Ioannina, northwest Greece. *J. Quat. Sci.* **2011**, *26*, 616–626. [CrossRef]
60. Bouyoucos, G.J. Hydrometer method improved for making particle size analysis of soils. *Agron. J.* **1962**, *54*, 464–465. [CrossRef]
61. Soil Survey Staff. Soil Survey Staff. Soil Survey Laboratory Methods Manual, Version No.4.0. USDA NRCS. In *Soil Survey Investigations Report 2004*; U.S. Govternment Print Office: Washington, DC, USA, 2004.
62. Iñigo, V.; Marín, A.; Andrades, M.; Jiménez-Ballesta, R. Evaluation of the copper and zinc contents of soils in the vineyards of La Rioja (Spain). *Environments* **2020**, *7*, 55. [CrossRef]
63. De Vos, B.; Lettens, S.; Muys, B.; Deckers, J.A. Walkley–Black analysis of forest soil organic carbon: Recovery, limitations and uncertainty. *Soil Use Manag.* **2007**, *23*, 221–229. [CrossRef]
64. Salminen, R.; Batista, M.J.; Bidovec, M.; Demetriades, A.; De Vivo, B.; De Vos, W.; Tarvainen, T.; Geochemical Atlas of Europe. Part 1–Background Information, Methodology and Maps. *Geological Survey of Finland, Espoo, Finland.* 2005. Available online: http://weppi.gtk.fi/publ/foregsatlas (accessed on 17 March 2021).
65. EPA (Environmental Protection Agency) Cleanup Regulations and Standards. Available online: http://www.epa.gov/cleanup/regs.htm (accessed on 20 March 2021).
66. CCME (Canadian Council of Ministers of the Environment). Available online: http://st-ts.ccme.ca (accessed on 20 March 2021).
67. DEC (Department of Environment and Conservation). Contaminated Sites Management Series: Assessment Levels for Soil, Sediment and Water. 2010. Available online: http://www.dec.wa.gov.au/contaminatedsites (accessed on 20 March 2021).
68. Papadopoulou-Vrynioti, K.; Mertzanis, A.; Vryniotis, D.; Vassiliades, E.; Karakitsios, V. The contribution of karstic rocks to soil quality, Ioannina plain (Epirus, Hellas). *J. Geochem. Explor.* **2015**, *154*, 224–237. [CrossRef]
69. Migiros, G.; Bathrellos, G.D.; Skilodimou, H.D.; Karamousalis, T. Pinios (Peneus) River (Central Greece): Hydrological—Geomorphological elements and changes during the Quaternary. *Cent. Eur. J. Geosci.* **2011**, *3*, 215–228. [CrossRef]
70. Kamberis, E.; Bathrellos, G.; Kokinou, E.; Skilodimou, H. Correlation between the structural pattern and the development of the hydrographic network in a portion of the Western Thessaly basin (Greece). *Cent. Eur. J. Geosci.* **2012**, *4*, 416–424. [CrossRef]
71. Kokinou, E.; Skilodimou, H.D.; Bathrellos, G.D.; Antonarakou, A.; Kamberis, E. Morphotectonic analysis, structural evolution/pattern of a contractional ridge: Giouchtas Mt., Central Crete, Greece. *J. Earth Syst. Sci.* **2015**, *124*, 587–602.
72. Rozos, D.; Skilodimou, H.D.; Loupasakis, C.; Bathrellos, G.D. Application of the revised universal soil loss equation model on landslide prevention. An example from N. Euboea (Evia) Island, Greece. *Environ. Earth Sci.* **2013**, *70*, 3255–3266. [CrossRef]

73. Skilodimou, H.D.; Livaditis, G.; Bathrellos, G.D.; Verikiou-Papaspiridakou, E. Investigating the flooding events of the urban regions of Glyfada and Voula, Attica, Greece: A contribution to Urban Geomorphology. *Geogr. Ann. A* **2003**, *85*, 197–204. [CrossRef]
74. Skilodimou, H.D.; Bathrellos, G.D.; Maroukian, H.; Gaki-Papanastassiou, K. Late Quaternary evolution of the lower reaches of Ziliana stream in south Mt. Olympus (Greece). *Geogr. Fis. Din. Quat.* **2014**, *37*, 43–50.
75. Bathrellos, G.D.; Skilodimou, H.D.; Chousianitis, K.; Youssef, A.M.; Pradhan, B. Suitability estimation for urban development using multi-hazard assessment map. *Sci. Total Environ.* **2017**, *575*, 119–134. [CrossRef]
76. Bathrellos, G.D.; Kalivas, D.P.; Skilodimou, H.D. Landslide Susceptibility Assessment Mapping: A Case Study in Central Greece. In *Remote Sensing of Hydrometeorological Hazards*; Petropoulos, G.P., Islam, T., Eds.; CRC Press; Taylor & Francis Group: London, UK, 2017; pp. 493–512.
77. Youssef, A.M.; Pradhan, B.; Al-Kathery, M.; Bathrellos, G.D.; Skilodimou, H.D. Assessment of rockfall hazard at Al-Noor Mountain, Makkah city (Saudi Arabia) using spatio-temporal remote sensing data and field investigation. *J. Afr. Earth Sci.* **2015**, *101*, 309–321. [CrossRef]
78. Chousianitis, K.; Del Gaudio, V.; Sabatakakis, N.; Kavoura, K.; Drakatos, G.; Bathrellos, G.D.; Skilodimou, H.D. Assessment of Earthquake-Induced Landslide Hazard in Greece: From Arias Intensity to Spatial Distribution of Slope Resistance Demand. B. *Seismol. Soc. Am.* **2016**, *106*, 174–188. [CrossRef]
79. Skilodimou, H.D.; Bathrellos, G.D.; Chousianitis, K.; Youssef, A.M.; Pradhan, B. Multi-hazard assessment modeling via multi-criteria analysis and GIS: A case study. *Environ. Earth Sci.* **2019**, *78*, 47. [CrossRef]
80. Li, G.; Lu, N.; Wei, Y.; Zhu, D. Relationship between heavy metal content in polluted soil and soil organic matter and pH in mining areas. *IOP Conf. Ser. Mater. Sci. Eng.* **2018**, *394*, 052081. [CrossRef]
81. Kabata-Pendias, A.; Pendias, H. *Trace Elements in Soils and Plants*; CRC Press Inc: Boca Raton, FL, USA, 1995.
82. Alexakis, D.; Gamvroula, D. Arsenic, chromium, and other potentially toxic elements in the rocks and sediments of Oropos-Kalamos basin, Attica, Greece. *Appl. Environ. Soil Sci.* **2014**, *2014*, 718534. [CrossRef]
83. Stamatis, G.; Alexakis, D.; Gamvroula, D.; Migiros, G. Groundwater quality assessment in Oropos-Kalamos basin, Attica, Greece. *Environ. Earth Sci.* **2011**, *64*, 973–988. [CrossRef]
84. Golfinopoulos, S.; Varnavas, S.; Alexakis, D. The status of arsenic pollution on the Greek and Cyprus environment: An Overview. *Water* **2021**, *13*, 224. [CrossRef]
85. Amato, F.; Moreno, T.; Pandolfi, M.; Querol, X.; Alastuey, A.; Delgado, A.; Pedrero, M.; Cots, N. Concentrations, sources and geochemistry of airborne particulate matter at a major European airport. *J. Environ. Monit.* **2010**, *12*, 854–862. [CrossRef]
86. Agrawal, H.; Sawanta, A.; Jansen, K.; Miller, J.W.; Cocker, D.R. Characterization of chemical and particulate emissions from aircraft engines. *Atmos. Environ.* **2008**, *42*, 4380–4392. [CrossRef]
87. Dousova, B.; Lhotka, M.; Buzek, F.; Cejkova, B.; Jackova, I.; Bednar, V.; Hajek, P. Environmental interaction of antimony and arsenic near busy traffic nodes. *Sci. Total Environ.* **2020**, *702*, 134642. [CrossRef]
88. Wiseman, C.L.S.; Zereini, F.; Püttmann, W. Traffic-related trace element fate and uptake by plants cultivated in roadside soils in Toronto, Canada. *Sci. Total Environ.* **2013**, *442*, 86–95. [CrossRef]
89. Bergqvist, C.; Herbert, R.; Persson, I.; Greger, M. Plants influence on arsenic availability and speciation in the rhizosphere, roots and shoots of three different vegetables. *Environ. Pollut.* **2014**, *184*, 540–546. [CrossRef]
90. Schroeder, H.A.; Balassa, J.J. Abnormal trace metals in man. Cadmium. *J. Chronic Dis.* **1961**, *14*, 236–258. [CrossRef]
91. Rahman, A.; Kumarathasan, P.; Gomes, J. Infant and mother related outcomes from exposure to metals with endocrine disrupting properties during pregnancy. *Sci. Total Environ.* **2016**, *56*, 1022–1031. [CrossRef]
92. Vryniotis, D. *Soil Geochemical—Environmental Survey of the N.W. Part of the Ioannina Basin*; I.G.M.E. Open File Report: Athens, Greece, 2010; p. 95.
93. Boyle, K.A. Evaluating particulate emissions from jet engines: Analysis of chemical and physical characteristics and potential impacts on coastal environments and human health. *Transport. Res. J. Transp. Res. Board* **1996**, *1517*, 1–9. [CrossRef]
94. Nunes, L.M.; Stigter, T.Y.; Teixeira, M.R. Environmental impacts on soil and groundwater at airports: Origin, contaminants of concern and environmental risks. *J. Environ. Monit.* **2011**, *13*, 3026–3039. [CrossRef] [PubMed]
95. Adamiec, E.; Jarosz-Krzeminska, E.; Wieszała, R. Heavy metals from non-exhaust vehicle emissions in urban and motorway road dusts. *Environ. Monit. Assess.* **2016**, *188*, 369. [CrossRef]
96. Apeagyei, E.; Bank, M.S.; Spengler, J.D. Distribution of heavy metals in road dust along an urban-rural gradient in Massachusetts. *Atmos. Environ.* **2011**, *45*, 2310–2323. [CrossRef]
97. Aravelli, K.; Heibel, A. Improved lifetime pressure drop management for robust cordierite (RC) filters with asymmetric cell technology (ACT). *SAE Tech. Paper Series* **2007**, *1*, 920. [CrossRef]
98. Spikes, H. Low—And zero-sulphated ash, phosphorus and sulphur anti-wear additives for engine oils. *Lubric. Sci.* **2008**, *20*, 103–136. [CrossRef]
99. Alexakis, D. Multielement contamination of land in the margin of highways. *Land* **2021**, *10*, 230. [CrossRef]
100. Jaradat, Q.M.; Masadeh, A.; Zaitoun, M.; Maitah, B. Heavy metal contamination of soil, plant and air of scrapyard of discarded vehicles at Zarqa city, Jordan. *Soil Sediment Contam. Int. J.* **2005**, *14*, 449–462. [CrossRef]
101. Doncheva, S.; Stoynova, Z.; Velikova, V. Influence of succinate on zinc toxicity of pea plants. *J. Plant Nutr.* **2001**, *24*, 789–804. [CrossRef]

102. Duruibe, J.O.; Ogwuegbu, M.O.C.; Egwurugwu, J.N. Heavy metal pollution and human biotoxic effects. *Int. J. Phys. Sci.* **2007**, *2*, 112–118.
103. Fosmire, G.J. Zinc toxicity. *Am. J. Clin Nutr.* **1990**, *51*, 225–227. [CrossRef] [PubMed]

Review

A Review of Underground Pipeline Leakage and Sinkhole Monitoring Methods Based on Wireless Sensor Networking

Haibat Ali and Jae-ho Choi *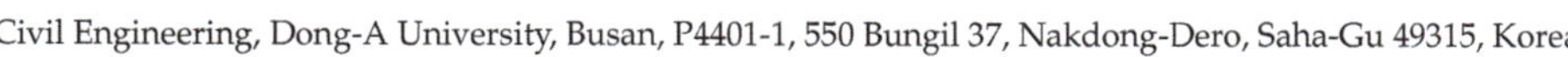

Civil Engineering, Dong-A University, Busan, P4401-1, 550 Bungil 37, Nakdong-Dero, Saha-Gu 49315, Korea
* Correspondence: jaehochoi@dau.ac.kr; Tel.: +82-51-200-762

Received: 9 July 2019; Accepted: 21 July 2019; Published: 24 July 2019

Abstract: Major metropolitan cities worldwide have extensively invested to secure utilities and build state-of-the-art infrastructure related to underground fluid transportation. Sewer and water pipelines make our lives extremely convenient when they function appropriately. However, leakages in underground pipe mains causes sinkholes and drinking-water scarcity. Sinkholes are the complex problems stemming from the interaction of leaked water and ground. The aim of this work is to review the existing methods for monitoring leakage in underground pipelines, the sinkholes caused by these leakages, and the viability of wireless sensor networking (WSN) for monitoring leakages and sinkholes. Herein, the authors have discussed the methods based on different objectives and their applicability via various approaches—(1) patent analysis; (2) web-of-science analysis; (3) WSN-based pipeline leakage and sinkhole monitoring. The study shows that the research on sinkholes due to leakages in sewer and water pipelines by using WSN is still in a premature stage and needs extensive investigation and research contributions. Additionally, the authors have suggested prospects for future research by comparing, analyzing, and classifying the reviewed methods. This study advocates collocating WSN, Internet of things, and artificial intelligence with pipeline monitoring methods to resolve the issues of the sinkhole occurrence.

Keywords: WSN; pipeline leakage; human-induced sinkhole; leakage detection; sewer pipeline; sensors

---

1. Introduction

Sewer and water leakages in underground pipelines have become a critical issue for water-management authorities in most countries—developed and developing alike—worldwide. Leakages in sewer and water pipelines may lead to several problems such as a shortage of drinking water, groundwater contamination, and ground subsidence [1]. Numerous countries are investing a considerable amount of their annual budget towards the prevention and control of the probable effects of sewer and water pipeline leakage. These issues further exacerbate infrastructure and environmental conditions that support human socioeconomic activities. In recent years, developed countries, such as the United Kingdom, Australia, France, Spain, and the United States of America, have experienced shortage in domestic water supply because of leaking pipe mains [2]. The after effects of these leakages in pipelines cause ground subsidence and sinkholes [3]. These sinkholes result in damage to infrastructure (roads, highways, railways, and underground fluid transportation networks).

A sinkhole refers to a cavity in the ground formed by underground erosion and the depression of the ground surface. In general, there exist two types of sinkholes—natural and human induced. Natural sinkholes are mainly observed in regions with large deposits of salt, limestone, and carbonate rocks. The accurate prediction of the location and time of the occurrence of these sinkholes is rather difficult [4]. Groundwater extraction, construction in adjoining areas, and leakage in underground pipelines are the leading causes of human-induced sinkhole formation in urban areas [5]. Among them, the presence of leakages, bursts, or blockages in sewer, drain, and/or water pipelines are the most frequently reported causes of sinkholes.

The issue of sinkhole creation has witnessed global escalation in recent years owing to ever-increasing urbanization and the continuous construction, development, and expansion of urban areas [6]. A cavity begins to develop as leaked water erodes the soil surrounding pipelines. This reduces the bearing capacity of the soil layer above the cavity, and hence, the ground collapses to form a sinkhole [7]. The size of sinkholes ranges from 2 m deep and 1.5 m wide [5] to a massive scale of up to 15 m deep and 30 m wide [8], as reported in Jeju, South Korea and Southwest Japan, respectively. Human-induced sinkholes have been reported in San Antonio, Texas [9]; Oakwood, Georgia; and Tracer, Colorado [10]. These incidences have reportedly caused considerable economic damage and loss of human lives. As reported in Fraser, USA [10], the abrupt collapse of a 44-year-old sewer pipeline destroyed 22 homes, and the reconstruction of the damaged road and sewer pipelines cost the city administration approximately $70 million. More than 20 sinkholes have been observed in USA alone owing to the failure of underground pipe mains [11]. Figure 1 illustrates the effects of underground pipeline leakage.

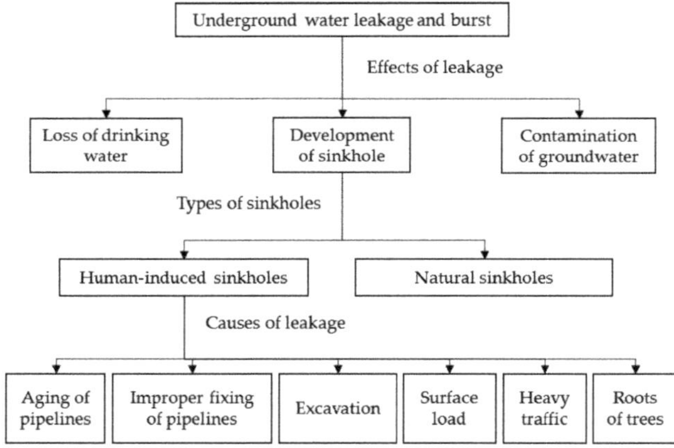

Figure 1. Causes and effects of leakage in underground water systems and sewer pipeline systems.

Various hardware, software, mathematical formulas, and algorithm-based methods have been proposed for monitoring, detecting, and preventing leakages in underground water pipelines and the occurrence of sinkholes. Conventional techniques, such as acoustic-based approaches, have been used for this purpose. These techniques require an expert who scans a suspected area by listening to leak sounds. However, such methods are extremely time consuming and their accuracy highly depends on the skills and experience of personnel [12]. Vibration analysis is another conventional technique utilized to locate leaks in pipelines [13]. However, in conventional vibration and acoustic-based techniques, hydrophones must be placed on both sides of the pipeline section under consideration for leak detection. Such conventional techniques require the exact length of the pipeline to accurately locate a leak. In such cases, the length of the pipeline is measured either by walking with a measurement wheel or by utilizing the recorded data from maps [14]. The lengths obtained using these methods are inappropriate for locating or detecting a leak and result in errors of up to 30% [15]. Various methods have been adopted to monitor and detect sinkholes. Over the years, wireless sensor networking (WSN) systems [16], Internet of things (IoT) [3], and image processing technology using artificial intelligence (AI) have been used for monitoring underground pipeline infrastructures [17]. WSN was first used for pipeline leakage monitoring and detection in 2004 [18]. However, most methods related to WSN, IoT, and image processing mainly focus on sewer and water pipeline monitoring to detect any defects (including leakage) in a more efficient manner in the context of solving the drinking water shortage problem. The after effects (sinkholes) of leakages in sewer and water pipelines have not been discussed and must be examined.

Sinkholes are the most complex civil engineering problems stemming from the interaction of water and the ground. The sinkholes induced by leaking underground pipelines are an important problem that must be extensively investigated. The authors of this paper highlight the major technology-based approaches to obtain a better understanding of the current trends in leakages in underground water and sewer pipelines and their after effects (such as sinkholes). This work aims to find an opportunity for extending existing smart technologies, such as WSN, IoT, and AI, to monitor and detect any signs of sinkholes due to leakages in underground water or sewer pipelines.

The authors aim to demonstrate the exceptional performance of WSN and its application to the detection of sinkholes and to find a suitable solution to optimize the leakage process. Human-induced sinkholes due to leakages in underground pipelines are examined in detail, and the main challenges, issues, and future research areas are elucidated. Furthermore, the authors highlight the WSN-based approaches used to resolve the issue of sinkhole occurrence. The use of smart technology can ensure the future success of underground construction infrastructure industries and create new and clear business objectives. It is important to emphasize that the focus of this study is only on sewer and water pipeline leakages and their after effects.

2. Review Method

2.1. Scope and Objective

This review article concentrated on leakages in sewer and water pipelines and their after effects. It focused on the status of the methods related to the use of WSN to address problems of leakage, burst, and/or blockage monitoring in underground sewer and water pipelines along with the monitoring of the sinkholes caused by these phenomena. The analysis of these methods can provide direction for future research in this area to reduce the occurrence of sinkholes due to leakages. The objective of this review was to provide a comprehensive overview of the state-of-the-art development in leakage and human-induced sinkhole detection and WSN-based monitoring methods.

2.2. Review Execution

This study applied two approaches to comprehensively review the damage caused by sinkholes and leakages in underground water and sewer pipe mains—patent analysis and extensive literature review.

Patent analysis—the patents filed by organizations in different regions worldwide were analyzed using relevant keywords to examine related trends concerning underground sewer and water pipeline leakages and sinkhole creation over the past 18 years. Relevant national and international bodies—Google Patent, United States Patent and Trademark Office, Korean Intellectual Property Rights Information Service (KIRIS), State Intellectual Property Office of the P.R.C, and so on—served as the sources for the patents mentioned in this section, as described in Table 1. Section 3 provides a detailed explanation of this approach.

Literature review—this includes journal articles and conference papers published in regard to the monitoring and detection of leakages in underground water and sewer pipelines and the sinkholes caused by these leakages using WSN systems. Different research article search databases, such as Science Direct, Web of Sciences, and Engineering Village, were used for finding relevant literature published over the past 18 years (2000 to 2018). The reason for reviewing the literature in between 2000 to 2018 was that WSN was first applied to leakage monitoring in 2004 [18]. Section 4 provides a detailed description of WSN-based and IoT-based sewer and water pipeline leakage and sinkhole monitoring and detection methods. Detailed web-of-science analysis was performed to provide statistical data on the reviewed literature. The aforementioned information was collected using keywords such as pipeline leakage, sinkhole, subsidence, sensors, and wireless sensor network.

Table 1. Results of patent search pertaining to sinkhole detection and underground pipeline leakage monitoring.

No.	Patent Database/Office	Total Search Results	Pipeline Leakage Monitoring	Sinkhole Detection	Organization/Country
1	Google Patent Database	251	10	9	Google
2	The United States Patent and Trademark Office	100	8	3	USA
3	Korean Intellectual Property Rights Information Service Database	18	4	2	South Korea
4	State Intellectual Property Office of the P.R.C	15	4	0	China
5	Canadian Intellectual Property Office	15	9	0	Canada
6	World Intellectual Property Organization Office	14	1	0	United Nation
7	Espacenet Patent Search Database	6	2	4	Europe
8	German Patent and Trade Mark Office	6	0	1	Germany
9	Taiwan Patent Search System Database	4	3	0	Taiwan
10	Australian Government Patent Office	2	2	0	Australia
	Total	431	43	19	

3. Patent Analysis

Patent analysis is a unique management tool that deals with a company's technology and strategic management of a product development or service development process [19]. By converting patent data into competitive intelligence, companies can monitor current technological advancements, predict technology trends, and plan for potential competition based on new technologies. This section discusses the patents closely related to the development of leakage detection techniques in underground sewer and water pipelines to prevent potential adverse effects such as drinking water shortage and sinkhole formation.

In this regard, the authors first performed a preliminary search using several keywords related to sinkhole detection and its primary cause—water leakage in sewer pipelines—by utilizing Espacenet Patent Search, which is mainly used in European countries. The authors rapidly realized that different patent-search databases must be used because of the limited number of patents published in this domain. To select patents most relevant to the subject of concern in this study, 10 patent search engines, including official websites, were accessed, as described in Table 1. The database search yielded 431 patents. Approximately 10% (43) of these patents were related to underground pipeline monitoring and leakage detection using WSN, whereas only 4% (19) were related to sinkhole monitoring and detection using WSN.

Table 2 describes the scope of the patents listed in Table 1 in terms of academic and industrial relevance with regard to investigations concerning underground pipeline leakage monitoring and sinkhole formation during 2000 to 2018. Relevant patents have been classified based on the leakage monitoring/detection of underground sewer and water pipelines and sinkhole formation. Out of 43 patents, only 16 patents for pipeline leakage monitoring are cited in Table 2, which are the most relevant to the subject of concern in this study, and 19 patents relevant to sinkhole occurrence are cited. Table 2 further classifies the patents based on whether they belong to the class of safety equipment or method, robotic devices or sensors, or experimental setup or method. As shown in Table 2, until 2010, there were no patents on the implementation of WSN and IoT. However, the application of WSN and IoT has increased over the past two decades. There have been no significant technical advancements in the development of equipment or devices to ensure safety against pipe bursts and sinkhole formation.

Among the patents (16 leakage monitoring and 19 sinkhole detection or monitoring patents) shown in Table 2, the patents most relevant to the subject of concern (patents that used WSN) were selected for further study. Among these filtered patents, six were related to natural sinkhole and water pipeline monitoring and leakage detection methods, while only a single patent was related to human-induced sinkhole detection or monitoring using WSN [39]. The remaining patents mentioned in Table 2 are relevant to leakage monitoring and sinkhole monitoring. However, they do not use a WSN system. The authors created Figure 2 in order to illustrate the concept of different patented methods, which used a WSN system.

Table 2. Patents published concerning underground water and sewer pipeline leakage and sinkhole formation.

Patent Subject Classification	Subcategories	References and Years of Patent Publication							
Leakage monitoring	Safety equipment or method	-	-	[20]	-	-	-	-	[21,22]
	Devices (robots/sensors)	-	-	-	[23,24]	[25,26]	[27,28]	[29–31]	-
	Experimental setup or methods	-	-	-	-	[32]	-	[33]	[34,35]
Sinkhole detection or monitoring	Safety equipment or method	-	-	-	[36]	[37]	-	-	-
	Devices (robots/sensors)	-	-	-	-	[38–40]	[41]	[42–45]	[46]
	Experimental setup or methods	-	-	-	-	[20,47,48]	[49]	[50,51]	[52,53]
Years		2000	2010	2013	2014	2015	2016	2017	2018

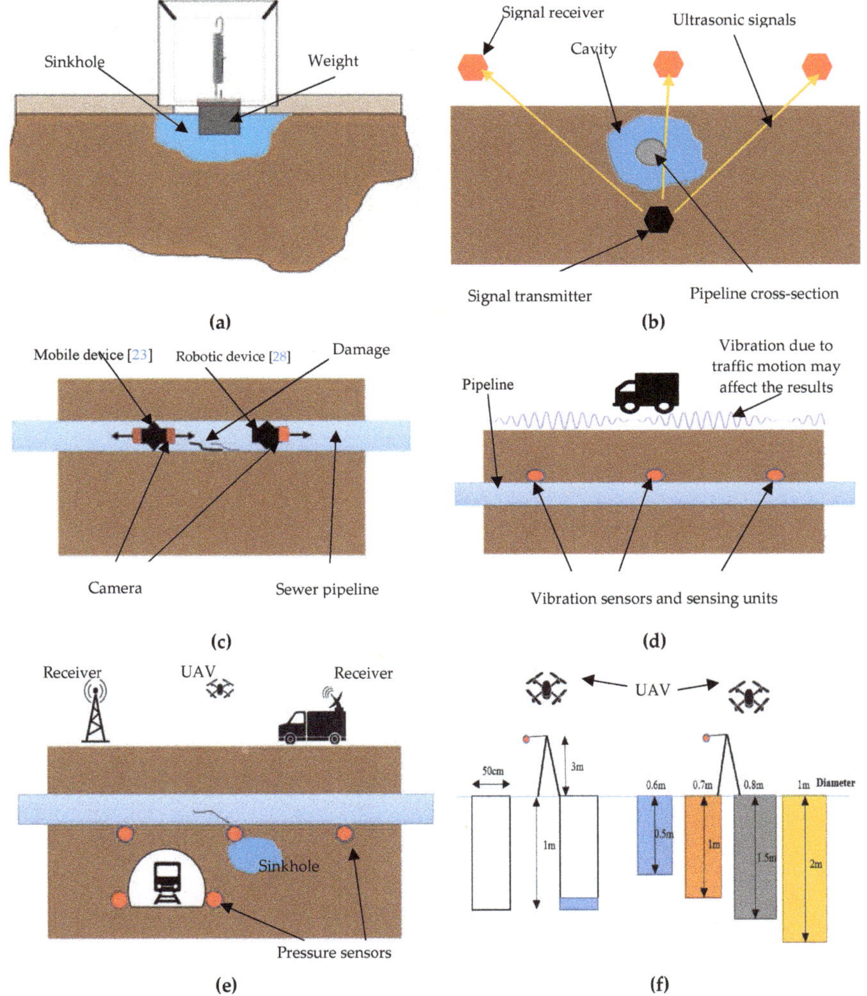

Figure 2. Visual representation of underground water pipeline leakage and sinkhole monitoring methods based on wireless sensor networking (WSN) systems; (**a**) sinkhole detector; (**b**) ultrasonic medium change detection; (**c**) remote pipeline inspection method and mobile-type monitoring device for pipelines; (**d**) pipeline safety and monitoring device; (**e**) sinkhole monitoring system based on pressure sensors; (**f**) simulated sinkhole to develop a neural network learning database for sinkhole detection.

3.1. Patents Related to Sinkhole Detection or Monitoring

First, the patents relevant to sinkhole detection and monitoring are overviewed with their limitations. There are various patents with various operating and functioning techniques such as safety methods to overcome the sudden collapse or sinkhole occurrence [36,37]. Likewise, some patents were invented for sinkhole monitoring via developing experimental setups in laboratories, such methods and experimental setup have been cited in Table 2 [47–53]. However, herein authors just considered patents with applications of WSN. As reported in relevant patents, a device for detecting underground sinkhole formation is partially buried in the ground. A typical sinkhole detector comprises of a cable, a control circuit, and a weight, as depicted in Figure 2a. If a sinkhole is created in the vicinity of the device diameter or under the detector, the device releases the attached weight. This activates a circuit that sends a notification to an administration center [45]. The application of the said device is limited in the sense that it can only detect sinkholes created under the detector or those in the vicinity of its diameter. As the device is of limited size and does not include a WSN system, it cannot be installed for sinkhole detection throughout a pipeline network. Similarly, a sinkhole monitoring system based on pressure sensors has been developed for monitoring human-induced sinkholes. In this method, pressure sensors are placed close to underground utilities such as pipelines and subway tunnels. These pressure sensors comprise of four springs fixed on an axle [39]. When pressure changes in the vicinity of a pressure sensor, the sensor detects the change and wirelessly sends alarm signals to an end user, as depicted in Figure 2e. The signals transmitted underground by the pressure sensors are received by an unmanned aerial vehicle (UAV) or a receiver mounted on a moving vehicle such as a train [39].

A patent describes the application of a neural network learning database for sinkhole detection. This patent is related to the development of a method for creating a small simulated sinkhole cavity characterized by the shape, texture, and complex background of an actual sinkhole. This method utilizes a drone fitted with a thermal imaging camera to detect sinkhole cavities at heights of the order of 10 m [49]. The images captured by the thermal drone camera can be used to construct an image database for the development of a neural network based sinkhole detection model. Figure 2f depicts the schematic of a simulated sinkhole with shapes similar to those of actual sinkholes. A similar device with abilities to detect changes in an underground medium has been developed. This device is used to detect any leakage or deterioration in underground liquid pipelines carrying oil or water [38].

As depicted in Figure 2b, a transmitter device is buried in the middle layer of soil, and several signal receivers are placed above the ground to receive wireless ultrasonic signals emitted from the transmitter. The changes in the paths of these signals while propagating towards the receiver end can be detected by measuring their travel time between the transmitter and receiver. Any difference in travel time indicates a change in the underground medium. A limitation of this device is that ultrasonic signal propagation can be affected by the vibrations and sound effects generated by vehicular traffic on the ground or any other source during practical use.

3.2. Patents Related to Leakage Monitoring

Similar to the above-mentioned devices for sinkhole detection and monitoring, mobile-type monitoring devices for pipeline inspection have been developed to overcome the leakage issues in sewer pipelines [23–31]. Meanwhile, other experimental setups [32–35] and safety equipment [20–22] have been invented for pipeline leakage monitoring. However, these experimental setups and devices did not used the WSN system for pipeline monitoring. Leakage in sewer pipelines that transport wastewater and raw sewage to wastewater treatment plants leads to the contamination of surrounding soil and groundwater and the creation of sinkholes. As depicted in Figure 2c, this problem can be resolved by utilizing modern technology in the form of a remote pipeline-inspection device [28]. This robotic device comprises a camera to capture the images of pipeline interiors, and this data can be transmitted via a communication network for detailed visual inspection, as depicted in Figure 2c. A similar mobile device capable of moving forward and backward with cameras installed on its front and rear sides, as depicted in Figure 2c, can be used to inspect the existence of any leakage, crack, or damage [23] within a pipeline. The said device can be placed inside a pipeline, moved forward up to a fixed point, and made to automatically return to its starting point. The inspection of captured images helps conclude whether the sewer pipe mains has been damaged or not.

These devices [23,28] are currently under use. However, automated image processing and analysis techniques are necessary for replacing human visual inspection, which requires time and effort. With regard to real-time application, the said devices are expensive, time consuming, and require human intervention. Moreover, the devices can only be used to inspect a pipeline section by section, and they cannot inspect the entire pipeline network at once. However, AI-based automation methods are being investigated to facilitate more efficient maintenance compared to visual-inspection methods.

The risk of sinkhole creation has increased because of various ground subsidence phenomena. The method shown in Figure 2d is related to a safety monitoring system for a pipeline, wherein a sensing unit attached to a pipeline network generates position change and vibration signals at regular intervals in accordance with fluid flow inside the pipe to detect leakage and/or rupture in the pipeline. Using the pipe network information already stored in a management server, it is possible to track the position of the damaged parts of a pipeline via the interpretation of received position-change and vibration signals [24]. Sensing units are attached to pipelines using a magnetic outer-bottom portion comprising two subunits—(1) position change sensing unit including acceleration and gyro sensors to generate a position change signal; (2) vibration sensing unit that converts the vibration signals of the installed weight into electrical signals. The device provides the advantage of being able to detect changes in the vibration pattern and position due to the changes in flow velocity when a pipeline is distorted or broken. However, the reliability of the said system may deteriorate because of the transmission of an abnormal signal in the event of strong vibrations caused by construction work or rail and road traffic load.

4. Literature Review

4.1. Web of Science Analysis

A web of science analysis was performed to provide relevant articles and other information such as citations and article categories. From the standpoint of applying the WSN system to monitor and prevent pipeline leakage and sinkhole occurrence, the following different sets of keywords have been used according to trial and error—(1) leakage, pipeline, wireless sensor, and monitoring; (2) leakage, pipeline, and wireless sensor; (3) leakage, water pipeline, and underground; (4) sinkhole and underground; (5) leakage, pipeline, and sinkhole; (6) pipeline, sinkhole, and wireless sensor; and (7) human-induced sinkholes.

Research articles published between 2000 and 2018 with different sets of keywords are summarized in Table 3. It can be observed that a majority of extant studies focused on either leakage monitoring and detection of underground fluid pipelines (numbers marked in red under the water pipeline monitoring category in Table 3) or natural sinkhole monitoring and detection (numbers marked in red under the natural sinkhole category in Table 3). The numbers in orange under the others category in Table 3 indicate the results that are not relevant to the scope of this study (i.e., pipeline corrosion, gas pipelines, underwater pipelines, and aboveground pipeline leakage monitoring). In other instances, search results with different keyword combinations yielded identical results. Only one article was found to be related to sinkhole detection and monitoring due to leakage in underground sewer/water pipelines, as described in Table 3 (the number marked in green).

In accordance with Table 3, after filtering the duplicated articles, 47 articles published in different journals related to sewer/water pipeline and human-made sinkhole monitoring using WSN systems were observed.

Among those 47 articles, majority of the contributions (i.e., 35 articles) were published in journals related to technical domains such as computer science and information systems, telecommunication, electronics, and WSN systems. However, a relatively smaller number of articles (i.e., 12 articles) were published in journals related to application domains such as civil engineering, water resources, soil mechanics, and environmental sciences.

Table 3. Web of science analysis of articles and contributions of different authors related to the field of interest (time span 2000–2018).

Search Trial	Keywords \ Category	Total Articles	Human-Induced Sinkholes	Water and Sewer Pipeline Monitoring	Natural Sinkhole	Others	No. of Citations
①	Leakage, pipeline, wireless sensor, monitoring	22	00	18	N.R [1]	04	146
②	Leakage, pipeline, wireless sensor	23	00	17	N.R [1]	05	154
③	Leakage, water pipeline, underground	17	00	09	N.R [1]	08	185
④	Sinkhole, underground	126	01	02	51	72	956
⑤	Leakage, pipeline, sinkhole	03	00	00	02	01	-
⑥	Pipeline, sinkhole, wireless sensor	00	00	00	00	00	00
⑦	Man-made, sinkholes	23	01	00	13	09	151
	Total	**213**	**01 [2]**	**46**	**66**	**99**	**1592**

[1] not related to the intended field of study, [2] total sum of the column is "01" because both articles are duplicate.

The histogram in Figure 3 depicts the above-mentioned 47 articles with the number of publications per year between 2000 and 2018 pertaining to the use of WSN for underground water and sewer pipeline leakage and burst monitoring, and man-made sinkholes.

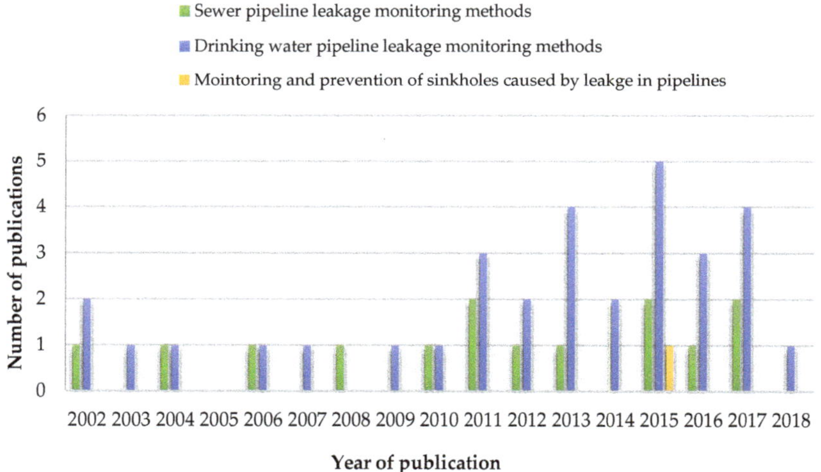

Figure 3. Distribution of number of relevant studies published between 2000 and 2018.

4.2. Overview of Previous Review Articles Related to Underground Water and Sewer Pipeline Monitoring

The WSN system has been tested by monitoring various types of faults in pipelines carrying different substances such as gas, oil, and water [54,55]. Various researchers have reviewed previous methods used for monitoring underground water and sewer pipelines, and they have suggested prospects for improvement. A hybrid method (combining any two different pipeline leakage monitoring methods) to form a new and more reliable solution was suggested in articles [56] and [57]. According to [58,59], there are two critical issues that need improvement in terms of development of a preferable underground WSN system—(1) efficiency of network-communication range and (2) energy efficiency of WSN.

Adedeji et al. [60], in their review article, discussed and suggested the importance of sensor-deployment strategy for reliable and scalable data propagation in the soil channel. Similarly, different methods for underground pipeline-leakage monitoring were compared in [61]. Acoustic reflectometry was found to be the most reliable method among all the reviewed methods, because it is applicable to various linear pipeline layouts (i.e., straight, short, long, and zigzag) while being one of the most energy efficient and reliable methods. Table 4 lists extant review articles published on monitoring underground water and sewer pipeline leakage using WSN-based systems in the reverse chronological order. The listed articles also discuss suggestions for future research and challenges encountered during each review.

Table 4. Summary of future research suggestions pertaining to underground water and sewer pipeline leakage and sinkhole monitoring using WSN-based systems, as discussed in extant review articles.

Article	Future Research Suggestions	Research Field	Publication Year
Abdelhafidh et al. [56]	Combination of several leak detection methods to form one hybrid system for better results.	Computer Sciences	2018
Adedeji et al. [60]	Sensors deployment strategies in WSN nodes need to be considered.	Electrical Engineering	2017
Datta et al. [61]	Acoustic reflectometry is most suitable for leakage and blockage in underground pipelines.	Mechanical Engineering	2016
Sheltami et al. [57]	Hybrids of different WSN-based pipeline monitoring techniques to enhance the detection and localization of leakage.	Computer Sciences	2016
Obeid et al. [62]	Integrated energy-aware system on chip solution for non-invasive pipeline monitoring.	Electronics and Communication	2016
BenSaleh et al. [58]	Developing energy-efficient nodes with adequate sleep and wake-up mechanisms.	Electronics and Communication	2015
BenSaleh et al. [63]	Develop a robust and reliable system, which is cost-effective, scalable, and customizable in future.	Electronics and Communication	2013
Tariq et al. [59]	Improvement in the communication radius efficiently.	Computer Sciences	2013

From Table 4, it can be concluded that all suggestions for future research are related to the fields of computer science, electrical, electronics, and telecommunication sectors. Therefore, it can be considered that all researchers, in their respective review articles, aimed to present the challenges encountered in the field of WSN, sensor-deployment strategies, and communication radius of the system. However, in the fields of soil mechanics, civil infrastructure, and geology, the after effects of leakage on the surrounding soil must be considered. As mentioned earlier, leakage can lead to soil erosion and ultimately results in sinkholes.

4.3. Architecture of Underground WSN Systems for Pipelines

Different networks and communication protocols have been used to monitor the underground and aboveground water pipelines for the transmission and propagation of sensor data to end users. A WSN consists of various units that work together to gather desired data in a specific environment where communication can be established over a wireless channel, including sensing, computing, and communication devices [64,65]. There are three tiers—(1) sensor tier; (2) master-node tier; and (3) end user—in any underground WSN system for pipelines, as depicted in Figure 4. In the sensor tier, sensors are placed along the pipeline network to wirelessly transmit assigned data to the master node, which subsequently transmits data to the end user(s). The main distinction between underground and terrestrial WSN systems is the communication medium (soil or air) [66].

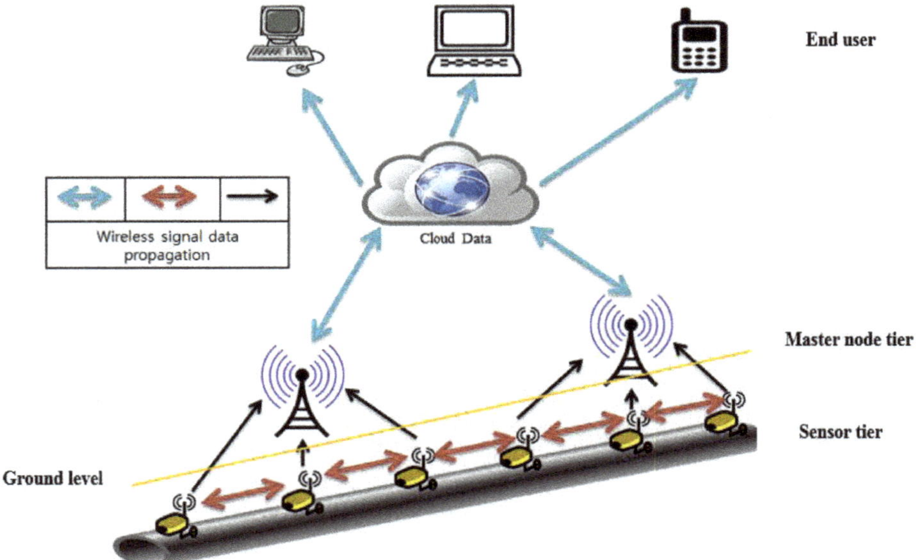

Figure 4. Typical WSN architecture for underground water pipeline monitoring.

Many wireless sensor communication protocols and interfaces, such as Wi-Fi, ZigBee, LoRa, and Bluetooth, have been used to overcome multiple transducer connectivity problems [67,68]. Table 5 explains the characteristics of commonly used communication protocols along with their properties such as speed, range, frequency, and limitations. Bluetooth is applicable for communication over a short range of 1–100 m with better data transmission speed in comparison to ZigBee and GPRS. Certain communication protocols such as GPRS and 3G possess low data transfer speed but their range is up to 10 km. Some communication protocols such as 5G, 4G, 3G, and LTE provide a somewhat longer communication range with high data-transmission speed, whilst requiring a large spectrum fee [69,70]. It is, therefore, essential to select a communication protocol based on the end application and availability of required resources to achieve desired results of cost-effectiveness, reliability, and low power consumption. The communication protocols listed in Table 5 are in order of increasing range of different protocols.

Table 5. Communication protocols and their properties [69,70].

Communication Protocol	Speed	Range	Frequency	Limitations
Bluetooth	3 Mbps	1–100 m	2.4–2.48 GHz	Short range
Wi-Fi	300 Mbps	100 m	2.4–5.4 GHz	Short range
ZigBee	250 Kbps	75 m	2.4 GHz	Low data rate and short range
Z-Wave	40 Mbps	30 m	868.42–908.42 GHz	Low data rate and short range
GPRS	Up to 170 Kbps	1–10 km	900–1800 MHz	Low data rate
3G	384 Kbps-2 Mbps	1–10 km	2.5, 3.5, 5.8 GHz	Costly spectrum fee
WiMAX	Up to 75 Mbps	10–50 km	1–30 MHz	Not widespread

4.4. Water Pipeline Leakage Monitoring Methods Based on WSN System

The loss of water is one of the most critical issues, particularly in urban localities. In an urban framework, the occurrence of any severe pipeline leakage issue such as water supply interruption and traffic flow interruption may intensify, thereby resulting in an increase in maintenance costs. Therefore, the implementation of a robust, reliable, and energy-efficient WSN-based system is imperative to overcome water loss issues and after effects of leakage [71]. Ng et al. [72] proposed a vibration sensor-based sound variation detection device that records suspicious water leakage sounds and compares them with normal pipeline sounds, thereby identifying the symptoms of water leakage. The device was used on two types of pipelines—metal and PVC—but not concrete ones. The said device has a disadvantage, that is it only functions properly when installed near a leak point, and it is immensely difficult to predict the exact location of leakage. It is also not clear to judge the utility of the device when on concrete pipes.

Four things—cost efficiency, reliability, power consumption, and sensor placement across the pipeline network—must be considered regarding monitoring underground sewer/water networks and sinkholes using WSN. Cattani et al. [73] proposed a method called ADIGE based on the long-range LoRa WSN technology to reduce the number of gateways uploading data gathered from wireless sensors. The proposed ADIGE system comprised two parts—(1) sensing, controlling, and collection of data; and (2) data management, including data fusion and analysis—as depicted in Figure 5a. This method was designed to achieve high reliability and energy efficiency along with wider coverage of water pipeline monitoring using a large number of low-cost sensors for collecting more real-time information at multiple locations instead of using lesser numbers of expensive sensors installed at fewer locations. However, using the WSN system to accurately identify leakage locations inside a long-running water pipeline and the effects caused by leakage remains a major challenge for researchers.

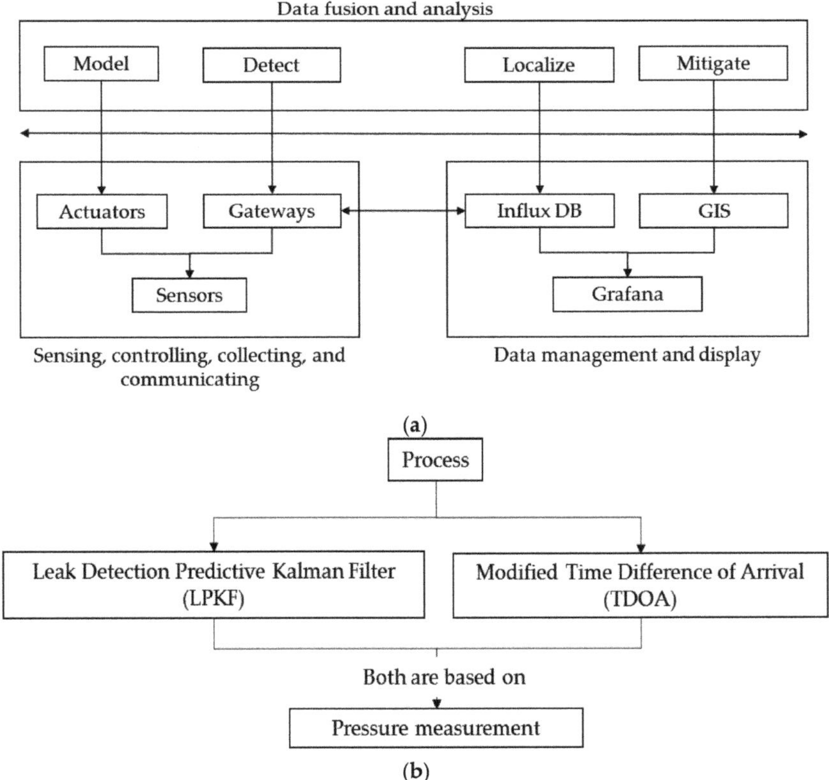

Figure 5. Leakage monitoring methods using different combinations and algorithms for leak detection and localization: (**a**) System combination of method based on the long-range LoRa WSN technology; (**b**) different types of methods and algorithms used for leak detection and localization.

Karray et al. [74] also proposed a method to identify leakage locations within a pipeline network. This method involved the use of a combination of leakage detection algorithms, localization algorithms, and system-on-chip (SoC) architecture. This solution was based on two methods, as described in Figure 5b. Instead of using multiple algorithms at once to compromise battery life, the author preferred using single algorithm—the Kalman Filter (KF)—to conserve energy. The method used for this purpose was referred to as energy-aware reconfigurable sensor node for water pipeline monitoring (EARNPIPE). A plastic pipeline capable of supporting pressures of up to 25 bar and 1000 m^3 volume of water was used to supply water using 1 hp power supplied by the laboratory test bed motor. A force-resistive sensor (FSR) was used to measure the pressure within the pipelines. The said method was proposed for use in long-distance pipelines installed on the ground but was not suitable for use with underground pipelines.

Sun et al. in [75] suggested a new solution for water pipeline monitoring, leakage, and burst detection. Real-time leakage and burst detection were set as the primary objectives of this method. Considering the propagation of sensor signal data in soil, a WSN system based on magnetic induction (MI) was proposed. The sensors were distributed into two layers—(1) the hub layer and (2) in-soil sensor layer. In the hub layer, pressure and acoustic sensors were deployed at checkpoints and pump stations inside pipelines to measure pressure and vibration changes caused by leakage, as depicted in Figure 6. After measuring the values of pressure and vibration, data were wirelessly sent to the administration center using MI channels. Whereas in the in-soil layer, sensors were deployed along the pipeline to measure different soil parameters such as temperature and humidity. This was accomplished by rolling MI relay coils around the pipes. Beyond the system architecture and framework presented in this article, more work regarding evaluating the system performance is required to be performed by deploying MI-based WSN for underground pipeline monitoring (MISE-PIPE) in real-life applications.

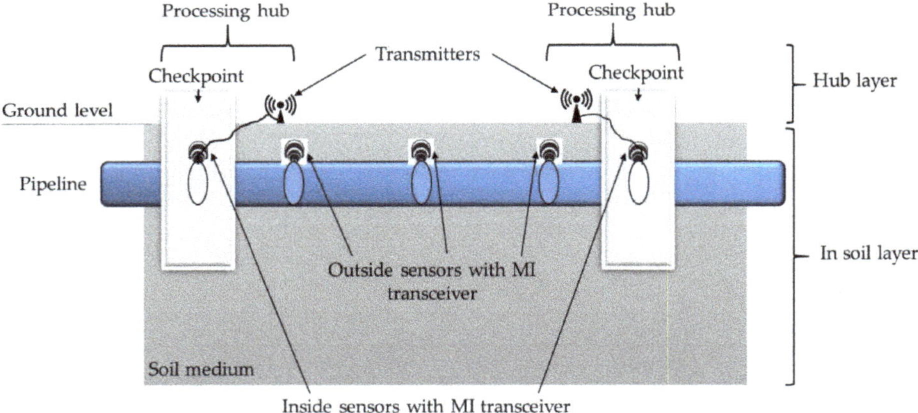

Figure 6. System architecture of magnetic induction (MI)-based WSN for underground pipeline monitoring (MISE-PIPE) [75].

4.5. Sewer Pipeline Leakage Monitoring Methods Based on WSN Systems

Leakage in sewer pipelines needs to be considered as a potential source of underground cavity creation and sinkholes in urban areas. According to Kuwano et al. [76], the sewer pipe mains buried underground for over 25 years demonstrate a remarkable increase in cracks and defects. Researchers and scientists have contributed toward the development of means to address issues related to cracking, leakage, and bursting of sewer pipelines. Kim et al. [77] proposed a novel RFID-based autonomous mobile device monitoring system for pipelines, called RAMP. The mobile robotic device comprising sensors (visual-, chemical-, pressure-, and SONAR-sensing) was capable of moving inside the pipelines along the direction of fluid flow whilst monitoring the presence of any defect on its way and localizing the same. The primary function of the robotic agent was pipeline inspection. However, the system requires a structure with improved fluid resistance and enhanced mobility of the robotic agent because at present its usage is only limited to inside straight pipelines. Additionally, to achieve better results, energy efficiency and sensor power must be improved. Such methods are equally applicable to water, sewer, and pipelines [77].

Similarly, PIPENET is a system proposed for pipelines with large diameters, such as sewer and drainage pipelines, and comprises a WSN-based system for underground pipeline leakage detection and localization [78]. For real-time monitoring, the system was deployed in collaboration with Boston Water and Sewer Communication (BWSC), USA. Their deployment primarily focused on two critical applications—(1) sewer collector's water level monitoring and (2) hydraulic and water quality monitoring [79]. The method aimed to detect and localize the presence of any leakage or burst within the pipeline by collecting the hydraulic and vibration data at a high sampling rate using different sensing parameters—flow, pH, vibration, and pressure—and data collection was performed for an extended period of over 22 months in the city of Boston, USA. The experiment was performed both inside a laboratory and in the field to compare the experimental results and develop an algorithm for leakage detection and localization. Some limitations of this method include false alarms and low energy efficiency. Considering sewer pipelines, robot-sensing devices have been preferred to monitor the condition and performance of pipelines [77].

Most robots used for the inspection of sewer pipelines moved along one direction (straight). In contrast, KANTARO was an innovative, fast, and robust sensing device intended for use in sewer-pipeline inspection. The device could move in a straight path as well as bend around the curves. For this, a particular patented mechanism called the "Sewer Inspection Robot (nSIR) Mechanism" was developed [80]. KANTARO was equipped with a fisheye camera mounted to detect any damage or blockage within the sewer pipeline network. It could only fit in pipelines with internal diameters in the range of 200–300 mm and included lithium-polymer batteries for power supply to motors, sensors, computer systems, and underground wireless communication modules.

4.6. Pipeline Leakage Monitoring Methods Based on Computer Vision and Image Processing

Other such approach involved the use of closed-circuit television (CCTV) image processing by concerned researchers for the detection of defects and damages in sewer pipelines. Manual interpretation of images and videos has previously been used, and those methods were observed to be more time consuming, labor intensive, and the results obtained were deemed to be less reliable and possibly inaccurate. In contrast, automated defect detection methods have been proposed by researchers using artificial intelligence (AI) and CCTV image and video processing techniques, such as the deep learning technique called the faster region based convolutional neural network (faster R-CNN) [81]. The faster R-CNN model has been demonstrated to detect defects in sewer pipelines much faster with higher accuracy.

However, the faster R-CNN approach only works on static images. Similarly, the authors in [17] used an automated approach to detect cracks, deformations, joint displacements, and settled deposits in sewer pipelines. Their approach was based on image processing and mathematical formulations to analyze the output obtained from CCTV images. To validate the performance of the proposed method, the authors in [17] constructed a confusion matrix. The results observed for the accuracy of crack, displaced-joint, and settled-deposit detection were found to 74%, 65%, and 54%, respectively, which could be improved by increasing the number of images captured. A drawback of the proposed methodology was that the method relied heavily on the expertise and experience of operators. From the literature overviewed in earlier sections it is evident that over time, technologies have developed from the visual inspection of sewer/water pipelines to the application of wired sensors, wireless sensors, IoT, big data, and AI technologies, as illustrated in Figure 7.

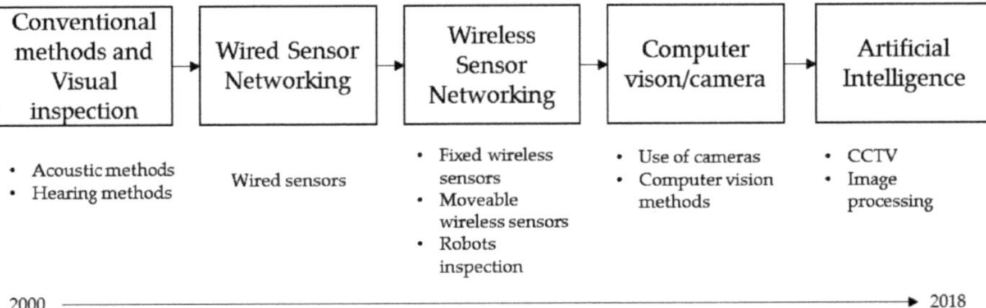

Figure 7. Illustration of technological developments in underground pipeline network inspection methods over time.

Using WSN-based systems, previously discussed sewer and water pipeline surveillance models (Sections 4.4 and 4.5) are summarized in Table 6. This table provides a detailed perspective of WSN-based sewer and water pipeline leakage monitoring methods in terms of the protocols used for wireless communication between the transmitter and receiver, type of wireless sensors used (non-invasive or invasive sensors, static or mobile sensors, etc.), field of application, and water or sewer pipeline network. Table 6 makes it easier to determine the most feasible method to monitor water and/or sewer pipelines under certain conditions. Meanwhile, it also helps to select the most feasible methods, which can be further modified for the application of sinkhole due to leakage. For example, Sun et al. [75] used sensors both inside and outside the pipelines to detect various factors related to leakage. In such methods, more soil property measuring sensors can be added to further collect data related to the factors (soil moisture, density, temperature, overburden, and porosity, among others) that contribute to the occurrence of sinkholes.

Table 6. Comparison of various methods for leakage detection in water and sewer pipelines.

Method (Ref.)	Communication Protocol	Types of Sensors	NIS [1]	IS [2]	Mobility	Application Field
[82]	-	-		•	Mobile	Sewer pipeline
[72]	3G	Acoustic sensors.	•		Static	Drinking water pipeline (PVC and steel pipes)
[83]	GSM	Ultrasonic and water sensors.	•		Static	Leak in home pipelines and the water level of the tank.
[73]	LoRa	Adige sensors.	•		Static	Drinking water pipeline
[74]	Bluetooth	Pressure sensors (FSR).	•		Mobile	Drinking water pipeline
[75]	Magnetic induction	Soil property, pressure, and acoustic sensors.	•	•	Static	Drinking water pipeline
[17]	3G	Temperature, pressure, flow rate, and hydrophone.		•	Mobile	Drinking water pipeline
[18]	RF signals	Acoustic sensors.			Static	Drinking water pipeline
[77]	RFID	Vibration sensors and robot agents.	•		Static	Drinking water and sewer pipeline
[78]	Bluetooth	Vibration, flow rate, and pressure sensors.	•		Static	Drinking water pipeline
[80]	RF signals	IR, tilt and image sensors, and laser scanner.		•	Mobile	Sewer pipeline
[3]	ZigBee	Pressure, piezoelectric sensor, gyroscope, and accelerometer.	•		Static	Drinking water pipeline
[84]	CAN (wired), X stream, X Bee, and Wi-Fi (wireless)	Vibration sensors and accelerometer.	•		Static	Drinking water pipeline
[85]	RF signals	Hydrophone.		•	Mobile	Drinking water pipeline
[86]	X stream, X Bee, and Wi-Fi	Accelerometer.	•		Static	Drinking water pipeline
[87]	RF signals	FSR sensors.	•		Static	Drinking water pipeline
[88]	ZigBee	Vibration sensors.	•		Static	Drinking water pipeline
[89]	-	-	•		Mobile	Sewer pipeline
[90]	WSN	Pressure and flow rate sensors.	•		Static	Sewer pipeline
[91]	-	-		•	Mobile	Oil and gas
[92]	-	-	•	•	Mobile and Static	Any pipelines

[1] NIS: Non-invasive sensors (sensors placed outside a pipeline), [2] IS: Invasive sensors (sensors placed inside a pipeline).

4.7. Sinkhole Monitoring and Detection Methods

A record of the past 60 years of Gauteng, South Africa, reports the occurrence of more than 3000 sinkholes, including natural and human-induced ones [93]. The Maryland geological survey reported a total of 139 sinkhole occurrences, 51 of which occurred naturally, while the remainder were the human-induced type [94]. According to the British Geological Survey (BGS) (a British government institute), 10% of all sinkholes occur as a result of leakage in underground pipelines [82]. As discussed in previous sections (Sections 4.4–4.6), various methods are adopted for leakage monitoring. However, sinkholes due to leakage require proper concentration as they result in considerable damage. According to the US Geological Survey, nearly $300 million per year is spent on the reconstruction of, and compensation for, damage caused by sinkholes to the infrastructure [95]. The frequency of sinkhole occurrence has increased considerably in South Korea, since 2010, mainly because of underground construction in urban areas. An account of the total occurrence of sinkholes in Seoul, South Korea, between January and July 2014 reported that ruptured sewer pipelines resulted in approximately 85% of all man-made sinkholes, 18% were triggered by excavation during construction activities, and the remaining 3% were caused by leakage in freshwater pipelines [96].

Sinkholes occurs on roads, highways, and railways owing to the infiltration of leaking underground pipelines or rainwater. These cavities can be detected beneath the surfaces using different monitoring and detection methods. Although many sinkhole detection methods have been developed, it is still difficult to predict the formation of sinkholes. There are alternative methods to predict when and where sinkholes will form. Researchers have drawn maps for sinkhole-prone regions that predict risk based on the ground composition or other local sinkholes; however, this still does not provide a definite answer.

4.7.1. Sinkhole Monitoring Methods Using Image Processing and Radar

Many researchers have contributed toward the prediction of the occurrence of natural sinkholes via the use of diverse technologies, such as the trenching method along with ground penetration radar (GPR), electrical resistivity tomography (ERT), high-precision leveling [97], Brillouin optical fiber sensor [98], high-precision differential leveling [99], laser imaging detection and ranging (LIDAR), and interferometric synthetic aperture radar (InSAR) [100]. The combined use of the geographic information system (GIS) and analytical hierarchical process (AHP) has previously been proposed to identify hazard zones in Kuala Lumpur and Am pang Jaya of Malaysia that are susceptible to the occurrence of natural sinkholes on the basis of five criteria—lithology, groundwater level decline, soil types, land use, and proximity to groundwater wells [4]. However, the human-made (owing to leakage) requires proper concentration.

As in the UK, the British Geological Survey used digital map data for geohazard monitoring in urban areas. This digital map data can be accessed from GIS [101]. High-resolution ground-based InSAR (GB-InSAR) and LiDAR are methods available for sinkhole monitoring. These technologies use GIS data for sinkhole mapping and are particularly effective for sinkhole tracking [100]. However, it is challenging to discern sinkholes or subsurface features based on GIS data or satellite data [102]. Soil erosion gradually leads to collapse, and the application of GB-InSAR, LiDAR, scanning, and photogrammetry methods have to scan and analyze different areas continuously to observe the changes over time, which can be expensive and time-consuming. Therefore, the current practices in urban areas for monitoring and detection of the sinkhole are limited.

Similarly, in some developed countries, ground penetration radar (GPR) is currently used for sinkhole monitoring and detection, especially in urban areas. The results of a GPR rely on the reflection of a high frequency (25–1000 MHz) electromagnetic (EM) pulse from subsurface contacts and other anomalies such as boulders and cavities. [103]. GPR uses radar pulsations to predict the changes in the subsurface. This method uses EM radiation of the radio spectrum and detects the reflected signals from the subsurface structures [104]. However, there are limitations associated with GPR, as noise effects the magnetic techniques, and the presence of clayey soils affects the transmission of EM signals from GPR [105]. Previously, the application of GPR was excluded from the investigation in Nigeria owing to the presence of clay [106].

Other methods include geomorphological mapping, borehole drilling, and air photo interpretation. Table 7 lists the methods currently available to address challenges and obtain solutions for sinkholes; their advantages and limitations are also discussed.

Table 7. Sinkhole monitoring and detection techniques.

Sinkhole Detection and Monitoring Techniques	Positive Aspects	Gaps	References
GIS with digital map data	Allows for spatial analysis, assessed specific problems using multi-criteria approach.	Scale-dependent, needs geo referencing, software, and manpower expertise.	[107]
GPR and InSAR	Subsurface detection and mapping.	Equipment (some expensive/specialist) and expertise required.	[108–110]
Leaser image detection and ranging system (LiDAR)	Provides high-resolution images for monitoring of sinkholes or subsidence.	Detection of subsurface changes are not possible; only limited to surface detection.	[111,112]
Electric resistivity tomography (ETR)	Subsurface detection and mapping.	Expensive equipment and expertise required to operate.	[109,110,113]
Drones (UAV) and photogrammetry	Spatio-temporal analysis is possible.	Limited to surface analysis and dependent on availability of drones and photogrammetry equipment.	[114–116]
WSN	Can be integrated with geological information; can integrate geophysical information.	Limited to surface-based assessment.	[3,117,118]

4.7.2. Human-Induced Sinkhole Monitoring Methods Using WSN

Soil strata and profile vary from for different regions and are characterized based on the unique geomorphological and hydrological conditions, and pipelines are not limited to areas with specific geology. If we consider the subsurface of any urban area, miles of water and sewer pipelines are passing through different soil profiles. Therefore, the effect of water leakage from sewer or water pipelines will be different for different subsurface soil strata.

A leakage in the sewer or water pipeline results in the change in underground soil and other various problems mentioned above. Over the past few years, numerous cases of sinkhole creation owing to the rapid deterioration of underground sewer and water pipelines have been reported. To overcome this issue, researchers have developed sensor-based pipe safety units to detect leakage by analyzing water leaks and pipeline behavior. Pipeline data are collected by the use of smart sensors and WSN [3]. However, the system is still being developed, and the primary purpose of the program is to develop a sinkhole risk index (SRI) via the real-time monitoring of underground construction activities. This is an individual research endeavor aimed at the prevention of sinkhole creation due to leakages or ruptures in underground sewer pipelines owing to human interference. In the said study, experiments were conducted on a buried pipeline, exposed pipeline, exposed acrylic pipeline, and test-object pipe. During the experiments, the concerned researchers used cables for the connection between sensors.

However, previously, researchers have suggested several methods for the study of the mechanism of natural sinkholes by considering different factors that influence the occurrence of natural sinkholes. Therefore, different numerical, mathematical, and experimental models have been developed. Each researcher in their experimental model considered different sinkhole mitigating factors and soil types to understand the mechanism.

Researchers considered sea side soil (mudflat, sea clay) as a soil type, and subsurface void growth, ground water level, salt dissolution, and overburden pressure as the factors responsible for sinkhole or subsidence in their experiments [119,120] as the researchers aimed to understand the mechanism of sinkhole along the seaside. Therefore, this sinkhole mechanism cannot be true for other cases, as other soil types need to be considered (clay, sand, bedrock, carbonates, etc.) and other sinkhole mitigating factors (size of cavity, subsurface soil properties, aquifer, etc.). Similarly, Tao et al. [121] also performed a numerical simulation to understand the mechanism of natural sinkholes owing to changes in ground water table. Sand and clay were used as the soil profiles for the sinkhole model. However, similar to previous contributions, this study was just limited to understanding the natural sinkholes mechanism. In addition, it must be improved to increase the simulation capabilities.

5. Discussion and Conclusions

Numerous conventional techniques have been used over the years for the accurate detection of water and sewer pipeline leakage and prevention of sinkhole creation. However, there are a few limitations in the execution of conventional techniques such as high cost, need for large workforce and experts, long execution times, and unreliable results. Over the years, conventional methods have been replaced by methods based on IoT, WSN, smart sensors, and AI; these methods are more manageable and reliable.

This article focuses on research contributions regarding the use of WSN for water and sewer pipeline leakage and sinkhole monitoring and detection, thereby providing methods to prevent the creation of sinkholes in urban areas. The proposed review was performed following a two-pronged methodology—patent analysis and literature review of methods used for monitoring underground pipeline leakage and sinkhole creation using WSN-based systems. Both patent analysis and literature review demonstrated a lack of research on the prevention or monitoring of sinkhole creation caused by leakage in water and/or sewer pipelines.

The development of modern technology has led to the development of improved water and sewer leakage monitoring systems. However, modern, technology-based monitoring systems require efficient communication technologies such as 5 G in combination with WSN to create an advanced infrastructure such as underground fluid transportation networks. Over the last decade, new wireless communication networks have been developed to serve a range of new applications and deployment scenarios. Similarly, smart cities are being developed, where WSN and 5 G are considered to be one of the key enabling technologies that will provide smart solutions to smart cities to improve the overall quality of life [122].

5.1. Review Findings

The results of a preliminary search of patent databases demonstrate that a majority of patents published in the past 18 years were related to overcoming leakage problems associated with water and sewer pipelines. However, they lack in terms of suggesting measures to counter the after effects—sinkholes and ground subsidence—of leaked water and sewer pipelines. Additionally, existing patents for sinkhole monitoring and detection face several limitations. For instance, the "sinkhole detector" can only detect sinkholes near the installed device [45]. Similarly, the patent named "device for detecting changes in underground medium" explored the use of ultrasonic sensors; however, the propagation of ultrasonic signals can be easily affected by the soil medium, traffic-induced vibrations, and sound effects generated by different means of transportations or other sources.

Similarly, the literature review conducted in this study revealed that majority of researchers are concerned with overcoming the water losses caused by leakages in sewer and water pipelines. Particularly, a burst or leak of a sewer pipeline can endanger human lives as well as damage nearby infrastructures such as railways and highways; this is because most cases of human-induced sinkholes and ground subsidence are caused by leakages in sewer pipelines [123]. Compared to naturally occurring sinkholes, human-induced sinkholes incur less damage to infrastructures and human lives and this is why they have not been considered a significant issue by researchers in the past. However, over the last few years, sinkholes caused by leakage in water and sewer pipelines have become a severe problem across the world. Natural, as well as human-induced, sinkholes cannot be directly detected, and in this regard, the development of SRI or a sinkhole risk model is considered likely to play the role of a bridge between leaky pipelines and sinkhole detection [3].

However, no prominent studies have yet been found concerning the monitoring, detection, prevention, and localization of human-induced sinkholes. There exist many factors—soil type, topography of the area under consideration for sinkhole evaluation, history of sinkhole occurrence, recharge-area category, thickness and depth of underground cavity, groundwater table, and public safety—that need to be considered during sinkhole risk or risk index evaluation [124]. Similarly, other factors such as temperature, moisture content, and porous water pressure of soil, which may also change owing to a leakage or rupture in underground water or sewer pipelines, require proper attention.

5.2. Future Research Directions

After a critical analysis of various methodologies used by researchers previously in the domain of water and sewer pipeline leakage and sinkhole induction, the authors believe that there are certain major challenges that must be addressed on priority. Sinkhole formation, or collapse of the ground caused by a rupture or leakage in underground sewer and water pipeline, requires considerable attention. Similarly, there exists an urgent need for the development of a human-induced SRI to determine the magnitude of the occurrence of sinkholes induced by underground sewer and water pipeline leakage.

The authors advocate toward utilizing a set of technological tools to assess the occurrence of sinkholes by prioritizing the critical factors associated with them. State-of-the-art technologies such as AI and WSN can be used in combination to monitor and access the geographical changes and the locations that are at a higher risk of sinkhole formation owing to leaked underground pipelines. Image processing techniques based on AI platforms can be implemented to monitor disruptions at both the ground surface level and the underground pipeline level. Combining these techniques with WSN, which uses sensors to determine soil properties such as moisture, density, porosity, pH, temperature, and bearing capacity, can help to gain a better understanding of the processes at this geographical location.

As WSN was previously used to analyze the physical properties (vibration, flow rate, sound, and so on) of pipelines and their flow in order to monitor leakage, it did not concentrate on the properties of the soil profile, which can change after the interaction of soil with water leaked from the sewer/water pipelines. Thus, the systematic approach proposed in this paper can pave the path for future research in this area. In the future, such technological tools can be used in smart cities to overcome the issues of pipeline leakage and sinkhole formation, where radio technologies such as 5 G enables smart city networks to support interconnected infrastructure elements and to manage big data from existing smart infrastructures [122].

An evaluation of previous research publications in this area clearly shows that a majority of the methods used are based on laboratory experiments and the findings and proposals appear to be difficult for practical application. Therefore, to achieve significant improvement, we suggest implementing collaborative exercises between research institutes and water supply agencies.

5.3. Conclusions

Based on the analysis reported in this study, it can be concluded that the key focus of research in existing studies has been concerned with (1) the use and enhancement of wireless sensor networking systems; (2) types and quality of sensors; and (3) improvement of hardware and software systems adopted for underground water and sewer pipeline monitoring. Although soil properties, such as bearing capacity, water content, soil density, air voids, pH level, ground subsidence, and others, change owing to pipeline leakage, these have seldom been considered in previous investigations. Nonetheless, these parameters are critical with regard to triggering sinkhole formation and ground subsidence owing to leakage in water and sewer pipelines. This article presents a state-of-the-art review of different methods for monitoring sinkhole and underground water and sewer pipeline leakage and their effects whilst considering the use of applications based on IoT and WSN systems over the past two decades.

This review would be of interest to researchers intending to work in this area as it serves as a platform to direct future studies whilst accounting for challenges likely to be encountered during the same.

Author Contributions: H.A. and J.C. wrote the original draft supervised by J.C.

Funding: This work was supported and funded by the Basic Science Research Program through the National Research Foundation of Korea (NRF) and funded by the Ministry of Education (2016R1A6A1A03012812 and 2017R1D1A1B03036200).

Conflicts of Interest: The authors declare no conflict of interest.

References

1. Karoui, T.; Jeong, S.Y.; Jeong, Y.H.; Kim, D.S.; Karoui, T.; Jeong, S.Y.; Jeong, Y.H.; Kim, D.S. Experimental study of ground subsidence mechanism caused by sewer pipe cracks. *Appl. Sci.* **2018**, *8*, 679. [CrossRef]
2. Oren, G.; Stroh, N. Antileaks: A device for detection and discontinuation of leakages in domestic water supply systems. *Eur. J. Young Sci. Eng.* **2012**, *1*, 10–13.
3. Kwak, P.J.; Park, S.H.; Choi, C.H.; Lee, H.D.; Kang, J.M.; Lee, I.H. IoT (Internet of Things)-based underground risk assessment system surrounding water pipes in Korea. *Int. J. Control Autom.* **2015**, *8*, 183–190. [CrossRef]
4. Rosdi, M.A.H.M.; Othman, A.N.; Zubir, M.A.M.; Latif, Z.A.; Yusoff, Z.M. Sinkhole susceptibility hazard zones using GIS and Analytical Hierarchical Process (AHP): A case study of Kuala Lumpur and Ampang Jaya. *ISPRS Int. Arch. Photogramm. Remote Sens. Spat. Inf. Sci.* **2017**, *XLII-4/W5*, 145–151. [CrossRef]
5. Jaya, P. Two Pedestrians Fall Into 3m-Deep Sinkhole in Seoul. *Star Online*. 25 February 2015. Available online: https://www.thestar.com.my/news/world/2015/02/25/sink-hole-seoul/#wGy2vrEth6cJBmhd.99 (accessed on 26 November 2018).
6. De Waele, J.; Gutiérrez, F.; Parise, M.; Plan, L. Geomorphology and natural hazards in karst areas: A review. *Geomprphology* **2011**, *134*, 1–8. [CrossRef]
7. Han, Y.; Hwang, H.W. Discussion on the sinkhole forming mechanism. *Int. J. Adv. Sci. Eng. Technol.* **2017**, *5*, 42–44.
8. Julian, R. Japan Giant Sinkhole in Fukuoka Repaired in a Week Begins to Sink Again. *Daily Mail Online*, 28 November 2016. Available online: https://www.dailymail.co.uk/news/article-3978468/Panic-Japan-giant-sinkhole-repaired-just-week-begins-sink-forcing-road-close-did-carry-repairs-quickly.html (accessed on 26 November 2018).
9. Steve, R.; Haworth, J. Police Officer Killed After Massive Sinkhole Opens Up in Road Swallowing Two Cars. *Mirror*, 6 December 2016. Available online: https://www.mirror.co.uk/news/world-news/man-killed-after-being-swallowed-9397110 (accessed on 26 November 2018).
10. Roger, S. Huge Sinkholes Are Now Appearing in The Wrong Places. *AP News*, 9 May 2017. Available online: https://apnews.com/41a17da407d241769e54085a8118ac7c (accessed on 26 November 2018).
11. Joel, B. Old Pipes, Big Problems: More Than 20 Sinkholes in Asheville. *Citizen Times*, 13 December 2016. Available online: https://www.citizen-times.com/story/news/local/2016/12/13/old-pipes-big-problems-more-than-20-sinkholes-asheville/95152436/ (accessed on 26 November 2018).
12. Hunaidi, O.; Chu, W.; Wang, A.; Guan, W. Detecting leaks in plastic pipes. *J. Am. Water Works Assoc.* **2000**, *92*, 82–94. [CrossRef]
13. Liston, D.A.; Liston, J.D. Leak detection techniques. *J. N. Engl. Water Work. Assoc.* **1992**, *106*, 103–108.
14. Gao, Y.; Brennan, M.J.; Joseph, P.F.; Muggleton, J.M.; Hunaidi, O. A model of the correlation function of leak noise in buried plastic pipes. *J. Sound Vib.* **2004**, *277*, 133–148. [CrossRef]
15. Farley, M.; Liemberger, R. Developing a non-revenue water reduction strategy: Planning and implementing the strategy. *Water Sci. Technol. Water Supply* **2005**, *5*, 41–50. [CrossRef]
16. Lin, M.; Wu, Y.; Wassell, I. Wireless sensor network: Water distribution monitoring system. In Proceedings of the 2008 IEEE Radio and Wireless Symposium, Orlando, FL, USA, 22–24 January 2008; pp. 775–778.
17. Hawari, A.; Alamin, M.; Alkadour, F.; Elmasry, M.; Zayed, T. Automated defect detection tool for closed circuit television (cctv) inspected sewer pipelines. *Autom. Constr.* **2018**, *89*, 99–109. [CrossRef]
18. Hunaidi, O.; Wang, A.; Bracken, M.; Gambino, T.; Fricke, C. Acoustic Methods for Locating Leaks in Municipal Water Pipe Networks. In Proceedings of the International Water Demand Management Conference, Dead Sea, Jordan, 30 May–3 June 2004; pp. 1–14.

19. Craig, F.; Babette, B. *Strategic and Competitive Analysis: Methods and Techniques for Analyzing Business Competition*, 1st ed.; Prentice Hall: Upper Saddle River, NJ, USA, 2002; 457p.
20. Ahn, K.S.; Cho, S.N. System for Detecting Leakage of Water in Conduit Line Installed Underground. WO Patent WO2018105764A1, 14 June 2018.
21. Kim, A. System for Detecting a Pipeline Leaks. K.R. Patent KR 20170075452A, 3 July 2018.
22. Ko, Y.M. Monitoring System for Leak Detection of Waterworks. K.R. Patent KR 101876730B1, 9 August 2018.
23. Merlo, S.A. Remote Pipe Inspection Using a Mounted Camera and Sensor. U.S. Patent US 8087311B2, 4 September 2008.
24. Oh, H.J.; Choi, D.S.; Lee, S.M. Safety and Monitoring Device of Pipeline. K.R. Patent KR 101571366B1, 25 November 2015.
25. Choi, R.G. Sewer Inspection System to Prevent Sinkhole. K.R. Patent KR 101564223B1, 30 October 2015.
26. Ko, Y.M. Monitoring System of Water Leakage and Sinkhole Using Water Leakage Sensor. K.R. Patent KR 101807739B1, 12 December 2017.
27. Yang, B.; Recane, M. Methods and Appratus for Pattern Match Filtering for Real Time Acoustic Pipeline Leak Detection and Location. U.S. Patent US 6389881B1, 21 May 2002.
28. Stephen, M. Remote Pipe Inspection. U.S. Patent US 8087311B2, 4 September 2008.
29. Ggorbel, F.H.; Dabney, J.B. Pipes and Sewer Inspection Means Maintenance Vehicle. U.S. Patent US 7182025B2, 27 Febuary 2018.
30. Kim, K.B.; Shin, S.H.; Kim, S.Y. Mobile Type Monitoring Device for Pipelines. K.R. Patent KR 101753707B1, 5 July 2017.
31. Kim, J.K. System for Detecting Water Leakage of Water Pipe. K.R. Patent KR 101967405B1, 8 March 2017.
32. Rye, D.W.; Yeom, B.W.; Bang, E.S.; Lee, H.J.; Jung, B.J.; Jung, S.W.; Lee, I.W.; Choi, C.H. Method for Evaluating Subsidence Risk in Urban Area. K.R. Patent KR 101808127B1, 12 May 2016.
33. Jung, S.W.; Yeom, B.W.; Ryu, D.W.; Lee, H.J.; Jeong, B.J.; Bang, E.S.; Lee, I.W. Apparatus for Measuring Shear Strength of Soil in Ground Subsidence Envirinment. K.R. Patent KR 101742107B1, 12 May 2015.
34. Gary, P. Sinkhole Detection Systems and Methods. U.S. Patent US 9638593B2, 2 May 2017.
35. Jung, Y.S.; Jung, S.H.; Cho, Y.G.; Park, J.J.; Kim, K.S.; Kang, H.H.; Kim, J.H. Injection Packer Restoration, Reclamation Method and Apparatus for Ground Subsidence. K.R. Patent KR 101849929B1, 18 March 2018.
36. Yang, G.H.; Yu, I.K.; Sim, J.I. Lightweight Soil for Preventing Sinkhole. K.R. Patent KR 101583257B1, 18 December 2014.
37. Kang, T.S.; Park, H.I.; Won, C.H. Paving Structure Capable of Preventing the Falling of the Pedestrian by Small Sinkhole. K.R. Patent KR 101699496B1, 1 April 2015.
38. Hwang, G.; Lee, Q.S.; Woo-Sub, Y. Device for Detecting Change in Underground Medium. U.S. Patent US 20160146760A1, 26 May 2016.
39. Lee, K.M.; Cho, J.H.; Shin, D.H. Sensor for Sensing Sinkhole, Sinkhole Monitoring System, and Method Using Sensor for Sensing Sinkhole. K.R. Patent KR 101680846B1, 13 December 2015.
40. Park, J.K.; Bong, T.G.; Kwon, S.W. System for Monitoring Groundwater with Function of Detecting Sinkhole. K.R. Patent KR 101528831B1, 16 June 2015.
41. Kang, H.J. A Sinkhole Detecting Device. K.R. Patent KR 101704527B1, 8 Febuary 2017.
42. Kang, K.Y.; Kim, C.H. Ground Penetrating Radar Survey Vehicle. K.R. Patent KR 101820805B1, 22 January 2018.
43. Kim, S.S. Switchboard System for Sensing Sinkhole. K.R. Patent KR 101904072B1, 14 November 2017.
44. Kim, Y.J. Sinkhole Exploration Device Using V_{rms}, the Inversion Process and Signal-to-Noise Ratio Upgrade of Permeable Underground Radar Exploration Using Common Reflection Surface Method. K.R. Patenet KR 101799813B1, 21 November 2017.
45. Jerry, S. Sinkhole Detector. U.S. Patent US 9552716B1, 24 January 2017.
46. Ko, B.C.; Jang, J.H.; Shin, S.Y.; Lee, E.J.; Lee, D.J. Sinkhole Detection System and Method Using a Drone-Based Thermal Camera and Image Processing. K.R. Patent KR 101857961B1, 4 October 2016.
47. Lee, H.J. Sinkholes Prevention Methods by Closing Settlement. K.R. Patent KR 101553599B1, 17 September 2015.
48. Ryu, D.W.; Kim, E.; Lee, K.; Yum, B.W.; Lee, I.; Lee, J.; Jeong, W.S.; Lee, H.; Jung, B.; Bnag, S.E.; et al. Method for Detecting Change in Underground Environment by Using Magnetic Induction, Detection Sensor and Detection System. U.S. Patent US 20180209787A1, 14 July 2015.
49. Ko, B.C.; Jang, J.H.; Shin, S.Y.; Joo, L.E. A Method for Construction of Simulation Mock Sinkhole. K.R. Patent KR 20180050045A, 14 May 2018.

50. Jung, S.W.; Yeom, B.W.; Ryu, D.W.; Lee, H.J.; Jung, B.J.; Bang, E.S.; Lee, I.H. Monitoring System and Method for Predicting Urban Sinkhole Measured by Suction Stress and Pore Water Pressure. K.R. Patent KR 101779017B1, 19 September 2017.
51. Jung, S.W.; Yeom, B.W.; Rye, D.W.; Lee, H.J.; Jung, B.J.; Bang, E.S.; Lee, I.H. System and Method for Predicting Urban Sinkhole Based on Shear Strength Measured by Vane Tester. K.R. Patent KR 101781409B1, 26 September 2017.
52. Wy, Y.H. Sinkhole Experimental Apparatus and Its Manufacturing Method. K.R. Patent KR 101890268B1, 29 September 2018.
53. McVay, M.; Tran, K.T. Detection of Sinkholes or Anomalies. U.S. Patent US 10088586B2, 2 October 2018.
54. Ali, S.; Ashraf, A.; Qaisar, S.B.; Afridi, M.K.; Saeed, H.; Rashid, S.; Felemban, E.A.; Sheikh, A.A. SimpliMote: A Wireless Sensor Network Monitoring Platform for Oil and Gas Pipelines. *IEEE Syst. J.* **2018**, *12*, 778–789. [CrossRef]
55. El-Darymli, K.; Khan, F.; Ahmed, M.H. Reliability Modeling of Wireless Sensor Network for Oil and Gas Pipelines Monitoring. *Sens. Transducers J.* **2009**, *106*, 6–26.
56. Abdelhafidh, M.; Fourati, M.; Fourati, L.C.; Laabidi, A. An investigation on wireless sensor networks pipeline monitoring system. *Int. J. Wirel. Mob. Comput.* **2018**, *14*, 25. [CrossRef]
57. Sheltami, T.R.; Bala, A.; Shakshuki, E.M. Wireless sensor networks for leak detection in pipelines: A survey. *J. Ambient Intell. Humaniz. Comput.* **2016**, *7*, 347–356. [CrossRef]
58. Bensaleh, M.S.; Qasim, S.M. Optimal Solutions for Water Pipeline Monitoring using Wireless Sensor Network [Extended Abstract]. In Proceedings of the SemiCon China-CSTIC Conference, Shanghai, China, 16–17 March 2014; pp. 1–4.
59. AL-Kadi, T.; AL-Tuwaijri, Z.; AL-Omran, A. Wireless sensor networks for leakage detection in underground pipelines: A Survey Paper. *Procedia Comput. Sci.* **2013**, *21*, 491–498. [CrossRef]
60. Adedeji, K.B.; Hamam, Y.; Abe, B.T.; Abu-Mahfouz, A.M. Towards achieving a reliable leakage detection and localization algorithm for application in water piping networks: An Overview. *IEEE Access* **2017**, *5*, 20272–20285. [CrossRef]
61. Datta, S.; Sarkar, S. A review on different pipeline fault detection methods. *J. Loss Prev. Process Ind.* **2016**, *41*, 97–106. [CrossRef]
62. BenSaleh, M.S.; Qasim, S.M.; Abid, M.; Jmal, M.W.; Karray, F.; Obeid, A.M. Towards realisation of wireless sensor network-based water pipeline monitoring systems: A comprehensive review of techniques and platforms. *IET Sci. Meas. Technol.* **2016**, *10*, 420–426.
63. BenSaleh, M.S.; Qasim, S.M.; Obeid, A.M.; Garcia-Ortiz, A. A review on wireless sensor network for water pipeline monitoring applications. In Proceedings of the 2013 IEEE International Conference on Collaboration Technologies and Systems (CTS), San Diego, CA, USA, 20–24 May 2013; pp. 128–131.
64. Sohraby, K.; Minoli, D.; Znati, T.F. *Wireless Sensor Networks: Technology, Protocols, and Applications*, 1st ed.; Wiley Publication: Hoboken, NJ, USA, 2007; 328p.
65. Yick, J.; Mukherjee, B.; Ghosal, D. Wireless sensor network survey. *Comput. Netw.* **2008**, *52*, 2292–2330. [CrossRef]
66. Vuran, M.C.; Akyildiz, I.F. Channel model and analysis for wireless underground sensor networks in soil medium. *Phys. Commun.* **2010**, *3*, 245–254. [CrossRef]
67. Ilyas, M. *The Handbook of Ad Hoc Wireless Networks*, 1st ed.; CRC Press: Boca Raton, FL, USA, 2003; 624p.
68. Labiod, H.; Afifi, H.; DeSantis, C. *Wi-Fi, Bluetooth, Zigbee and WiMAX, 1st ed*; Springer: Berlin, Germany, 2007; ISBN 1402053975.
69. Lee, J.S.; Su, Y.W.; Shen, C.C. A Comparative Study of Wireless Protocols: Bluetooth, UWB, ZigBee, and Wi-Fi. In Proceedings of the 33rd Annual Conference of the IEEE Industrial Electronics Society (IECON), Taipei, Taiwan, 5–8 November 2007; pp. 46–51.
70. Pothuganti, K.; Chitneni, A. A Comparative Study of Wireless Protocols: Bluetooth, UWB, ZigBee, and Wi-Fi. *Adv. Electron. Electr. Eng.* **2014**, *4*, 655–662.
71. Almazyad, A.; Seddiq, Y.; Alotaibi, A.; Al-Nasheri, A.; BenSaleh, M.; Obeid, A.; Qasim, S.; Almazyad, A.S.; Seddiq, Y.M.; Alotaibi, A.M.; et al. A Proposed Scalable Design and Simulation of Wireless Sensor Network-Based Long-Distance Water Pipeline Leakage Monitoring System. *Sensors* **2014**, *14*, 3557–3577. [CrossRef] [PubMed]

72. Ng, K.S.; Chen, P.Y.; Tseng, Y.C. A design of automatic water leak detection device. In Proceedings of the 2017 IEEE 2nd International Conference on Opto-Electronic Information Processing (ICOIP), Singapore, 7–9 July 2017; pp. 70–73.
73. Cattani, M.; Boano, C.A.; Steffelbauer, D.; Kaltenbacher, S.; Günther, M.; Römer, K.; Fuchs-Hanusch, D.; Horn, M. Adige: An efficient smart water network based on long-range wireless technology. In Proceedings of the 3rd International Workshop on Cyber-Physical Systems for Smart Water Networks—CySWATER '17, Pittsburgh, PA, USA, 21 April 2017; pp. 3–6.
74. Karray, F.; Garcia-Ortiz, A.; Jmal, M.W.; Obeid, A.M.; Abid, M. EARNPIPE: A Testbed for Smart Water Pipeline Monitoring Using Wireless Sensor Network. *Procedia Comput. Sci.* **2016**, *96*, 285–294. [CrossRef]
75. Sun, Z.; Wang, P.; Vuran, M.C.; Al-Rodhaan, M.A.; Al-Dhelaan, A.M.; Akyildiz, I.F. MISE-PIPE: Magnetic induction-based wireless sensor networks for underground pipeline monitoring. *Ad Hoc Netw.* **2011**, *9*, 218–227. [CrossRef]
76. Kuwano, R.; Horii, T.; Kohashi, H.; Yamauchi, K. Defects of sewer pipes causing cave-in's in the road. In Proceedings of the 5th International Symposium on New Technologies for Urban Safety of Mega Cities in Asia, Phuket, Thailand, 16–17 November 2006; pp. 347–353.
77. Kim, J.H.; Sharma, G.; Boudriga, N.; Iyengar, S.S.; Prabakar, N. Autonomous pipeline monitoring and maintenance system: A RFID-based approach. *EURASIP J. Wirel. Commun. Netw.* **2015**, *2015*, 262. [CrossRef]
78. Stoianov, I.; Nachman, L.; Madden, S.; Tokmouline, T. PIPENETa wireless sensor network for pipeline monitoring. In Proceedings of the 6th International Conference on Information Processing in Sensor Networks—IPSN '07, Cambridge, MA, USA, 25–27 April 2007; ACM Press: New York, NY, USA, 2007; pp. 264–273.
79. Stoianov, I.; Nachman, L.; Whittle, A.; Madden, S.; Kling, R. Sensor Networks for Monitoring Water Supply and Sewer Systems: Lessons from Boston. In Proceedings of the 8th Annual Water Distribution Systems Analysis Symposium (WDSA), Cincinnati, OH, USA, 27–30 August 2006; pp. 1–17.
80. Nassiraei, A.A.F.; Kawamura, Y.; Ahrary, A.; Mikuriya, Y.; Ishii, K. A New Approach to the Sewer Pipe Inspection: Fully Autonomous Mobile Robot "KANTARO". In Proceedings of the IECON 2006—32nd Annual Conference on IEEE Industrial Electronics, Paris, France, 6–10 November 2006; pp. 4088–4093.
81. Cheng, J.C.P.; Wang, M. Automated detection of sewer pipe defects in closed-circuit television images using deep learning techniques. *Autom. Constr.* **2018**, *95*, 155–171. [CrossRef]
82. Lai, T.T.T.; Chen, W.J.; Li, K.H.; Huang, P.; Chu, H.H. TriopusNet: Automating wireless sensor network deployment and replacement in pipeline monitoring. In Proceedings of the 2012 ACM/IEEE 11th International Conference on Information Processing in Sensor Networks (IPSN), Beijing, China, 16–20 April 2012; pp. 61–71.
83. Daadoo, M.; Daraghmi, Y. Smart Water Leakage Detection Using Wireless Sensor Networks (SWLD). *Int. J. Netw. Commun.* **2017**, *7*, 1–16.
84. Shinozuka, M.; Chou, P.H.; Kim, S.; Kim, H.R.; Yoon, E.; Mustafa, H.; Karmakar, D.; Pul, S. Nondestructive monitoring of a pipe network using a MEMS-based wireless network. In Proceedings of the SPIE Smart Structures and Matrials and Non Distructive Evaluation and Health Monitoring, San Diego, CA, USA, 12 April 2010; p. 76490P.
85. Kadri, A.; Abu-Dayya, A.; Trinchero, D.; Stefanelli, R. Autonomous sensing for leakage detection in underground water pipelines. In Proceedings of the 2011 Fifth IEEE International Conference on Sensing Technology, Palmerston North, New Zealand, 28 November–1 December 2011; pp. 639–643.
86. Mustafa, H.; Chou, P.H. Embedded Damage Detection in Water Pipelines Using Wireless Sensor Networks. In Proceedings of the 2012 IEEE 14th International Conference on High Performance Computing and Communication & 2012 IEEE 9th International Conference on Embedded Software and Systems, Liverpool, UK, 25–27 June 2012; pp. 1578–1586.
87. Sadeghioon, A.; Metje, N.; Chapman, D.; Anthony, C.; Sadeghioon, A.M.; Metje, N.; Chapman, D.N.; Anthony, C.J. SmartPipes: Smart Wireless Sensor Networks for Leak Detection in Water Pipelines. *J. Sens. Actuator Netw.* **2014**, *3*, 64–78. [CrossRef]
88. Ismail, M.I.M.; Dziyauddin, R.A.; Samad, N.A.A. Water pipeline monitoring system using vibration sensor. In Proceedings of the 2014 IEEE Conference on Wireless Sensors (ICWiSE), Subang, Malaysia, 26–28 October 2014; pp. 79–84.

89. Kim, J.H.; Sharma, G.; Boudriga, N.; Iyengar, S.S. SPAMMS: A sensor-based pipeline autonomous monitoring and maintenance system. In Proceedings of the 2010 Second International Conference on Communication Systems and NETworks (COMSNETS 2010), Bangalore, India, 5–9 January 2010; pp. 1–10.
90. Yoon, S.H.; Ye, W.; John, H.; Littlefield, B.; Hahabi, S.C. SWATS: Wireless Sensor Networks for Steamflood and Waterflood Pipeline Monitoring. *IEEE Netw.* **2011**, *25*, 50–56. [CrossRef]
91. Lim, J.S.; Kim, J.; Friedman, J.; Lee, U.; Vieira, L.; Rosso, D.; Gerla, M.; Srivastava, M.B. SewerSnort: A drifting sensor for in situ Wastewater Collection System gas monitoring. *Ad Hoc Netw.* **2013**, *11*, 1456–1471. [CrossRef]
92. Metje, N.; Chapman, D.N.; Cheneler, D.; Ward, M.; Thomas, A.M.; Metje, N.; Chapman, D.N.; Cheneler, D.; Ward, M.; Thomas, A.M. Smart Pipes—Instrumented Water Pipes, Can This Be Made a Reality? *Sensors* **2011**, *11*, 7455–7475. [CrossRef] [PubMed]
93. Constantinou, S.; Van Rooy, J.L. Sinkhole and subsidence size distribution across dolomitic land in Gauteng. *J. S. Afr. Inst. Civ. Eng.* **2018**, *60*, 2–8. [CrossRef]
94. James, R. Foundation Engineering Problems and Hazards in Karst Terranes. Maryland Department of Natural Resources, 2015. Available online: http://www.mgs.md.gov/geology/geohazards/engineering_problems_in_karst.html (accessed on 26 November 2018).
95. Stop Sinkholes Before They Start. Municipal Sewer and Water. 2017. Available online: https://www.mswmag.com/online_exclusives/2017/10/stop_sinkholes_before_they_start_sc_00125 (accessed on 26 November 2018).
96. Road Cave-in and Special Countermeasure. Seoul Metropolitan Government, 2015. Available online: http://english.seoul.go.kr/#none (accessed on 3 May 2019).
97. Sevil, J.; Gutiérrez, F.; Zarroca, M.; Desir, G.; Carbonel, D.; Guerrero, J.; Linares, R.; Roqué, C.; Fabregat, I. Sinkhole investigation in an urban area by trenching in combination with GPR, ERT and high-precision leveling. Mantled evaporite karst of Zaragoza city, NE Spain. *Eng. Geol.* **2017**, *231*, 9–20. [CrossRef]
98. Xu, J.; He, J.; Zhang, L. Collapse prediction of karst sinkhole via distributed Brillouin optical fiber sensor. *Measurement* **2017**, *100*, 68–71. [CrossRef]
99. Desir, G.; Gutiérrez, F.; Merino, J.; Carbonel, D.; Benito-Calvo, A.; Guerrero, J.; Fabregat, I. Rapid subsidence in damaging sinkholes: Measurement by high-precision leveling and the role of salt dissolution. *Geomorphology* **2018**, *303*, 393–409. [CrossRef]
100. Intrieri, E.; Gigli, G.; Nocentini, M.; Lombardi, L.; Mugnai, F.; Fidolini, F.; Casagli, N. Sinkhole monitoring and early warning: An experimental and successful GB-InSAR application. *Geomorphology* **2015**, *241*, 304–314. [CrossRef]
101. Cooper, A.H.; Farrant, A.R.; Price, S.J. The use of karst geomorphology for planning, hazard avoidance and development in Great Britain. *Geomorphology* **2011**, *134*, 118–131. [CrossRef]
102. Comerci, V.; Vittori, E.; Cipolloni, C.; Di Manna, P.; Guerrieri, L.; Nisio, S.; Succhiarelli, C.; Ciuffreda, M.; Bertoletti, E. Geohazards Monitoring in Roma from InSAR and In Situ Data: Outcomes of the PanGeo Project. *Pure Appl. Geophys.* **2015**, *172*, 2997–3028. [CrossRef]
103. Nouioua, I.; Laid Boukelloul, M.; Fehdi, C.; Baali, F. Detecting Sinkholes Using Ground Penetrating Radar in Drâa Douamis, Cherea Algeria: A Case Study. *Electron. J. Geotech. Eng.* **2013**, *18*, 1337–1349.
104. Al-fares, W.; Bakalowicz, M.; Guérin, R.; Dukhan, M. Analysis of the karst aquifer structure of the Lamalou area (Hérault, France) with ground penetrating radar. *J. Appl. Geophys.* **2002**, *51*, 97–106. [CrossRef]
105. Pueyo Anchuela, Ó.; Casas Sainz, A.M.; Pocoví Juan, A.; Gil Garbí, H. Assessing karst hazards in urbanized areas. Case study and methodological considerations in the mantle karst from Zaragoza city (NE Spain). *Eng. Geol.* **2015**, *184*, 29–42. [CrossRef]
106. Yacine, A.; Ridha, M.M.; Laid, H.M.; Abderahmane, B. Karst Sinkholes Stability Assessment in Cheria Area, NE Algeria. *Geotech. Geol. Eng.* **2014**, *32*, 363–374. [CrossRef]
107. Al-kouri, O.; Al Fugara, A.; Al-Rawashdeh, S.; Sadoun, B.; Pradhan, B. Geospatial modeling for sinkholes hazard map based on GIS & RS data. *J. Geogr. Inf. Syst.* **2013**, *5*, 584–592.
108. Galve, J.P.; Lucha, P.; Castañeda, C.; Bonachea, J.; Guerrero, J. Integrating geomorphological mapping, trenching, InSAR and GPR for the identification and characterization of sinkholes: A review and application in the mantled evaporite karst of the Ebro Valley (NE Spain). *Geomorphology* **2011**, *134*, 144–156.
109. Carbonel, D.; Rodríguez, V.; Gutiérrez, F.; McCalpin, J.P.; Linares, R.; Roqué, C.; Zarroca, M.; Guerrero, J.; Sasowsky, I. Evaluation of trenching, ground penetrating radar (GPR) and electrical resistivity tomography (ERT) for sinkhole characterization. *Earth Surf. Process. Landf.* **2014**, *39*, 214–227. [CrossRef]

110. Nouioua, I.; Rouabhia, A.; Fehdi, C.; Boukelloul, M.L.; Gadri, L.; Chabou, D.; Mouici, R. The application of GPR and electrical resistivity tomography as useful tools in detection of sinkholes in the Cheria Basin (northeast of Algeria). *Environ. Earth Sci.* **2013**, *68*, 1661–1672. [CrossRef]
111. Joyce, K.E.; Samsonov, S.V.; Levick, S.R.; Engelbrecht, J.; Belliss, S. Mapping and monitoring geological hazards using optical, LiDAR, and synthetic aperture RADAR image data. *Nat. Hazards* **2014**, *73*, 137–163. [CrossRef]
112. Kobal, M.; Bertoncelj, I.; Pirotti, F.; Dakskobler, I.; Kutnar, L. Using Lidar Data to Analyse Sinkhole Characteristics Relevant for Understory Vegetation under Forest Cover—Case Study of a High Karst Area in the Dinaric Mountains. *PLoS ONE* **2015**, *10*, e0122070. [CrossRef]
113. VanSchoor, M. Detection of sinkholes using 2D electrical resistivity imaging. *J. Appl. Geophys.* **2002**, *50*, 393–399. [CrossRef]
114. Ge, L.; Chang, H.-C.; Rizos, C. Mine Subsidence Monitoring Using Multi-source Satellite SAR Images. *Photogramm. Eng. Remote Sens.* **2007**, *73*, 259–266. [CrossRef]
115. Filin, S.; Baruch, A. Detection of Sinkhole Hazards using Airborne Laser Scanning Data. *Photogramm. Eng. Remote Sens.* **2010**, *76*, 577–587. [CrossRef]
116. Suh, J.; Choi, Y. Mapping hazardous mining-induced sinkhole subsidence using unmanned aerial vehicle (drone) photogrammetry. *Environ. Earth Sci.* **2017**, *76*, 144. [CrossRef]
117. Kim, K.; Park, D.-H.; Lee, J.; Jin, S. UGS middleware for monitoring state of underground utilities. In Proceedings of the 2015 IEEE International Conference on Information and Communication Technology Convergence (ICTC), Jeju, Korea, 28–30 October 2015; pp. 1183–1185.
118. Kwak, P.J.; Park, S.H.; Choi, C.H.; Lee, H.D. Safety Monitoring Sensor for Underground Subsidence Risk Assessment Surrounding Water Pipeline. *J. Sens. Sci. Technol.* **2015**, *24*, 306–310. [CrossRef]
119. Al-Halbouni, D.; Holohan, E.P.; Taheri, A.; Schöpfer, M.P.J.; Emam, S.; Dahm, T. Geomechanical modelling of sinkhole development using distinct elements: Model verification for a single void space and application to the Dead Sea area. *Solid Earth* **2018**, *9*, 1341–1373. [CrossRef]
120. Oz, I.; Eyal, S.; Yoseph, Y.; Ittai, G.; Elad, L.; Haim, G. Salt dissolution and sinkhole formation: Results of laboratory experiments. *JGR Earth Surf.* **2016**, *121*, 1746–1762. [CrossRef]
121. Tao, X.; Ye, M.; Wang, X.; Wang, D.; Castro, R.; Zhao, J. Experimental and Numerical Investigation of Sinkhole Development and Collapse in Central Florida. In Proceedings of the Fourteenth Multidisciplinary Conference on Sinkholes and the Engineering and Environmental Impacts of Karst; University of South Florida Tampa Library: Tampa, FL, USA, 2015; pp. 501–506.
122. Dimitrakopoulos, G. Sustainable mobility leveraging on 5G mobile communication infrastructures in the context of smart city operations. *Evol. Syst.* **2017**, *8*, 157–166. [CrossRef]
123. Kang, J.; Choi, C.; Kwak, P.J.; Kim, J.; Park, S.H.; Park, Y. Ground subsidence monitoring and risk assessment around water supply and sewerage. *J. Korean Soc. Civ. Eng.* **2017**, *65*, 22–27.
124. Gray, K.M. Central Florida Sinkhole Evaluation. In *District Geotechnical Engineer*; Florida Department of Transportation: Tallahassee, FL, USA, 2014.

Article

Research on the Construction of a Natural Hazard Emergency Relief Alliance Based on the Public Participation Degree

Yingxin Chen [1], Jing Zhang [2,3,*], Zhaoguo Wang [4,*] and Pandu R. Tadikamalla [5]

1. School of Economics and Management, Harbin Engineering University, Harbin 150001, China; chenyxdingdang@hrbeu.edu.cn
2. School of Information Science and Engineering, University of Jinan, Jinan 250022, China
3. Shandong Provincial Key Laboratory of Network-based Intelligent Computing, University of Jinan, Jinan 250022, China
4. School of Marxism Studies, Northeast Agricultural University, Harbin 150030, China
5. Joseph M. Katz Graduate School of Business, University of Pittsburgh, Pittsburgh, PA 15260, USA; Pandu@katz.pitt.edu

* Correspondence: ise_zhangjing@ujn.edu.cn (J.Z.); wangzhaoguo@neau.edu.cn (Z.W.); Tel.: +86-0531-8276-7069 (J.Z.); +86-0451-5519-1994 (Z.W.)

Received: 16 February 2020; Accepted: 18 March 2020; Published: 25 March 2020

Abstract: At present, in light of new situations and the new task of natural hazard response, effective public participation in emergency relief has become an urgent task that can reduce economic losses and casualties. The purpose of this paper is to construct a natural hazard emergency relief alliance and analyze the mechanisms and dynamics of public participation. In this study, methods based on a multi-agent system were adopted, and we used different participants as heterogeneous agents with different attitudes and resources. Using four different processes, namely participation proposals, negotiation interval, negotiation decision-making function, and participation strategy, we comprehensively construct an emergency relief alliance for natural hazards. In addition, the dynamic public interaction process is analyzed and a construction algorithm is given. The experimental results show that the proposed method has better performance in alliance formation efficiency, negotiation efficiency, and agent utility. The research results illustrate that the public's attitudes and resources influence the construction of emergency relief alliances; a greater degree of public participation contributes to a more efficient alliance formation. The findings of this study contribute to the promotion of public cooperation and improvement in the efficiency of natural hazard emergency relief.

Keywords: natural hazards; multi-agent system; emergency relief; public participation

---

1. Introduction

As a result of its vast territory, complex environmental conditions, large population, low disaster prevention, and management capabilities, China has suffered serious losses and used a great deal of human and material resources for natural hazard prevention and relief. The government plays an important role in emergency management, but it is difficult to achieve high efficiency by only relying on the strength of the government [1]; therefore, determining how the public can effectively participate in natural hazard emergency relief has become an urgent task. Emergency relief refers to the money, goods, rescue, and services provided for victims after natural hazards [2]. Practice has proven that due to traffic inconveniences, secondary disasters, and an imperfect system, public participation has caused a certain degree of traffic congestion and disorder. Therefore, determining how to organize the public to participate in emergency relief in an orderly and effective way has become a popular topic.

In 2016, the new "National Natural Hazard Emergency Relief Plan" was released in China, but the subject of public participation in emergency relief has concerned scholars.

Questions to clarify this emergency plan are listed below.

a. What technology should be used to formalize the emergency relief alliance?
b. How is the emergency relief alliance constructed? What about its performance?
c. Does the participation degree affect emergency alliance construction?

This paper will address these questions. The remainder of this paper is structured as follows. In Section 2, we review relevant literature and identify the research gap that this paper focuses on. In Section 3, we give a formal expression based on a multi-agent system negotiation interval, the participation decision function is presented, and the interaction process of the participants is analyzed. Section 4 gives the formal definition of the participation degree and participation strategy function, and an algorithm for constructing an emergency relief alliance is proposed. In Section 5, we compare our approach and the competing algorithms and summarize our results. Conclusions and an outlook on future research directions are given in Section 6.

Our contributions are as follows: We propose the establishment of a method for creating a natural hazard emergency relief alliance. At present, research regarding the field of public participation in emergency management mainly focuses on macroscopic, qualitative research, such as participation status and problems, influencing factors, and participation strategy; however, quantitative research is mainly concerned with resource allocation, personnel evacuation and path planning, and a lack of quantitative analysis of participation models. This paper presents a model for an emergency relief alliance and proposes a corresponding algorithm.

Moreover, taking into account that different participants have different resources and participation attitudes, we put forward the concept of public participation degree. The adoption of multi-agent technology reflects autonomy, and a natural hazard emergency relief alliance based on the public participation degree is proposed. The experiment shows that our method is reliable and applicable.

2. Literature Review

2.1. Disaster Risk Reduction

Many scholars have done large amounts of research in the field of disaster risk reduction. Adnan et al. [3] selected eleven precedent (model) cities to study their various initiatives for reducing coastal flood risks. Their findings showed that protecting cities from flooding and reducing exposure to floods are two different but interrelated approaches to disaster risk reduction (DRR). Few et al. [4] discussed findings from a two-year research project on DRM (Disaster Risk Management) capacity development. They pointed out that adaptability, ownership, sustainability, the inclusion of actors, and scales reflect the capacity of DRR. Sim et al. [5] explored the way that social workers can apply transdisciplinary strategies to work with other professional practitioners and stakeholders in meeting disaster risk reduction needs in a remote Chinese village. Marlowe et al. [6] emphasized the importance of incorporating the concepts of reach, relevance, receptiveness, and relationships into a DRR approach; these concepts were presented as an embedded guiding framework that can aid in disaster communication. Carrao et al. [7] assessed drought hazard (DH) changes for the mid-century (2021–2050) and late-century (2071–2099). The results showed that a major challenge for risk management is for human populations or their activities to not adapt to DH changes; instead, they should continue developing global initiatives that mitigate their impact on the entire carbon cycle by the late-century. Migliorini et al. [8] conducted surveys and interviews with National Sendai Focal Points and stakeholders in science and research, governmental agencies, non-governmental organizations, and industry, enhancing data interoperability for disaster risk reduction. Marchal et al. [9] pointed out that reinsurance and insurance industries play a role in helping to manage risks and improve disaster risk reduction (DRR) as well as loss prevention. Webb [10] stated that early tropical cyclone warnings are necessary but not sufficient to ensure that households and communities are prepared. There is a strong case for investments in mid- to long-term DRR focused on community and household capacity, prioritizing women's active and equal participation as community leaders, and disability inclusion. Totaro et al. [11] drew maps of monothematic and synthetic indices to describe the hazard status of metropolitan areas; a hazard hotspot map was also elaborated to identify areas with high hazards.

2.2. Emergency Relief for Natural Hazards

Many scholars have researched in the field of natural hazard emergency relief. Narayanan et al. [12] built a disaster recovery network that provides food, water, shelter, medical services, and other relief resources for the affected people through reliable communication network coordination to avoid social chaos caused by natural hazards. Wachinger [13] pointed out that personal experience, government investment, experts' awareness of disaster risk, and media reports have a significant impact on natural hazard emergency relief. Wex et al. [14] proposed a decision support model of disaster emergency relief based on the heuristic algorithm of Monte Carlo, which can effectively allocate and dispatch rescue units, greatly reducing the casualties and economic losses. Oral et al. [15] assessed the impact of earthquake experience on earthquake preparedness, and the results showed that disaster experience, residential area, and disaster relief work have an obvious relationship, and reasonable suggestions were put forward for disaster relief. Berariu et al. [16] analyzed the cascade effects of natural hazards and investigated their impact on relief operations concerning critical infrastructure; in particular, the transport infrastructure, electricity, and human health.

Alem et al. [17] took into account practical characteristics, such as budget allocation and fleet sizing of multiple types of vehicles, and developed a new two-stage stochastic network flow model to help decide how to rapidly supply humanitarian aid to victims of a disaster. Toyasaki et al. [18] focused on horizontal cooperation in inventory management, which is currently implemented in the United Nations Humanitarian Response Depot (UNHRD) network, and proposed a policy priority for the first-best system of optimal inventory management. Deo [19] used integrated disaster management technology, a quantitative method, and big data analysis to create disaster and early warning models to reduce the impact of these disasters and provide a comprehensive method for disaster management systems. Nassereddine et al. [20] presented a multi-criteria decision-making approach to evaluate emergency response systems by taking into account the interaction synergy. Borowska-Stefanska et al. [21] developed an optimization pattern for the process of a preventive evacuation of people from flood-risk areas aimed at mitigating the negative effects of the flood.

Some scholars have researched natural disaster management based on multi-agent system theory. Izida et al. [22] pointed out that communication in natural disasters can be simulated through the multi-agent paradigm, which provides better assistance in emergencies of natural disasters. Yue et al. [23] built a disaster emergency collaborative decision-making framework based on a multi-agent system, designed a collaborative decision-making model, and achieved a natural disaster dynamic emergency decision-making process. Naqvi et al. [24] proposed an agent-based model of a stylized low-income region to study the impact of natural disasters on population displacement, income, prices, and consumption with a focus on low-income groups. Na et al. [25] developed an agent-based discrete-event simulation (ABDES) modeling framework based on an embedded GIS module for making no-notice natural disaster evacuation planning. Lee et al. [26] presented a novel and efficient rescue system with a multi-agent simultaneous localization and mapping (SLAM) framework, which was proposed to reduce the rescue time while rescuing the people trapped inside a burning building.

2.3. Public Participation in Natural Hazard Emergency Relief

Since Sherry Arnstein published his famous paper "A Ladder of Citizen Participation" in 1969 and John Clayton Thomas proposed an effective decision model of citizen participation, more and more scholars have studied public participation. The research on public participation focuses on community development, e-government, social governance, environmental science, and democratic construction.

Chen et al. [27] developed an Environmental Community Consultative Group (ECCG), a small local group that can promote public environmental awareness, improve public environmental behavior, and facilitate public engagement in environmental management. Werner [28] pointed out that policymakers' dependencies, motivations, and decision-making processes lead them to evaluate firms by using a sociopolitical reputation as a differentiating heuristic. Say et al. [29] shed light on the basic principles related to the process of public participation and how effective participation can be attained in SEA (Strategic Environmental Assessment) practices at the policy, plan, and program levels. Park et al. [30] aimed to investigate the legal opportunities of public participation in managing state forests in the case of the Republic of Korea (ROK), which provide legal insights on promoting public participation in managing state forests. Xi [31] analyzed the problems of public participation in environmental protection and put forward three ways to optimize the system of public participation in environmental protection. Li [32] indicated that it is not necessary to impose a kind of external force on residents through empowerment, then make them engage in tourism. Self-empowerment means consciousness, awareness, and promotion of participation ability; these are fundamental and critical points for community residents participating in tourism. Xiao et al. [33] established a structural equation model to identify key factors influencing citizen environmental willingness to participate in waste management, and this indicated that the most important influencing factor was citizen knowledge, followed by social motivation, while institutional factors had the smallest positive effect. Ruiz-Villaverde et al. [34] identified several advantages of public and stakeholder participation in water management, such as the better use of knowledge and experiences from different stakeholders, increases in public acceptance, and reduced litigation, delays, and inefficiencies in implementation. Chen et al. [35] pointed out that government supervision has a positive effect on environmental governance and can urge enterprises to actively perform pollution control. The effect of government supervision is constrained by the income and costs of enterprises, and the penalties for passive pollution control should be raised.

In summary, scholars have done a lot of research in the field of natural hazard emergency relief and public participation. However, most of the studies are focused on issues such as the positive role of public participation, the influencing factors, and the ways of public participation. Some scholars have put forward quantitative models and methods of emergency management, but multi-agent system technology is rarely used, and it is mostly used for research of material distribution and personnel evacuation. The agent has the characteristics of autonomy, sociality, and initiative. It also has some human characteristics, such as knowledge, belief, obligation, intention, and so on. So, we can regard the public as agents, and management of such emergency relief is improved by combining human intelligence with efficiencies of multi-agent systems, and we adopt agent automatic negotiation coalition technology [36] to carry out our research.

3. The Establishment of the Natural Hazard Emergency Relief Alliance

3.1. The Formalization of the Natural Hazard Emergency Relief Alliance

The public refers to the masses, nongovernmental organizations (NGO), profit organizations, and individuals. The participants of natural hazard relief include governments and the public. We define the government, social masses, NGO, profit organizations, and individuals as agents. The natural hazard alliance, composed of multiple agents, is a connected graph with the agent as the node. $Agents = \{At_1, At_2, \ldots, At_n\}$, where *Agents* is set of agents. The agent that communicates directly with At_i is defined as the immediate neighbor of At_i, expressed as $DN(i) = \{i|(i,j) \in \mathrm{E}, i \neq j\}$, where E represents the set of edges that are connected between agents. R_i is the resource that At_i can provide when carrying out natural hazard relief. The multiple agents may form alliances of $RA_1, RA_2, \ldots, RA_n \in 2^{\mathrm{I}}$, where each member At_i in the alliance RA_i has the resource R_i, so that the resources that the alliance RA_i owns are described as $R_{RA} \in \prod_{i \in RA} R_i$. After the formation of alliance RA_i, the overall utility can be obtained as $P(R_{RAi}) \in Z^+$, which represents the aid given to the victims after the alliance was established. Based on the theory of stakeholders, the public, whether organizations or individuals, gains corresponding utility from participating in natural hazard relief activities; for example, an enterprise provides excavators and transportation vehicles, and gets rewards of social recognition and good reputation. The masses and volunteers get psychological satisfaction and a sense of accomplishment through donations and rescue services, which belong to the invisible utility. Thus, each member At_i can get a utility value, which is $p_i \in Z^+$, $\sum_{i \in RA} p_i = P(R_{RA})$.

3.2. Participation Proposal

At_i can invite direct neighbor At_j to join an alliance by sending a proposal when At_j agrees with the proposal; At_i and At_j will combine their resources to complete a disaster relief mission and receive corresponding utility. At_i sending a proposal to At_j defined as follows.

$$S_{ij}(R_i) = ((R_{RA-i}, R_i), p_i^{R_i}) \tag{1}$$

where $S_{ij}(R_i)$ means that At_i sends a proposal to At_j. $R_{RA\text{-}i}$ is the total resources owned by alliance RA before joining At_i; $RA\text{-}i$ stands for the alliance without At_i; RA is the alliance after At_i joins; $p_i^{R_i}$ is the utility that At_i gained after At_i joins the alliance RA and provides resource R_i. When At_i sends a proposal to neighbor At_j, At_j will evaluate the utility before deciding whether to accept the proposal or not. If At_i's proposal does not meet At_j's minimum requirement, At_j will send a counter-proposal S_{oppose}^{ji} and will require higher utility. At_i might not immediately agree to the counter-proposal, and will continue to send a new anti-proposal S_{oppose}^{ij} to At_j, assigning a new utility value to At_j at the same time. In this way, both participants in the alliance will conduct several rounds of negotiation.

3.3. The Negotiation Interval

The negotiation interval is the fluctuation range of the utility distribution between the two participants, that is, the minimum utility and the maximum utility. Suppose that At_i has formed an alliance RA with other agents; when At_j joins the alliance RA, the added utility is given in Formula (2).

$$\Delta P(R_{RA}, R_j) = P(R_{RA}, R_j) - P(R_{RA}) \tag{2}$$

The maximum utility that At_i can obtain is given in Formula (3).

$$p_i^{\max}(R_{RA}, R_j) = \Delta P(R_{RA}, R_j) \tag{3}$$

The optimal participation proposal that At_i received is $S^* = \max_{s \in proposali} p(S)$, where $proposal_i$ represents the participation proposal set that At_i received. The At_i's minimum utility is given in Formula (4).

$$p^{\min}(R_{RA}, R_j) = \max(0, p_i(S^*)) \tag{4}$$

Therefore, the negotiation interval is given as follows.

$$\left[p^{\min}(R_{RA}, R_j), p^{\max}(R_{RA}, R_j)\right] \tag{5}$$

3.4. The Negotiation Decision-Making Function

The negotiation decision-making function reflects that agents choose their behavior with the principle of maximizing their own utility after receiving the participation proposal. S_j represents the proposal that At_i received from At_j, S_{oppose} represents the counter-proposal; S_{oppose} is generated according to the participation strategy. $S^*_{m\_join}$ represents the participation proposal sent to the neighbor At_m that has the highest participation degree. $\overline{S}$ represents the history of the participation proposal; it is a set of participation proposals that At_i had proposed. The participation strategy of the inviter agent is defined as a Formula (6). The participation strategy of the invited agent is defined as Formula (7).

$$T_i([p_i^{\min}, p_i^{\max}], \overline{S}) = p_i^{\min} + f_{\partial_{ij}^n}(x)(p_i^{\max} - p_i^{\min}) \tag{6}$$

$$T_i([p_i^{\min}, p_i^{\max}], \overline{S}) = p_i^{\max} - f_{\partial_{ij}^n}(x)(p_i^{\max} - p_i^{\min}) \tag{7}$$

where $f_{\partial_{ij}^n}(x)$ represents the participation strategy function; the participation strategy can be divided into three types according to the different values of participation attitude values ∂_{ij}:

(1) Negative type ($\partial_{ij}^n < 0$). This means that At_i makes a quick concession at the start stage of the participation negotiations and quickly reaches its maximum concession state.
(2) Neutral type ($\partial_{ij}^n \approx 0$). This means that the concession process of At_i is approximately linear, i.e., the utility of agents varies linearly with the number of participation negotiation rounds.
(3) Positive type ($\partial_{ij}^n > 0$). This means that At_i makes a slow concession at the starting stage of the participation negotiations, only makes a quick concession when x = *round/MAX_Round*→1, where *round* is the number of negotiation rounds, with an initial value of 0; *MAX_Round* is the maximum number of negotiation rounds.

$S^*_k(k \in proposal_i)$ represents the participation proposal sent by At_k that belongs to the best alliance, that is, At_k can provide the highest utility for At_i in the neighborhood.

$$S^*_k = \max_{k \in DN(i) \cup \{i\}} p_i(T_i([p_i^{\min}, p_i^{\max}], \overline{S_k})) \tag{8}$$

The decision function for At_i that participates in the alliance is given in Formula (9):

$$S_{Decisioni} = \arg\max(S_j, S_{oppose} S^*_k, S^*_{m\_jon}) \tag{9}$$

The negotiation process of agent participation in the alliance is to produce different types of participation proposals: Acceptance proposal, counter-proposal, new proposal, and rejection proposal, so that decisions that can be made by At_i as follows. (1) At_i proposes a counter-proposal according to the participation strategy. (2) At_i sends a proposal to the agent who belongs to the optimal alliance and invites At_i to join the alliance, and At_i joins the optimal alliance. (3) At_i creates a new alliance, namely At_i sends a participation proposal to its neighbor At_m and invites At_m to participate in the alliance.

3.5. The Agent Interaction Process

The state of agents is defined as *State* = {Sleep, Wait, Done}, where the Sleep indicates the initial state of the agents and when waiting for other agents to send participation proposals; the Wait indicates that the agent has sent a participation proposal to another agent and wait for its response; the Done indicates that the agent has accepted a participation proposal and the participation negotiation has been completed.

The agent interaction process of public participation in the alliance is given as follows.

A. Initialization: All agents are in the Sleep state when they begin to establish alliances.
B. The interaction process: (1) When At_j sends a participation proposal to At_i, At_j sets $State_j$ = Wait; in order to avoid deadlock, all agents in the $DN\{j\}\backslash\{i\}$. $DN\{j\}\backslash\{i\}$ means that all agents cannot send a new proposal to At_j. At this time, only At_i can communicate with At_j. (2) At_i will send to At_j a counter-proposal, acceptance proposal, or rejection proposal. (3) If At_i accepts At_j's proposal, then At_i sets $State_i$ = Done. At this time, At_i will send an acceptance proposal to At_j and send a rejection proposal to the other agents; then, At_i will no longer receive any participation proposals.

At_j adds At_i to the alliance set *proposal set$_j$* and sets $State_j$ = Sleep, and then At_j can receive the new participation proposal. (4) If At_i sends a counter-proposal to At_j, then At_j sets $State_j$ = Sleep, and At_i sets $State_i$ = Wait. (5) If At_i sends a rejection proposal to At_j, then At_j sets $State_j$ = Sleep.

The agent interaction process is given in Figure 1.

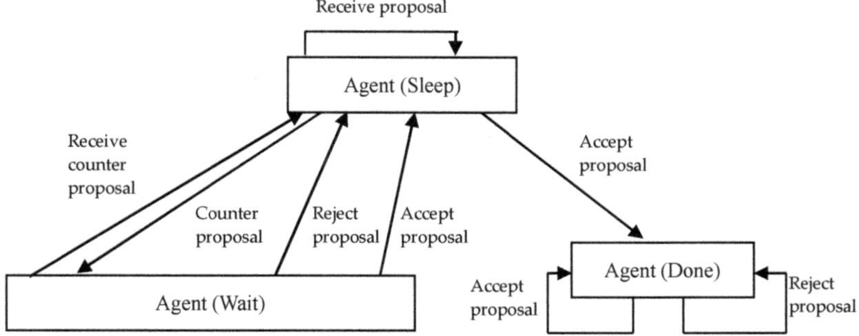

Figure 1. The agent interaction process.

4. The Participation Strategy Based on Participation Degree

4.1. The Participation Strategy Function

The participation strategy function given in Formula (10) uses an exponential function to gradually produce new proposals after concessions.

$$f_{\partial}(x) = e^{\partial_{ij}^{n} x} - 1/e^{\partial_{ij}^{n}} - 1 \quad (10)$$

The domain and range of $f_{\partial}(x)$ are [0, 1]; they are monotonically increasing, but with different values of ∂_{ij}^{n}, the concave and convex of the function will change. The agent concession degree can be adjusted by changing the value of ∂_{ij}^{n}; ∂_{ij}^{n} is At_i 's participation attitude value when At_j invites At_i to participate in the alliance in round n. The independent variable x is the ratio between the actual number of negotiation rounds and the maximum assignment negotiation rounds. That is, *x=round/MAX_Round*, *round* changes from 0 to *MAX_Round*, so x is guaranteed to change from 0 to 1.

4.2. The Participation Degree

The concept of the participation degree is proposed to reflect that different participants have different resources and different participation histories, and the participation process, guided by different strategies, is realized through the influence of participation degree on participation attitude value. The different values of the participation degree lead to different values of participation attitude, then lead to different participation strategies, and then lead to construction of different alliances. The details are given as follows.

Definition 1. *The historical participation degree of At_j relative to At_i*

$$pd_{ij-ADV} = \sum_{negotiation}^{participation\_success} (i \to j) \Big/ \sum_{negotiation}^{participation\_all} (i \to j) \quad (11)$$

where $pd_{ij-ADV} \in [0, 1]$. The denominator is the number of historical mutual proposals proposed by At_i and At_j, the molecule is the number of successful historical negotiations between At_i and At_j. Formula (11) reflects the probability of At_j accepting At_i 's participation proposal in historical negotiations, which is called the historical *participation degree of At_j relative to At_i. The higher the historical participation degree is, the higher the agent's participation consciousness is, and the higher the probability of At_i and At_j building the alliance is.*

Definition 2. *The current participation degree of At_j relative to At_i*

$$pd_{ij-current} = |DN(i)| \Big/ \sum_{n=\text{neighbor}_i} |DN(n)| \tag{12}$$

where $pd_{ij-current} \in [0, 1]$. The denominator is the sum of resources owned by At_i's neighbors, and the molecule is the number of At_j 's resources who are At_i's neighbors. When At_j has more resources relative to At_i's other neighbors, we think that At_j's current participation degree is high relative to At_i. The current participation degree reflects the resources owned by an agent. The more resources an agent has, the higher the current participation degree.

Definition 3. *The participation degree of At_j relative to At_i*

$$pd_{ij} = a \times pd_{ij-ADV} + b \times pd_{ij-current} \tag{13}$$

where $pd_{ij} \in [0, 1]$, $a + b = 1$. Definition 3 takes into account the historical participation degree in Definition 1 and the current participation degree in Definition 2. The a represents the weight of historical participation situations, and the b represents the weight of the current resource occupancy.

Based on the above definition of the participation degree, the participation attitude value of At_i relative to At_j is given as follows.

$$\partial_{ij}^{n} = \begin{cases} \partial_{ij}^{n} - \beta_{\partial}(pd_{ij} - \Delta T), \partial \geq \partial_{\min} \\ \partial_{\min}, \partial < \partial_{\min} \end{cases} \tag{14}$$

where the initial participation attitude value ∂_{ij}^{0} can be negative, neutral, or positive. β_{∂} is the influence factor of the participation degree. The bigger the β_{∂} value is, the bigger the impact of participation degree on participation strategy. ΔT represents the threshold of the participation strategy. $\partial_{\min}$ represents the minimum value of the participation attitude.

By proposing the concept of the participation degree, the participation process presents the following three characteristics.

(1) For participant At_j with a bigger participation degree ($pd_{ij} > \Delta T$), At_i will reduce the participation attitude value relative to At_j, which will make the participation strategy become passive.
(2) For participant At_j with a lower participation degree ($pd_{ij} < \Delta T$), At_i will increase the participation attitude value relative to At_j, which will make the participation strategy become positive.
(3) In the process of public participation in the disaster relief alliance, agents always preferred to negotiate with higher participation attitude values.

From these characteristics, we can see that the neighbor's participation attitude value is always referenced in the process of disaster relief alliance formation. The participation attitude values of the neighbors with low participation degree are reduced and the participation attitude values of the neighbors with high participation degree are increased. Agents prefer to negotiate with the neighbor with the best cession strategy and gradually eliminate the neighbor with the lowest participation degree, which makes the participation strategy intelligent and more efficient.

4.3. An Algorithm for Establishing the Emergency Relief Alliance

We give an Algorithm 1 for establishing the emergency relief alliance as follows:

Algorithm 1: The establishment of the emergency relief alliance

```
Input: Randomly generated agent networks
Output: The structure of the emergency relief alliance
{
Graph=graphGen. getGg ( );
Participation=participationInitialization ( ); //Randomly generate agent network, initialize the current participation degree and historical participation degree.
participationTemporary; //participationTemporary is used to save the historical participation degree.
for times:1→MaxADVTime //Construction of alliances for MaxADVTime rounds by dynamically updating the participation degree.
{
Agentall=initialiser. createAgents ( ); //Initialization of the agent set
Participation=participationTemporary; //Assign the value of the last round of participation degree to this agent
For clock:1→MAX_Round; //The maximum number of negotiation rounds shall not exceed MAX_Round
{
for agent: Agentall
{
if (agent. Issleep && !nonwaiting. isEmpty) //If the agent is not convergent, and some neighbors of the agent are in the Wait state.
{
PtDecision=agent. getPtDecision( ); //According to the negotiation decision-making function, the agent carries out negotiation and makes a decision
updateParticipation (PtDecision, Parameters); //Update parameters for calculating participation degree
updateState(PtDecision); //Update the state of agents based on the result of the participation decision
sortAgentneighbor(Participation); //Sort agent's neighbors according to values of participation degree; in the next round of negotiation, it will be preferred to choose the agent with the bigger participation degree.
}
End if
End for
}
clock=clock+1
End for
}
CalculateParticipation (participationTemporary, parameters);//Calculate the participation degree
End for
}
}
```

The algorithm flow chart is shown in Figure 2.

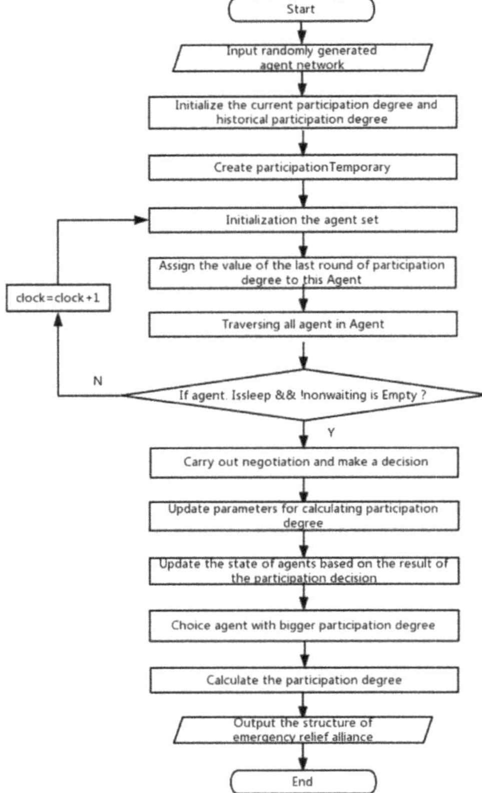

Figure 2. The algorithm flow chart.

5. Experiments and Analysis of Results

5.1. Experiments and Results

The experiments adopt an Eclipse platform to simulate the formation process of a multi-agent alliance; we verify the reliability and validity of our proposed model and Algorithm 1 by comparing with References [37,38].

We did three experiments: (1) Comparison and analysis of the influence of different parameters on the participation strategy and negotiation result. (2) By comparing and analyzing Reference [37] and our proposed method, we show the concession change caused by the dynamic change of participation degree. (3) As for the performance of alliance establishment, the differences among the References [37,38] and our proposed method are compared.

The initial parameters are given in Table 1.

Table 1. Parameter assignment.

Parameter Name	Parameter Values
Agent Number	10, 50, 100, 200, 500, 800
Single Agent Cost	C_i~$U(0,1)$
Single Agent Utility	p_i~$U(1.5C_i,3C_i)$
Repeat Times	100
Confidence Level	95%
∂_{ij}^{0}	15 (Positive), 0.001 (neutral), −5 (negative)
a	0.5
b	0.5
β_0	5
ΔT	0.5
Alliance Number	10

To validate the adaptability of different alliance construction methods for large-scale agent sets, the agent set scales are 10, 50, 100, 200, 500, and 800; single-agent overhead is a uniform distribution of [0, 1] and the single-agent utility is greater than the overhead. To remove the effect of a series of random numbers on the experimental results, a single alliance construction is repeated 100 times; the confidence level is 95%.

The first experiment is the comparative analysis of the effects of the parameters (a, b), β_0, and ΔT on the experimental results. The agent set scale is 100. The ∂_{ij}^{0} value is a random selection of 15, 0.001, and −5. The experimental results are shown in Tables 2–4.

Table 2. The effect of (a,b) on the alliance construction ($\beta_0 = 5$, $\Delta T = 0.5$).

(a,b)	**Average Negotiation Rounds (times)**	**Average Agent Utility**	**Negotiation Efficiency**	**Average Negotiation Success Rate (%)**
(0.1,0.9)	10.30 ± 0.86	28.96 ± 4.29	2.81 ± 0.04	58 ± 4
(0.5,0.5)	7.41 ± 0.87	34.03 ± 3.21	4.59 ± 0.12	65 ± 5
(0.9,0.1)	6.11 ± 0.34	39.11 ± 4.52	6.40 ± 0.15	70 ± 6

Table 3. The effect of β_0 on the alliance construction ($(a,b) = (0.5,0.5)$, $\Delta T = 0.5$).

β_0	**Average Negotiation Rounds (times)**	**Average Agent Utility**	**Negotiation Efficiency**	**Average Negotiation Success Rate (%)**
2	15.41 ± 0.75	30.22 ± 3.25	1.96 ± 0.06	56 ± 3
5	7.39 ± 0.88	33.08 ± 4.16	4.48 ± 0.13	65 ± 5
9	6.18 ± 0.77	39.47 ± 6.41	6.39 ± 0.14	69 ± 6

Table 4. The effect of ΔT on the alliance construction ($(a,b) = (0.5,0.5)$, $\beta_0 = 5$).

ΔT	**Average Negotiation Rounds (times)**	**Average Agent Utility**	**Negotiation Efficiency**	**Average Negotiation Success Rate (%)**
0.2	6.71 ± 0.82	38.21 ± 3.21	5.69 ± 0.06	69 ± 7
0.5	7.38 ± 0.89	33.09 ± 4.15	4.48 ± 0.14	65 ± 5
0.9	16.40 ± 0.84	31.31 ± 2.23	1.91 ± 0.06	58 ± 4

We can see from Table 2 that the higher the percentage of a in (a,b) is, the better the negotiation performance; this is because we do not consider different network topologies so that the resource owned by the agent does not change when an alliance is formed every time, while the participation attitude towards each other is dynamically changed with the process of alliance construction. The higher the proportion of a, the higher the performances of negotiation success rate, average agent utility, and negotiation efficiency are. As can be seen in Table 3, the bigger the β_0 value, the better the performance of the alliance construction; this is because when the β_0 value increases, the participation degree will play a greater role in the participation strategy. As shown in Table 4, the performance of the alliance construction becomes worse when the ΔT value increases; this is because ΔT is a threshold of the participation strategy—only when the participation degree is greater than ΔT will the participation attitude tend to be moderate. When the ΔT value increases, the agent requires that negotiation partners have higher participation degrees, which makes the moderate participation attitude more difficult to achieve.

As can be seen from the above experiment, the different sets of parameters have a significant influence on the performance of the participation alliance. In order to make good comparisons among our method and the References [37,38], in the following experiments, the parameters are set compromise values, namely a = 0.5, b = 0.5, $\beta_0 = 5$, $\Delta T = 0.5$.

In the second experiment, in order to obtain an empirical set of participation degrees, the 10 alliances are constructed and recorded as Alliance1, Alliance2, ... , Alliance10. The parameters are given in Table 1; the alliances are constructed by Reference [38] and our method, respectively. The agent set scale is 100, and positive, neutral, and passive participation strategies are adopted to construct alliances. The comparison between Reference [38] and our method (Alliance5, Alliance8, Alliance9, Alliance10) is given in Figure 3.

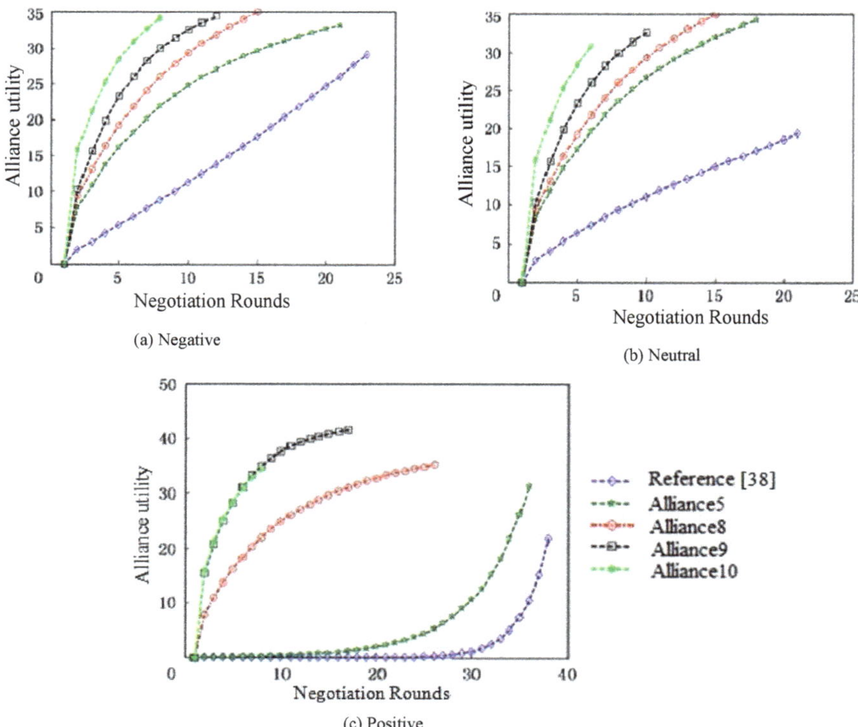

Figure 3. Experiment results.

As we can see from Figure 3, our proposed method can achieve better negotiation results by dynamically changing participation attitude values and selecting neighbors with better participation attitude values. The participation attitude becomes more and more moderate and achieves convergence after fewer rounds of negotiation.

In Reference [38], the parameter ∂_{ij}^{n} is assigned in advance and cannot change in the negotiation process, and the concession extent is relatively fixed. The ability of negotiation participants cannot be fully reflected in Reference [38], while the parameter ∂_{ij}^{n} is not assigned in advance in our proposed method. Once each alliance is formed, the negotiation data will be saved. The alliance formed in the next round will use this series of data to calculate the agent's participation degree, followed by the dynamic assignment of ∂_{ij}^{n+1} according to the ∂_{ij}^{n} value. This participation strategy based on participation degree fully reflects the difference in negotiation ability and negotiation situation.

The third experiment is a comparative analysis of References [37,38] and our proposed method. The parameters are given in Table 1. The agent set scales are 10, 50, 100, 200, 500, and 800; a single alliance construction is repeated 100 times, and the confidence level is 95%. The tenth-time experimental results of alliance construction (Alliance10) are adopted for comparison. The comparison results are given in Table 5.

Table 5. The performance comparison for alliance construction.

Type	Agent Number	Average Negotiation Rounds (times)	Average Agent Utility	Negotiation Efficiency	Average Negotiation Success Rate (%)
Reference [37]	10	69.88 ± 3.32	10.19 ± 1.65	0.15 ± 0.04	91 ± 4
Reference [38]		50.17 ± 2.63	16.74 ± 2.28	0.33 ± 0.07	51 ± 1
Alliance10		17.46 ± 2.04	19.23 ± 3.24	1.10 ± 0.10	61 ± 2
Reference [37]	50	82.17 ± 3.64	15.71 ± 2.08	0.19 ± 0.03	84 ± 3
Reference [38]		47.45 ± 3.31	18.82 ± 2.59	0.39 ± 0.06	51 ± 2
Alliance10		15.56 ± 2.18	21.61 ± 2.68	1.39 ± 0.08	61 ± 4
Reference [37]	100	85.16 ± 4.92	18.51 ± 3.12	0.22 ± 0.05	79 ± 5
Reference [38]		27.47 ± 5.26	25.42 ± 3.54	0.93 ± 0.10	56 ± 4
Alliance10		7.31 ± 0.96	33.21 ± 4.03	4.54 ± 0.13	63 ± 5
Reference [37]	200	86.28 ± 5.06	18.19 ± 3.09	0.21 ± 0.04	80 ± 3
Reference [38]		25.59 ± 5.12	26.38 ± 3.22	1.03 ± 0.15	60 ± 3
Alliance10		7.19 ± 0.88	35.17 ± 4.96	4.89 ± 0.07	68 ± 5
Reference [37]	500	89.52 ± 5.11	19.19 ± 3.95	0.21 ± 0.02	74 ± 5
Reference [38]		20.78 ± 4.30	29.68 ± 3.92	1.43 ± 0.14	62 ± 8
Alliance10		6.99 ± 0.78	39.17 ± 5.13	5.60 ± 0.27	70 ± 9
Reference [37]	800	91.14 ± 6.90	15.93 ± 2.03	0.17 ± 0.04	71 ± 2
Reference [38]		19.19 ± 4.45	31.32 ± 4.92	1.63 ± 0.15	67 ± 9
Alliance10		6.89 ± 0.62	42.44 ± 6.14	6.16 ± 0.41	80 ± 8

As can be seen from Table 5, our proposed method has obvious advantages in the negotiation rounds; because of the lack of an intelligent negotiation strategy in Reference [37], the negotiation rounds are greater than in the latter two methods, and our method reduces the negotiation rounds by more than 50% compared with the Reference [38], which is without the participation degree. In terms of the negotiation success rate, the negotiation success rate of Reference [37] is much larger than the latter two methods because Reference [37] is based on the candidate union set, and the negotiation success rate of our method is 10% higher than that of Reference [38]. As for average agent utility, Reference [38] is significantly higher than Reference [37] and our method, which shows that Reference [38] achieves higher individual utility. As for the negotiation efficiency, our method is much better than the other two methods, which shows that our method can achieve higher utility in a few rounds of negotiation, so the negotiation efficiency is high.

As for the different scales of the agent set, we can see that the larger the agent scale, the higher the negotiation success rate in our method; this is because the resources among agents have increased. However, in Reference [37], the negotiation success rate decreases with the increase of agent scale. From the perspective of negotiation efficiency, our method has the absolute advantage compared with the other two methods; this advantage is more obvious with the increase of agent scale, and shows that our method is suitable for large-scale agent systems. This is because the heterogeneity among agents is more pronounced when the agent scale increases. Our method, based on the participation degree, can well reflect this heterogeneity to improve the negotiation efficiency and make it easier to reach agreement.

5.2. Results Discussion

5.2.1. Experiment 1

As can be seen from Table 2, the bigger the proportion of *a* in (*a*,*b*), the fewer the negotiation rounds, and the higher the negotiation success rate and negotiation efficiency. However, the bigger the proportion of *b* in (*a*,*b*), the more the negotiation rounds, and the lower the negotiation success rate and negotiation efficiency. The reason is that for the public, such as enterprises, NGOs, and volunteers, if they have higher historical participation degrees, that means that they have a sense of participation and positive participation attitude, and it is easy to build alliances in the face of natural hazards, so the negotiation efficiency and negotiation success rate are high. If the public provides more resources and gets less utility from participating in natural hazard relief, they will refuse to build an alliance. Therefore, more rounds of negotiation are needed, and the negotiation efficiency and success rate are low.

As can be seen from Table 3, the bigger the β_0, the fewer the negotiation rounds, and the higher the negotiation success rate and negotiation efficiency. The reason is that the bigger the β_0 is, the higher the participation degree is, and, the more positive the public's participation attitude is, the easier it is to build alliances in natural hazard relief, so the negotiation efficiency and success rate are high.

As can be seen from Table 4, the bigger the ΔT, the more negotiation rounds there will be, and the lower the negotiation success rate and negotiation efficiency. The reason is that if a certain type of public continuously requires others to provide more resources and services, but does not provide more resources and services itself, it will be difficult to build an alliance, so the negotiation efficiency and success rate are low.

5.2.2. Experiment 2

Experiment 2 verified that our proposed method can achieve better negotiation efficiency by dynamically changing participation attitude values. The reason is that the traditional negotiation-based alliance formation method does not take into account the heterogeneity of agents, but makes consistent assumptions about the negotiation attitudes and resources of all agents. The negotiation is conducted in the order of agent ID number.

Our method proposes that the participation degree will be dynamically calculated in real time according to the agent's historical participation performance and the resources owned by the agent. The agent dynamically adjusts the participation degree, and chooses agents with higher participation degrees to negotiate and build alliances. This is also in line with the actual situation of emergency relief for natural hazards. The public can dynamically decide whether to join or leave an alliance according to their participation attitude and resources.

5.2.3. Experiment 3

Compared with Reference [37], our method is only slightly lower in the negotiation success rate, and has better performance than that of the Reference [38]. This shows that our method has better negotiation efficiency and higher agent utility than the other two methods, and is more suitable for large-scale multi-agent systems. This coincides with the fact that many public subjects participate in natural hazard relief. With the increase of the public scale, the number of interactive objects also increases, which greatly improves the possibility of forming alliances. Therefore, our method is practical.

6. Conclusions

Although the government is the main body of emergency management, our era is an unprecedented stage in which natural hazards frequently happen, so relying solely on the government's efforts to deal with natural hazards is almost impossible. We must actively encourage the public to participate in natural hazard relief. Through this research, we draw following conclusions.

(1) A natural hazard emergency relief alliance can be formalized based on the multi-agent system theory. It takes the participants as agents and makes use of a multi-agent system to construct the emergency relief alliance model. According to the paradigm of agent dynamic negotiation alliance, the emergency relief alliance is composed of two parts: ① Negotiation elements. Negotiation elements include the participation proposal, negotiation interval, and negotiation decision function. Negotiation elements are the premise and foundation of emergency relief alliances. ② Dynamic negotiation protocol. Dynamic negotiation protocol includes the agent interaction process and participation strategy. The dynamic negotiation protocol guarantees the convergence, non-deadlock, and consistency of an emergency relief alliance.

(2) Our proposed model and algorithm have good performance. Computational experiments were performed to evaluate and compare our proposed model and algorithm. Through the Eclipse platform, we made three experiments to prove that our model and algorithm have better performance in negotiation success rate, negotiation efficiency, and agent utility than other methods, and our model is more in line with the actual situation of natural hazard emergency relief.

(3) The participation degree influences the construction of an emergency relief alliance. The historical participation degree reflects the participation consciousness of the public, and the current participation degree reflects the resources owned by the public. The findings of the study indicate that the historical participation degree has a positive impact on emergency relief alliance formation; the more positive the public attitude is, the easier it is to build alliance. The current participation degree has no direct positive impact on emergency relief alliance formation. It is not that the more resources the public has, the easier it is to build alliance.

Based on this, to encourage the public to participate in natural hazard relief, we put forward the following countermeasures and suggestions.

On the one hand, foster a sense of public participation. The government makes use of traditional and modern media, such as radio, television, portal site, government websites, mobile phone short messages, WeChat, official accounts, and short videos, to widely publicize the knowledge of crisis management. Meanwhile, disaster relief courses, lectures, training, and professional schools should be set up to cultivate the awareness of public participation, improve the attitude of participation, and enhance public participation.

On the other hand, incentives should be increased. Through preferential tax policies, financial discount policies, commercial publicity, material rewards, and other incentive measures, the government guides NGOs, enterprises, institutions, and the masses to actively participate in emergency relief. At the same time, spiritual incentives are given to the public, such as the honorary titles of "philanthropist" and "good citizen". When the utility for the public increases, the public is willing to build an emergency relief alliance.

In our future research, we will consider a different network topology and a dynamic agent resource change process in each instance of the construction of an alliance. More factors will be taken account, such as rescue time, participants' characteristics, and so on.

Author Contributions: Methodology, Y.C.; software, Y.C.; validation, Y.C.; formal analysis, Y.C.; data curation, Z.W.; writing—original draft preparation, Y.C.; writing—review and editing, Y.C., J.Z., Z.W., and P.R.T.; funding acquisition, Y.C. and J.Z. All authors have read and agreed to the published version of the manuscript.

Funding: This research was funded by "National Social Science Fund of China, grant number 17BGL181", "National Natural Science Foundation of China (NSFC), grant number 51679058", and "Shandong Natural Science Foundation in China, grant number ZR2019LZH005".

Conflicts of Interest: The authors declare no conflict of interest.

References

1. Yu, X.B.; Yu, X.R.; Li, C.L.; Ji, Z.H. Information diffusion-based risk assessment of natural hazards along the Silk Road Economic Belt in China. *J. Clean. Prod.* **2020**, *244*, 118744.
2. Ren, Z.Y. General concept, current situation and development of emergency relief. *China Soc. Welf.* **2008**, *10*, 15–17. (In Chinese)
3. Adnan, S.G.; Kreibich, H. An evaluation of disaster risk reduction (DRR) approaches for coastal delta cities, a comparative analysis. *Nat. Hazards* **2016**, *83*, 1257–1278. [CrossRef]
4. Few, R.; Scott, Z.; Wooster, K.; Avila, M.F.; Tarazona, M. Strengthening capacities for disaster risk management II, Lessons for effective support. *Int. J. Disaster Risk Reduct.* **2016**, *20*, 154–162. [CrossRef]
5. Sim, T.; Liu, Y.; Li, S.J. Working together, Developing disaster risk reduction first aid training in a post-earthquake Chinese context. *J. Soc. Work.* **2017**, *17*, 491–497. [CrossRef]
6. Marlowe, J.; Neef, A.; Tevaga, C.R.; Tevaga, C. A New Guiding Framework for Engaging Diverse Populations in Disaster Risk Reduction, Reach, Relevance, Receptiveness, and Relationships. *Int. J. Disaster Risk Sci.* **2018**, *9*, 507–518. [CrossRef]
7. Carrao, H.; Naumann, G.; Barbosa, P. Global projections of drought hazard in a warming climate, a prime for disaster risk management. *Clim. Dyn.* **2018**, *50*, 2137–2155. [CrossRef]
8. Migliorini, M.; Hagen, J.S.; Mihaljevic, J.; Mysiak, J.; Rossi, J.L.; Siegmund, A.; Meliksetian, K.; Sapir, D.G. Data interoperability for disaster risk reduction in Europe. *Disaster Prev. Manag. Int. J.* **2019**, *28*, 796–808. [CrossRef]
9. Marchal, R.; Piton, G.; Gunn, L.-; Miras, Z.-; Der Keur, V.; Dartée, K.; Pengal, P.; Matthews, J.; Tacnet, J.-M.; Graveline, N.; et al. The (Re) Insurance Industry's Roles in the Integration of Nature-Based Solutions for Prevention in Disaster Risk Reduction-Insights from a European Survey. *Sustainability* **2019**, *11*, 6212. [CrossRef]
10. Webb, J. What difference does disaster risk reduction make? Insights from Vanuatu and tropical cyclone Pam. *Reg. Environ. Chang.* **2020**, *20*, 20. [CrossRef]
11. Totaro, F.; Alberico, I.; Di Martire, D.; Nunziata, C.; Petrosino, P. The key role of hazard indices and hotspot in disaster risk management, the case study of Napoli and Pozzuoli municipalities (Southern Italy). *J. Maps* **2020**, *16*, 68–78. [CrossRef]
12. Narayanan, R.G.L.; Ibe, O.C. A joint network for disaster recovery and search and rescue operations. *Comput. Netw.* **2012**, *56*, 3347–3373. [CrossRef]
13. Wachinger, G.; Renn, O.; Begg, C.; Kuhlicke, C. The Risk Perception Paradox—Implications for Governance and Communication of Natural Hazards. *Risk Anal.* **2013**, *33*, 1049–1065. [CrossRef]
14. Wex, F.; Schryen, G.; Feuerriegel, S.; Neumann, D. Emergency response in natural hazards management, Allocation and scheduling of rescue units. *Eur. J. Oper. Res.* **2014**, *235*, 697–708. [CrossRef]
15. Oral, M.; Yenel, A.; Oral, E.; Aydin, N.; Tuncay, T. Earthquake experience and preparedness in Turkey. *Disaster Prev. Manag. Int. J.* **2015**, *24*, 21–37. [CrossRef]
16. Berariu, R.; Fikar, C.; Gronalt, M. Hirsch, P. Understanding the impact of cascade effects of natural hazards on disaster relief operations. *Int. J. Disaster Risk Reduct.* **2015**, *12*, 350–356. [CrossRef]
17. Alem, D.; Clark, A.; Moreno, A. Stochastic network models for logistics planning in disaster relief. *Eur. J. Oper. Res.* **2016**, *255*, 187–206. [CrossRef]
18. Toyasaki, F.; Arikan, E.; Silbermayr, L.; Sigala, I.F. Disaster Relief Inventory Management, Horizontal Cooperation between Humanitarian Organizations. *Prod. Oper. Manag.* **2016**, *26*, 1221–1237. [CrossRef]

19. Deo, R.C.; Salcedosanz, S.; Carrocalvo, L. BeatrizSaavedra, M. Drought prediction with standardized precipitation and evapotranspiration index and support vector regression models. *Integr. Disaster Sci. Manag.* **2018**, *2818*, 151–174.
20. Nassereddine, M.; Azar, A.; Rajabzadeh, A.; Afsar, A. Decision making application in collaborative emergency response, A new PROMETHEE preference function. *Int. J. Disaster Risk Reduct.* **2019**, *38*, 101221. [CrossRef]
21. Borowska-Stefańska, M.; Kowalski, M.; Turoboś, F.; Wiśniewski, S. Optimisation patterns for the process of a planned evacuation in the event of a flood. *Environ. Hazards* **2019**, *18*, 335–360. [CrossRef]
22. Izida, A.; Tedrus, T.D.; Marietto, M.D.B.; Steinberger-Elias, M.; Botelho, W.T.; Franca, R.D. Emergency Care in Situations of Natural Disaster, A Multi-Agent Approach. In Proceedings of the 2012 International Conference for Internet Technology and Secured Transactions, London, UK, 10–12 December 2012; pp. 237–242.
23. Yue, E.; Zhu, Y.P. Study on Emergency Decision-making of Natural Disaster Based on collaboration of Multi-Agent. In Proceedings of the 2nd International Conference on Information Technology and Management Innovation (ICITMI 2013), Zhuhai, China, 23–24 July 2013; pp. 2684–2693.
24. Naqvi, A.A.; Rehm, M. A multi-agent model of a low income economy, simulating the distributional effects of natural disasters. *J. Econ. Interact. Co-Ord.* **2014**, *9*, 275–309. [CrossRef]
25. Na, H.S.; Banerjee, A. Agent-based discrete-event simulation model for no-notice natural disaster evacuation planning. *Comput. Ind. Eng.* **2019**, *129*, 44–55. [CrossRef]
26. Lee, S.; Kim, H.; Lee, B. An Efficient Rescue System with Online Multi-Agent SLAM Framework. *Sensors* **2020**, *20*, 235. [CrossRef]
27. Chen, M.; Qian, X.; Zhang, L.J. Public Participation in Environmental Management in China, Status Quo and Mode Innovation. *Environ. Manag.* **2015**, *55*, 523–535. [CrossRef]
28. Werner, T. Gaining Access by Doing Good, the Effect of Sociopolitical Reputation on Firm Participation in Public Policy Making. *Manag. Sci.* **2015**, *61*, 1989–2011. [CrossRef]
29. Say, N.; Herberg, A. The Contribution of Public Participation in a Good-Quality Strategic Environmental Assessment (Sea). *Fresenius Environ. Bull.* **2016**, *25*, 5751–5757.
30. Park, M.S.; Lee, H. Legal Opportunities for Public Participation in Forest Management in the Republic of Korea. *Sustainability* **2016**, *8*, 369. [CrossRef]
31. Xi, W. A Study on China's System of Public Participation in Environmental Protection-Taking Haikou's "Two Constructions" for Example. In Proceedings of the International Conference on Environmental Engineering and Sustainable Development (CEESD), Sanya, China, 3–6 December 2016.
32. Li, J. Public Participation in Tourism Development, an Analysis of Community Development in Underdevelopment Areas. In Proceedings of the 3rd Asian Pacific Conference on Energy, Environment and Sustainable Development (APEESD), Singapore, 21–22 January 2017; pp. 237–240.
33. Xiao, L.S.; Zhang, G.Q.; Zhu, Y.; Lin, T. Promoting public participation in household waste management, A survey based method and case study in Xiamen city, China. *J. Clean. Prod.* **2017**, *144*, 313–322. [CrossRef]
34. Ruiz-Villaverde, A.; Garcia-Rubio, M.A. Public Participation in European Water Management, from Theory to Practice. *Water Resour. Manag.* **2016**, *31*, 2479–2495. [CrossRef]
35. Chen, Y.X.; Zhang, J.; Tadikamalla, P.R.; Gao, X.T. The Relationship among Government, Enterprise and Public in Environmental Governance from the Perspective of Multi-player Evolutionary Game. *Int. J. Environ. Res. Public Health* **2019**, *16*, 3351. [CrossRef]
36. Soh, I.K. *Negotiation-Based Coalition Formation Model for Agents with Incomplete Information and Time Constraints*; Department of Computer Science and Engineering, University of Nebraska-Lincoln: Lincoln, NE, USA, 2002.
37. Goradia, H.; Vidal, J. *Negotiation-Based Coalition Formation among Selfish Agents [R/OL]*; Department of Computer Science and Engineering, University of South Carolina: Columbia, SC, USA, 2007.
38. Ramchurn, S.; Gerding, E.; Jennings, N.R.; Hu, J. Practical distributed coalition formation via heuristic negotiation in social networks [C/OL]. In Proceedings of the 5th International Workshop on Optimization in Multi-Agent Systems (OPTMAS), Valincia, Spain, 4 June 2012.

Article

Evaluation of Vegetation Restoration along an Expressway in a Cold, Arid, and Desertified Area of China

Chunfeng Jia [1,2,3], Bao-ping Sun [1,*], Xinxiao Yu [1] and Xiaohui Yang [4]

1 School of Soil and Water Conservation Beijing Forestry University, Beijing 100083, China; jiachunfeng@bjfu.edu.cn (C.J.); yuxinxiao111@bjfu.edu.cn (X.Y.)
2 Shanxi Environmental Protection Institute of Transport (co., LTD.), Taiyuan 030006, China
3 Shanxi Transportation Research Institute, Taiyuan 030006, China
4 Chinese Academy of Forestry, Beijing 100091, China; xiaohuiyang@caf.ac.cn
* Correspondence: sunbaoping@bjfu.edu.cn; Tel.: +86-010-6233-8068

Received: 28 March 2019; Accepted: 13 April 2019; Published: 17 April 2019

Abstract: Vegetation restoration plays a significant role in the restoration of expressways in the arid zone of China, but we still do not know which soil and vegetation types are most effective. We investigated soil particle size (SPZ), volume weight of the soil (VWS), soil water content (SWC), total porosity of soil (TP), soil organic matter (SOM), water erosion (W_rE), and wind erosion (W_dE) of eight sites (S1–S8) and evaluated them using the gray correlation method (GCM). Based on our results, the average SWC of the treatments ranged from 9.6% to 18.8%, following the order S4 > S5 > S8 > S6 > S3 > S7 > S1 > S2. The average SPZ of soils in S1, S2, S4, S5, S6, and S8 was larger, ranging from 0.23 to 0.68 mm, while that of soils in S3 and S7 was smaller, ranging from 0.01 to 0.09 mm. The TP in different treatment areas ranged from 50% to 60%, which is not conducive to soil and water conservation. The SOM levels varied widely among the different soils and were always below the threshold levels established by the second National Soil Census, rendering the soils not suitable for plant growth. The W_rE (36–80 t/ha) was greater than the W_dE (7–24 t/ha). In general, to achieve high soil and water conservation outcomes in this area, S1 and S7 offered the best protection benefits in terms of soil and water conservation.

Keywords: benefit evaluation; gray correlation method; soil erosion; soil physical properties

---

1. Introduction

The construction of expressways inevitably destroys the surface vegetation, leading to significant ecological problems such as soil erosion and landscape degradation [1–3]. In this sense, vegetation restoration along expressways is crucial to stabilize the roadbed, improve the road environment, and enhance the aesthetic properties of such areas [4–6]. During construction, the soil structure is largely destroyed; the surface soil in the cutting slope was mixed with gravel, parent material, and, in some cases, even rocks [7]. As a result, the surface temperature difference between day and night is relatively high, physical weathering is extreme, and evaporation is high; consequently, precipitation easily initiates runoffs [8,9].

In this sense, vegetation restoration along expressways is challenging. Previous studies have taken a series of measures to restore expressway slopes. The traditional forms of slope vegetation protection include grass planting [10], three-dimensional vegetation networks [11], masonry stone wall establishment [12], skeleton grass planting [13], vine plant protection [14], and geogrid grass planting [15]. Slope grass planting is characterized by a low survival rate, high maintenance costs, and the frequent degradation of grass seeds, while shrub planting is restricted by the water and nutrient conditions of the slope [14]. Inevitably, such approaches led to a low protection effect of the slope. Some expressway greening projects were only possible with the use of engineering work [15], with large investments. Also, over time, with rock weathering, concrete aging, and steel bar erosion, the protection effects become weaker [16,17]. In the northern regions of China, cutting slopes are also highly vulnerable to wind erosion [18,19]. Especially in the greening of expressways in arid, cold, and desertified areas, this problem is more prominent and mainly manifests in the following aspects: Firstly, the project implementers emphasize short-term effects while neglecting long-term effects [20], which directly leads to the phenomenon of "1 year green, 2 years yellow, 3 years withered, and 4 years dead". Secondly, project implementers overemphasize the quantity of greening. For example, in some arid areas, the emphasis was also on vegetation coverage, and especially in the south, evergreen trees have been planted extensively, often with unsatisfactory results. Thirdly, plant species selection and allocation were unreasonable [21], without any specific norms and standards, mainly copying the model of landscape greening and ignoring the natural laws of ecosystem and plant growth, which often resulted in unsatisfactory outcomes.

To reduce the environmental impacts of expressway construction and to improve the restoration of the damaged areas, it is crucial to employ new technologies, especially in cold, dry, and desertified areas. In this context, we included seven indices, namely volume soil weight, total porosity, soil moisture content, soil particle size, soil organic matter, water erosion, and wind erosion, from eight different treatment areas to evaluate the soil and water conservation benefits, using the gray correlation method (GCM) [22]. In this study, we summarize a set of technical methods for the ecological restoration of expressway areas in cold, dry, and desertified loess areas. The results of our study are of great significance for the restoration of vegetation along expressways, providing a reference for greening technology in other construction projects in cold, arid, and desertified areas and for technological progress of the vegetation restoration industry in general.

2. Study Sites

The study area is located along the Dahu Expressway (113.275° E and 39.936° N) in Datong City, Shanxi Province, a typical loess plateau area in China (Figure 1), at an elevation range from 1320 to 1170 m. The climate is north-temperate semi-arid continental monsoon climate. Due to the monsoon and high pressure on the Siberian and Mongolian plateau, there is little snow; the area experiences mild temperatures in winter, drought and windiness in spring, hot and rainy summers, and cool autumns. Average annual precipitation is 399 mm, of which more than 60% fall from July to September. Average annual temperature is 6.1 °C, with minimum and maximum temperatures of −29.5 and 34.5 °C, respectively. The daily temperature difference is 13.0 °C, with a frost-free period of 125 days. The study area can be divided into five geomorphic units, including loess hilly area, intermountain valley plain area, erosion and denudation high hilly area, valley plain area, and alluvial and flood plain area. The soil is mainly weathered rock soil, coarse silty soil, collapsible loess, and liquefaction soil of saline soil with sandy soil (Reference to China Soil Classification System). The main tree species are *Pinus tabulaeformis*, *Larix gmelinii*, and *Pinus sylvestris var. mongolica*. The shrub species mainly include *Hippophae rhamnoides Linn.* and *Caragana korshinskii*, while the most abundant herb genera are *Carex* and *Artemisia*. The above information was obtained by consulting local historical meteorological data and the respective yearbook.

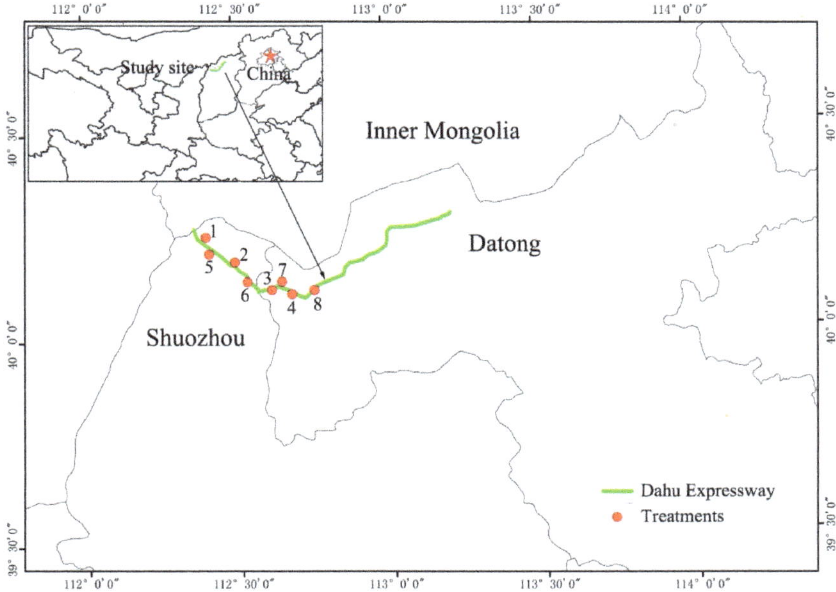

Figure 1. Geographic map of the study area. The numbers 1–8 represent the treatments S1–S8.

3. Materials and Methods

3.1. Experimental Design

Because the soil and vegetation were completely destroyed during the construction of roads and slopes, the developers filled the slopes with soil and planted shrubs to restore the ecological environment. During the implementation of the project, four kinds of soils, including weathered rock soil, coarse silty soil, collapsible loess, and liquefaction soil of saline soil with sandy soil were filled (Figure 2), and two plant species (*H. rhamnoides Linn.* and *C. korshinskii*) were planted. To determine which soil and vegetation allocation provided better soil and water conservation benefits, eight representative treatment areas were selected according to the actual situation of the project area. Specific information on the soil and vegetation allocation in each treatment area is shown in Table 1.

Table 1. Basic information of the treatments along the expressway in Datong City, Shanxi Province, China.

Number	Treatments	Specification (m)	Soil Type	Soil Thickness (cm)	Plant Species	Slope Ratio
1	S1	10 × 20	Weathered rock soil	43	*C. korshinskii*	1:1.5
2	S2	10 × 20	Coarse silty soil	40		
3	S3	10 × 20	Collapsible loess	45		
4	S4	10 × 20	Liquefaction soil of saline soil with sandy soil	39		
5	S5	10 × 20	Weathered rock soil	43	*H. rhamnoides Linn.*	
6	S6	10 × 20	Coarse silty soil	40		
7	S7	10 × 20	Collapsible loess	45		
8	S8	10 × 20	Liquefaction soil of saline soil with sandy soil	39		

Figure 2. Filled soil in the different treatments: S1 and S5 were the weathered rock soil; S2 and S6 were the coarse silty soil; S3 and S7 were the collapsible loess; S4 and S8 were the liquefaction soil of saline soil with sandy soil.

3.2. *Soil Sample Collection*

All samplings and erosion observations were performed in 2013. For the eight treatments along the expressway, nine sampling points were selected for each treatment. At each sampling point, soil samples were collected from the undisturbed original slope surface using a ring knife at depths of 0–20 cm, 20–40 cm, 40–60 cm, and 60–80 cm. The samples were brought to the laboratory to determine the volume weight of soil (VWS) and total porosity (TP). Additional soil samples were collected using an aluminum box to measure the soil moisture content (SWC) at the four soil depths, using the dry weighing method [23], and to determine soil organic matter levels via the potassium dichromate method [24]. Three soil samples were collected from each soil layer for each analysis.

3.3. *Determination of Soil Physical Properties*

3.3.1. Total Soil Porosity

To measure soil porosity, we opened the lower cover of the ring knife (with the mesh end), put it onto filter paper, and immediately weighed the sample (m_1). Subsequently, we removed the upper cover of the ring cutter and placed the mesh end of the ring cutter into the basin. The height of the water layer in the bowl was up to the edge of the knife (not submerged). After 12 hours of water absorption, we removed the ring knife, covered the lid, and weighed the sample again (m_2). After weighing, the ring cutter (bottom end of the mesh) was placed in a flat pan with dry sand and weighed (m_3) after 12 h. After weighing, the ring knife (mesh side down) was placed in a flat pan again and weighed (m_4) immediately after infiltration for 12 h. The weighed ring knife cover was opened, placed in an oven at 105 °C for 24 h, covered, and weighed (m_5). Subsequently, soil porosity was calculated as follows:

$$P = \left(\frac{m_2 - m_5}{v}\right) \times 100\% \quad (1)$$

where P is total porosity (g/cm^3) and v is the volume of the ring cutter.

3.3.2. Volume Weight of The Soil

The VWS was measured using the ring knife method [25]. We ensured that the soil in the ring cutter was not disturbed during each operational step. The ring cutter with the soil sample was placed in an oven at 105 °C and dried to constant weight; afterwards, the ring knife with soil samples (m_{1j}) and the soil samples (m_{2j}) were weighed separately, and the VWS was calculated as follows:

$$VWS = \frac{\sum_{j=1}^{3} (m_{1j} - m_{2j})}{3v} \quad (2)$$

where the v is the volume of the ring cutter, and J = 1, 2, and 3 is the repeat number of the same soil layer.

3.3.3. Soil Organic Matter

We used the potassium dichromate oxidation-external heating method to determine soil organic matter (SOM). Briefly, the soil mixture is boiled with potassium dichromate oxidizer and sulfuric acid (95%) in an oil bath at 170–180 °C for 5 min. The carbon in the SOM was oxidized to carbon dioxide by potassium dichromate, while the hexavalent chromium in the potassium dichromate was reduced to trivalent chromium. The remaining potassium dichromate was then titrated with standard solution of ferrous oxide. According to the amount of ferrous sulfate consumed by potassium dichromate before and after oxidation of organic carbon, the content of organic carbon was calculated and converted into SOM.

3.3.4. Soil Erosion

During the observation period, runoff barrels were used to collect runoff and sediment samples in the eight treatment plots. Weighing–sedimentation–drying–weighing and sampling–drying–weighing were used to correct the runoff and sediment yields, respectively. Hydraulic erosion per hectare was obtained through the area of the plot [26]. Secondly, wind erosion was measured via the cutting method [27]. The first choice is to record the depth of cutting, measure the change depth of cutting after the observation period, and, finally, calculate the wind erosion.

3.3.5. Evaluation of Benefits of Soil and Water Conservation of Different Treatments by the GCM

Here, X_0 is the reference series and X_i the comparison series. The value of the reference series represents the maximum value of each test indicator. First, we calculated the absolute value of the difference between each corresponding point of comparison sequence X_i and reference sequence X_0, and the GCM was evaluated as follows:

$$\Delta i(k) = \left|X_0(k) - X_i(k)\right| \tag{3}$$

$$X_0 = \{X_0(1), X_0(2), X_0(3), \ldots\ldots, X_0(n)\} \tag{4}$$

$$X_i = \{X_i(1), X_i(2), X_i(3), \ldots\ldots X_i(n)\}, i = 1, 2, 3, \ldots\ldots, m. \tag{5}$$

When using this approach, the actual value of each index must be converted into an evaluation value, and the original data should be processed without a dimension, according to the reference series data. On the one hand, the influence brought by each index dimension should be removed [28]; however, on the other hand, this indicator can also reflect the relative dominance of the community.

We used the formula $X_i(k)$ = original data sequence/reference data sequence to perform non-dimensionalization, reducing all data return to the [0,1] interval.

Subsequently, we determined the second-order maximum difference and the second-level minimum difference and calculated the correlation coefficient:

$$\xi_i(k) = rX_0(k), X_i(k) = \frac{\text{minmin}\Delta_i(k) + \rho\text{maxmax}\Delta_i(k)}{\Delta_i(i) + \rho\text{maxmax}\Delta_i(k)} = \frac{\Delta_{\min} + \rho\Delta_{\max}}{\Delta_{0,i}(k) + \rho\Delta_{\max}} \tag{6}$$

where $\xi_i(k)$ represents the relative value of the k point comparison curve (X_i) and the reference curve (X_0); that is, the X_i's correlation coefficient of X_o at point k. When $\Delta 0$, i(k) represents the absolute value of the X_0 sequence and the X_i sequence at the k point, $1 \leq i \leq m$, m was positive; Δmin and Δmax, respectively, represent the minimum and maximum values of the absolute difference of all comparison points in each point; ρ is the resolution coefficient ranging from 0–1. Here, ρ was artificially set to 0.5 (the artificial coefficient for qualitative analysis) to weaken the distortion effect caused by the excessively large maximum value to increase the significance of the difference between the correlation coefficients:

$$r_i = \frac{1}{n}\sum_{k=1}^{n} \xi_i(k) \tag{7}$$

where r_i indicates the degree of gray correlation; ξ indicates the gray correlation coefficient.

To measure the importance of each index more realistically and objectively, the coefficient of variation W_i was calculated using the coefficient of variation method:

$$W_i = \frac{r_i}{\sum r_i}. \tag{8}$$

Subsequently, we calculated the gray comprehensive evaluation value G (k) with the correlation degree value:

$$G_k = \sum_{k=1}^{n} \xi_i(k) W_i. \tag{9}$$

3.4. Statistical Analysis

Statistical analyses were performed using the software package SPSS 16.0. Descriptive statistics was used to calculate the mean and standard deviations for each set of replicates. First, a two-way ANOVA was used to analyze the differences in the SOM and SWC (n = 27 for each soil stratum), with treatment and soil depth as the independent factors. Two-way ANOVA was also used to analyze the differences in the SPT (n = 27 for each soil stratum), with treatment and soil particle size as independent factors. One-way ANOVA was also performed to test the effects of VWS, TP, W_rE, and W_dE on all treatments. All data were tested for normal distribution and homogeneity of variance analysis, meeting the requirements of variance analysis.

4. Results and Discussion

4.1. Basic Meteorological Characteristics

The study area is a temperate continental monsoon climate region, with an annual average precipitation of 384.02 mm. In 2013, average precipitation reached 418.7 mm, which was 9.03% higher than annual average precipitation (Figure 3). Precipitation from June to September was 296.1 mm, accounting for 70.71% of the total precipitation. Daily average precipitation was 1.14 mm, and maximum precipitation was 45.9 mm on September 9. During 2013, average daily temperature was 2.1 °C, with a single-peak "convex" pattern with seasonal changes. This gradually increased from January to July, reaching a maximum of 26.7 °C on July 30. After this peak, temperatures gradually decreased. The average daily wind speed was 1.04 m/s, with distinct differences between the seasons.

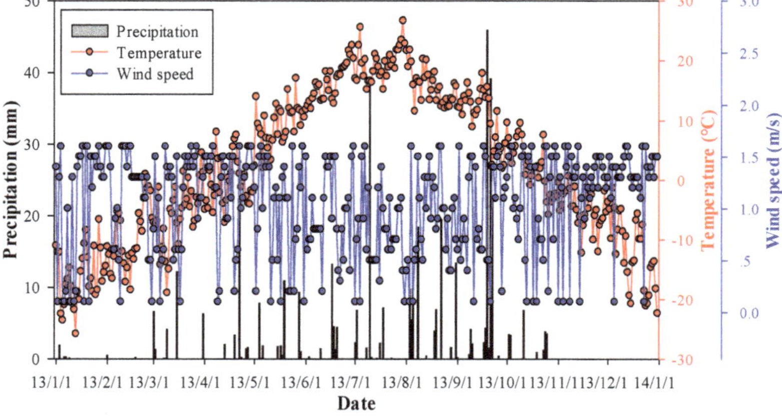

Figure 3. Precipitation, temperature, and wind speed in the study area along the expressway in Datong City, Shanxi Province, China, in 2013.

4.2. Weight Percentage of Soil Particle Size Under Different Treatments

Previous studies have stated that with decreasing soil particle diameter, the soil cohesiveness gradually increased [29–31]. According to the international classification standard of particles, when a particle of a sample has a diameter of 2–64 mm, it is called "gravel" [32]. In our study, the distribution of the particle size of the samples greatly varied among the different treatments (Figure 4). Site S2 contained the largest amount of gravel among all treatments, accounting for 25.35%. In the other treatments, the proportion of gravel was smaller, or gravel was completely absent. Particles with a diameter of 0.05–1 mm were most abundant in S8, with a significant difference when compared to the other treatments ($p < 0.05$). The weight percentages of d = 0.05–1 mm were 48.86%, 35.18%, 41.35%, 50.28%, and 61.72%, respectively, for the treatments S1, S2, S3, S7, and S8, which were significantly higher than those for the other particle size compositions ($p < 0.05$). The proportions of particles with a diameter of <0.005 mm were 39.52% and 38.65%, respectively, in the treatments S4 and S5, which were significantly higher than those in the other treatments ($p < 0.05$). Lu et al. also found that the size of the soil particles in the soil matrix differed; this was not only the case for the surface soil layers, but also for soil porosity in general [33]. Sandy soil contains coarse gravel and has a high soil porosity. On the contrary, the permeability of loam or clayey soil is lower than that of sandy soil, facilitating surface runoff [34]. Soil particle size and runoff are closely related, and a favorable soil particle size composition can effectively maintain water, nutrients, and organic matter [35]. According to our results, most of the particles with a diameter of more than 2 mm retain the original mineral composition of the parent rock. There are few available mineral nutrients, and the ability to absorb water was also poor. When the content of gravel in the soil exceeds 20% of the total volume of sample, changes in the temperature of the sample will be aggravated, and the water-holding capacity of the soil will be reduced [36].

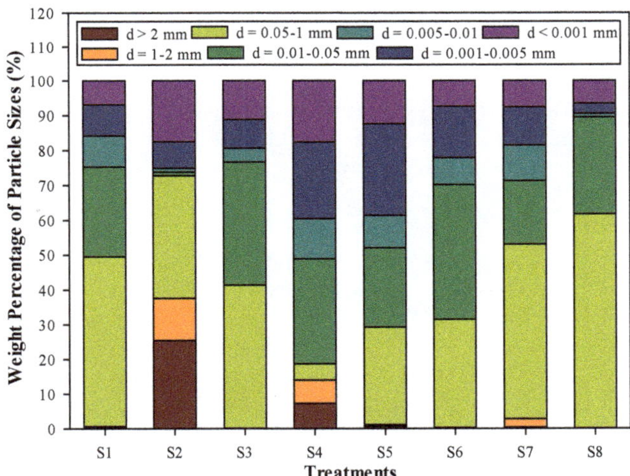

Figure 4. Variation in weight percentage of particle size of the samples under different treatments along the expressway in Datong City, Shanxi Province, China, in 2013. d represents soil particle diameter.

4.3. Variations in Volume Weight of the Soil and Total Porosity under Different Treatments

The volume weight of the soil (VWS) can be used as an indicator of soil solidity under certain conditions. At the same soil texture, soil with a low volume weight is relatively loose, while soil with a high VWS tends to be firm [36]. Generally, VWS varies greatly with soil texture, structure, and tightness [37]. Total porosity (TP) represents the percentage of soil porosity of the total soil volume. The amount of soil pores is related to the water permeability, air permeability, thermal conductivity, and compactness of the soil [38]. In our study, VWS and total porosity differed among the different treatments (Figure 5). The VWS was highest in S8, reaching 1.60 g/cm^3, and lowest in S3, with a value of 1.34 g/cm^3, with a difference of 19.4%. The VWS of S1 was 1.54 g/cm^3, with no significant difference between S1 and S7 ($p > 0.05$). The VWS values of S2 and S4 were 1.47 and 1.46 g/cm^3, respectively, also without a significant difference ($p > 0.05$). The VWS values of S1 and S8 were lager, while those of S3 and S5 were smaller, indicating a low soil compactness of S3 and S5, which is more suitable for plant growth. The TP values of S3, S5, and S8 were relatively high with 59.12%, 58.97%, and 58.97%, respectively, while that of S4 was 53.12% and therefore smaller than the values found for S3, S5, and S8. Generally, the TP ranged between 50% and 60%, and the texture was loose, which may not be conducive to water and fertilizer conservation.

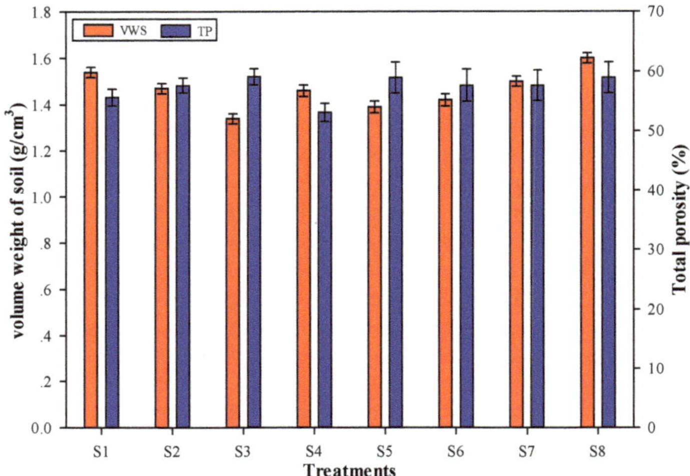

Figure 5. Variations (± SD) in volume weight of the soil and total porosity under different treatments along the expressway in Datong City, Shanxi Province, China, in 2013. VWS and TP represent volume weight of the soil and total porosity, respectively.

4.4. Variations in Soil and Water Content and Soil Organic Matter under Different Treatments

The soil water content (SWC) is mainly affected by both precipitation and evaporation [39]. In our study area, SWC and SOM levels differed greatly among the different treatments, most likely because of the inherent soil characteristics. The SWC levels increased with increasing soil depth, which might be explained by the high evaporation of the surface soil, resulting in low water content. However, SOM levels decreased with increasing soil depth (Figure 6). This, owing to these surface soil layers, can easily be supplemented with organic matter. The changes in SWC among treatments ranged from 9.6% of the average SWC of S2 to 18.8% of the average SWC of S4. The average SWC of S4 was twice the average SWC of S2. The average SWC levels of S1, S3, S5, S6, S7, and S8 were 10.3%, 14.5%, 17.9%, 14.8%, 12.7%, and 16.3%, respectively, following the order S4 > S5 > S8 > S6 > S3 > S7 > S1 > S2. The average SOM level of S8 was 0.40%, which was highest than in the other treatments, but still relatively low when compared with the lowest Grade 6 of the second National Soil Census and related standards (<0.6%) [40]. The average SOM content of S6 was 0.17%, which was 57.5% lower than that of S8. The average SOM levels of S1, S2, S3, S4, S5, and S7 were 0.23%, 0.25%, 0.23%, 0.20%, 0.19%, and 0.22%, respectively, following the order S8 > S2 > S1 > S3 > S7 > S4 > S5 > S6. The nutrient contents of the treatments were relatively low, without SOM, and the soils are therefore not suitable for the growth of newly transplanted. Most shrub species, when transplanted into a new environment, require adequate SOM levels to adapt and grow, in contrast to naturally growing shrubs, which have adapted to these environments through natural selection.

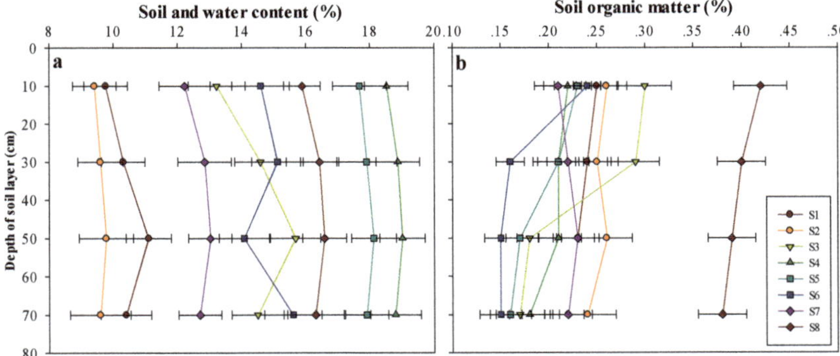

Figure 6. Variations (± SD) in soil and water content and soil organic matter under different treatments along the expressway in Datong City, Shanxi Province, China, in 2013. **a** and **b** represent the variations in soil and water content and soil organic matter, respectively.

4.5. Variation in Soil Erosion under Different Treatments

During the observation period, the water erosion (W_rE) of the different treatments varied because of the different soil properties. Throughout the study area, water erosion was greater than wind erosion (W_dE), ranging between 36 and 80 t/ha, while W_dE was between 7 and 24 t/ha (Figure 7). According to the classification of soil erosion intensity in China (SL190–2007), this area is moderately affected by soil erosion. The average W_rE of S4 was the largest (69.70 t/ha), while that of S3 was the smallest (39.70 t/ha). The W_rE of S3 was 43.03% lower than that of S4. The WE values of S2, S6, and S7 were relatively similar and reached about 50 t/ha. In S3 and S4, wind erosion was lower (8.10 and 8.63 t/ha, respectively), while at S7 and S8, it was higher (21.91 and 21.75 t/ha, respectively). The W_dE values of S7 and S8 were 2.70 and 2.52 times higher than those of S3 and S4, respectively. We found no significant differences in W_dE among S1, S2, S5, and S6 ($p > 0.05$). With increasing soil porosity, W_rE gradually decreased; generally, these two factors are not correlated [41]. The larger porosity of the surface soil facilitated the infiltration of runoff on the slope, thereby reducing runoff on the slope and the scouring force of runoff on the surface soil [42]. At the same time, the infiltration water also increased the erosion resistance of the surface soil [43]. Although there were numerous factors affecting slope surface erosion, and the interaction among them is more complex, in general, the pore size of the surface soil plays a role in inhibiting surface erosion.

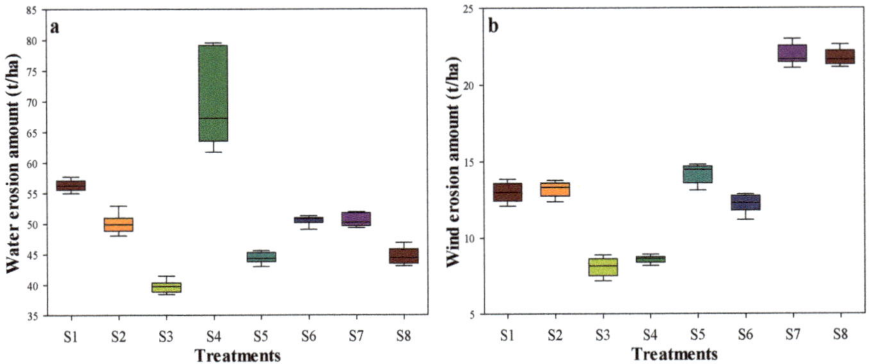

Figure 7. Variation (± SD) in soil erosion amount under different treatments along the expressway in Datong City, Shanxi Province, China, in 2013. **a** and **b** represent the variations in water erosion and wind erosion during the experimental period.

4.6. Evaluation of Different Soil and Water Conservation Measures Using the GCM

The results of comprehensive evaluations can directly reflect the effects of soil and water conservation measures [22,44,45]. The order of gray comprehensive evaluation values in the eight conditions was as follows: S1 (0.7593) > S7 (0.6995) > S3 (0.6656) > S8 (0.6516) > S5 (0.6485) > S2 (0.6385) > S4 (0.6377) > S6 (0.6194) (Table 2). This indicates that in S1, the soil and water conservation benefits were highest, suggesting the use of this soil in restoration programs in this area. Basically, these results lead us to infer that weathered rock soil and collapsible loess can be used when repairing slopes. In terms of plant species, *C. korshinskii* and *H. rhamnoides* Linn. were planted separately to achieve the best soil and water conservation benefits. The gray comprehensive evaluation values of each index were VWS (0.7376) > TP (0.7284) > SPZ (0.7233) > SWC (0.5777) > SOM (0.466) > W_rE (0.4294) > W_dE (0.369). The relationship between soil bulk density and total porosity was the most significant one, indicating that soil porosity and soil bulk density have the greatest impact on the effect of soil and water conservation measures. When repairing slopes, organic fertilizer can be applied to enhance plant growth and improve soil conditions. Additionally, adequate irrigation is necessary to support plant development.

Table 2. Correlation degrees and correlation coefficients of each evaluation index. SPZ, VWS, TP, SWC, SOMC, W_rE, W_dE, and G (k) represent soil particle size, volume weight of the soil, soil water content, total soil porosity, soil organic matter, water erosion, wind erosion, and gray comprehensive evaluation value, respectively. The parameter ri and Wi indicate the degree of gray correlation and the coefficient of variation, respectively.

Treatments	SPZ	VWS	TP	SWC	SOMC	W_rE	W_dE	G (k)
S1	0.7489	0.8513	0.8321	0.5286	0.3427	0.3388	0.341	0.7593
S2	0.5475	0.7981	0.6162	0.5955	0.3414	0.3538	0.336	0.6385
S3	0.8005	0.8142	0.784	0.5115	0.3576	0.4026	0.3492	0.6656
S4	0.6653	0.6316	0.7106	0.8678	0.5932	0.6929	0.3376	0.6377
S5	0.6872	0.7469	0.7106	0.8714	0.6494	0.4797	0.3347	0.6485
S6	0.6872	0.6498	0.7106	0.5496	0.5772	0.487	0.4074	0.6194
S7	0.8595	0.6427	0.7251	0.555	0.6653	0.3745	0.7557	0.6995
S8	0.8477	0.8864	0.7988	0.5079	0.6818	0.3576	0.7233	0.6516
ri	0.7233	0.7376	0.7284	0.5777	0.466	0.4294	0.369	0.5768
Wi	0.1975	0.1014	0.1389	0.1577	0.1272	0.1173	0.16	0.5661

5. Conclusions

We observed significant differences in soil and water conservation benefits among different soil types and two shrubs species. When the gravel content of the soil exceeded 20% of the total soil volume, the changes in soil temperature are aggravated, resulting in a reduced water holding capacity. Based on the soil porosity in the area, water and organic matter conservation are not optimal, and in some sites, the soil condition was not suitable for plant growth. Although we found relatively different organic matter levels in the experimental sites, all levels were beyond the threshold values established by the second National Soil Census. Based on the poor water and organic matter status, plant growth is severely limited. Soil erosion resistance varied greatly among the different sites, and slope protection is crucial to ensure soil and water conservation in this area. The results of the comprehensive evaluation value indicate that the sites of S1 and S7 provide the best soil and water conservation benefits. This also shows that the region can achieve high soil and water conservation benefits through the two allocation modes for slope protection. Weathered rock soil and collapsible loess can be filled in to protect slopes in this area, and subsequently, the two shrub species *C. korshinskii* and *H. rhamnoides* Linn. can be planted separately to achieve optimum protection and to conserve soil and water.

Author Contributions: C.J. wrote the paper; B.-p.S. conceived and designed the experiments; X.Y. (Xinxiao Yu) and X.Y. (Xiaohui Yang) analyzed the data.

Acknowledgments: This research was supported by the National Natural Science Foundation of China (Grant No. 41430747), the National Science and Technology Support Project (2015BA0804219).

Conflicts of Interest: The authors declare no conflict of interest.

References

1. Yang, J.; Liu, S.; Dong, S.; Zhao, Q.; Zhang, Z.M. Spatial analysis of three vegetation types in xishuangbanna on a road network using the network k-function. *Procedia Environ. Sci.* **2010**, *2*, 1534–1539. [CrossRef]
2. Zhang, L.T.; Gao, Z.L.; Yang, S.W.; Li, Y.H.; Tian, H.W. Dynamic processes of soil erosion by runoff on engineered landforms derived from expressway construction: A case study of typical steep spoil heap. *Catena* **2015**, *128*, 108–121. [CrossRef]
3. Nikomborirak, D. Private sector participation in infrastructure: The case of Thailand. *J. Organ. Chem.* **2015**, *45*. [CrossRef]
4. Roovers, P.; Bossuyt, B.; Igodt, B.; Hermy, M. May seed banks contribute to vegetation restoration on paths in temperate deciduous forest? *Plant Ecol.* **2006**, *187*, 25–38. [CrossRef]
5. Tian, H.; Cao, C.; Wei, C.; Bao, S.; Yang, B.; Myneni, R.B. Response of vegetation activity dynamic to climatic change and ecological restoration programs in inner mongolia from 2000 to 2012. *Ecol. Eng.* **2015**, *82*, 276–289. [CrossRef]
6. Azhari, A. Polycyclic aromatic hydrocarbons (pahs) in air and vegetation: Case study at three selected toll stations along north south expressway in Johor, Malaysia. *Eur. J. Cancer* **2012**, *29*, 1252–1256. [CrossRef]
7. Simões, M.D.S.; Rocha, J.V.; Lamparelli, R.A.C. Orbital spectral variables, growth analysis and sugarcane yield variáveis espectrais orbitais, indicadoras de desenvolvimento e produtividade da cana-de-açúcar. *Sci. Agricola* **2009**, *66*, 451–461. [CrossRef]
8. Normaniza, O.; Barakbah, S.S. Parameters to predict slope stability-soil water and root profiles. *Ecol. Eng.* **2006**, *28*, 90–95. [CrossRef]
9. Sangode, S.J.; Vhatkar, K.; Patil, S.K.; Meshram, D.C.; Pawar, N.J.; Gudadhe, S.S.; Badekar, A.G.; Kumaravel, V. Magnetic susceptibility distribution in the soils of pune metropolitan region: Implications to soil magnetometry of anthropogenic loading. *Curr. Sci.* **2010**, *98*, 516–527. [CrossRef]
10. Yang, Y.; Yang, J.; Zhao, T.; Huang, X.; Ping, Z. Ecological restoration of highway slope by covering with straw-mat and seeding with grass–legume mixture. *Ecol. Eng.* **2016**, *90*, 68–76. [CrossRef]
11. Wei, Z.; Wang, J.G.; Ling, W.; Jin, L. Stability analysis and supporting system design of a high-steep cut soil slope on an ancient l andslide during highway construction of tehran–chalus. *Environ. Earth Sci.* **2012**, *67*, 1651–1662. [CrossRef]
12. Burt, J.W. Developing restoration planting mixes for active ski slopes: A multi-site reference community approach. *Environ. Manag.* **2012**, *49*, 636–648. [CrossRef] [PubMed]
13. Wei, H.; Shao, L.; Zhang, Z. Shrubs increase soil resources heterogeneity along semiarid grass slopes in the loess plateau. *J. Arid Environ.* **2013**, *88*, 175–183. [CrossRef]
14. Maiti, S.K.; Maiti, D. Ecological restoration of waste dumps by topsoil blanketing, coir-matting and seeding with grass–legume mixture. *Ecol. Eng.* **2015**, *77*, 74–84. [CrossRef]
15. Farnhill, T. Union renewal and workplace greening—three case studies. *Br. J. Ind. Relat.* **2017**. [CrossRef]
16. Li, C.; Pan, C. The relative importance of different grass components in controlling runoff and erosion on a hillslope under simulated rainfall. *J. Hydrol.* **2018**, *558*, 76. [CrossRef]
17. Fore, S.; Mbohwa, C. Greening manufacturing practices in a continuous process industry. *J. Eng. Des. Technol.* **2015**, *13*, 94–122. [CrossRef]
18. Hüne, M.; González-Wevar, C.; Poulin, E.; Mansilla, A.; Fernández, D.A.; Barrera-Oro, E. Low level of genetic divergence between harpagifer fish species (perciformes: Notothenioidei) suggests a quaternary colonization of patagonia from the antarctic peninsula. *Polar Biol.* **2015**, *38*, 607–617. [CrossRef]
19. Graber, H.L.; Al, A.R.; Xu, Y.; Asarian, A.P.; Pappas, P.J.; Dresner, L.; Patel, N.; Jagarlamundi, K.; Solomon, W.B.; Barbour, R.L. Enhanced resting-state dynamics of the hemoglobin signal as a novel biomarker for detection of breast cancer. *Med. Phys.* **2015**, *42*, 6406. [CrossRef]

20. Androff, D.; Fike, C.; Rorke, J. Greening social work education: Teaching environmental rights and sustainability in community practice. *J. Soc. Work Educ.* **2017**, *53*, 1–15. [CrossRef]
21. Wardle, G.M.; Pavey, C.R.; Dickman, C.R. Greening of arid australia: New insights from extreme years. *Austral Ecol.* **2013**, *38*, 731–740. [CrossRef]
22. Lorenz, A.J. Resource allocation for maximizing prediction accuracy and genetic gain of genomic selection in plant breeding: A simulation experiment. *G3 Genes Genomes Genet.* **2013**, *3*, 481–491. [CrossRef]
23. Zuo, Q.; Zhang, D.; Jiaqiang, E.; Gong, J. Comprehensive analysis on influencing factors of composite regeneration performance of a diesel particulate filter. *Environ. Progress Sustain. Energy* **2016**, *35*, 882–890. [CrossRef]
24. Shigeno, T.; Brock, M.; Shigeno, S.; Fritschka, E.; Cervósnavarro, J. The determination of brain water content: Microgravimetry versus drying-weighing method. *J. Neurosurg.* **1982**, *57*, 99–107. [CrossRef]
25. Walinga, I.; Kithome, M.; Novozamsky, I.; Houba, V.J.G.; Lee, J.J.V.D. Spectrophotometric determination of organic carbon in soil. *Commun. Soil Sci. Plant Anal.* **1992**, *23*, 10. [CrossRef]
26. Hacisalihoglu, S. Determination of soil erosion in a steep hill slope with different land-use types: A case study in Mertesdorf (Ruwertal/Germany). *J. Environ. Biol.* **2007**, *28*, 433–438.
27. Visser, S.M.; Sterk, G.; Ribolzi, O. Techniques for simultaneous quantification of wind and water erosion in semi-arid regions. *J. Arid Environ.* **2004**, *59*, 699–717. [CrossRef]
28. Sarkar, U.K.; Deepak, P.K.; Negi, R.S. Length–weight relationship of clown knifefish chitala chitala (hamilton 1822) from the river ganga basin, India. *J. Appl. Ichthyol.* **2009**, *25*, 232–233. [CrossRef]
29. Xia, X.; Yu, S.; Kai, W.; Jiang, Q. Optimization of a straw ring-die briquetting process combined analytic hierarchy process and grey correlation analysis method. *Fuel Process. Technol.* **2016**, *152*, 303–309. [CrossRef]
30. Martin, C.M.; Randolph, M.F. Upper-bound analysis of lateral pile capacity in cohesive soil. *Geotechnique* **2006**, *56*, 141–146. [CrossRef]
31. Biscontin, G.; Pestana, J.M.; Nadim, F. Seismic triggering of submarine slides in soft cohesive soil deposits. *Marine Geol.* **2004**, *203*, 341–354. [CrossRef]
32. Alaoui, A.; Lipiec, J.; Gerke, H.H. A review of the changes in the soil pore system due to soil deformation: A hydrodynamic perspective. *Soil Till. Res.* **2011**, *115*, 1–15. [CrossRef]
33. Lu, Q.; Wang, E.; Chen, X. Effect of mechanical compaction on soil micro-aggregate composition and stability of black soil. *Trans. Chin. Soc. Agric. Eng.* **2015**, *31*, 54–59. [CrossRef]
34. Yahya, Z.; Mohammed, A.T.; Harun, M.H.; Shuib, A.R. Oil palm adaptation to compacted alluvial soil (typic endoaquepts) in malaysia. *J. Oil Palm Res.* **2013**, *24*, 1533–1541. [CrossRef]
35. Tang, A.M.; Cui, Y.J.; Richard, G.; Défossez, P. A study on the air permeability as affected by compression of three french soils. *Geoderma* **2011**, *162*, 171–181. [CrossRef]
36. Creamer, R.E.; Brennan, F.; Fenton, O.; Healy, M.G.; Lalor, S.T.J.; Lanigan, G.J.; Regan, J.T.; Griffiths, B.S. Implications of the proposed soil framework directive on agricultural systems in atlantic europe—A review. *Soil Use Manag.* **2010**, *26*, 198–211. [CrossRef]
37. Abdu, H.; Robinson, D.A.; Seyfried, M.; Jones, S.B. Geophysical imaging of watershed subsurface patterns and prediction of soil texture and water holding capacity. *Water Resour. Res.* **2008**, *44*, 5121–5127. [CrossRef]
38. Ashraf, M.A.; Maah, M.J.; Yusoff, I. Chemical speciation and potential mobility of heavy metals in the soil of former tin mining catchment. *Sci. World J.* **2012**, *2012*, 125608. [CrossRef]
39. Ross, D.S.; Bailey, S.W.; Lawrence, G.B.; Shanley, J.B.; Fredriksen, G.; Jamison, A.E.; Brousseau, P.A. Near-surface soil carbon, carbon/nitrogen ratio, and tree species are tightly linked across northeastern united states watersheds. *For. Sci.* **2011**, *57*, 460–469. [CrossRef]
40. Shirani, H.; Rizabandi, E.; Mosaddeghi, M.R.; Dashti, H. Impact of pistachio residues on compactibility, and permeability for water and air of two aridic soils from southeast of Iran. *Arid Soil Res. Rehabil.* **2010**, *24*, 20. [CrossRef]
41. Wei, J.; Knoche, H.R.; Kunstmann, H. Contribution of transpiration and evaporation to precipitation: An et-tagging study for the poyang lake region in southeast China. *J. Geophys. Res. Atmos.* **2015**, *120*, 6845–6864. [CrossRef]
42. Zhen, L. The national census for soil erosion and dynamic analysis in china. *Int. Soil Water Conserv. Res.* **2013**, *1*, 12–18. [CrossRef]

43. Nunes, A.N.; Lourenço, L.; Vieira, A.; Bento-Gonçalves, A. Precipitation and erosivity in southern portugal: Seasonal variability and trends (1950–2008). *Land Degrad. Dev.* **2016**, *27*, 211–222. [CrossRef]
44. Valinski, N.A.; Chandler, D.G. Infiltration performance of engineered surfaces commonly used for distributed stormwater management. *J. Environ. Manag.* **2015**. [CrossRef]
45. Jouquet, P.; Bottinelli, N.; Kerneis, G.; Henry-des-Tureaux, T.; Doan, T.T.; Planchon, O.; Tran, T.D. Surface casting of the tropical metaphire posthuma increases soil erosion and nitrate leaching in a laboratory experiment. *Geoderma* **2013**, *204–205*, 10–14. [CrossRef]

MDPI
St. Alban-Anlage 66
4052 Basel
Switzerland
Tel. +41 61 683 77 34
Fax +41 61 302 89 18
www.mdpi.com

Sustainability Editorial Office
E-mail: sustainability@mdpi.com
www.mdpi.com/journal/sustainability

www.ingramcontent.com/pod-product-compliance
Lightning Source LLC
LaVergne TN
LVHW070131170726
843515LV00007B/1778

* 9 7 8 3 0 3 6 5 6 3 9 7 8 *